Richard Aubert

MICROBIOLOGY

MICROBIOLOGY

By

LOUIS P. GEBHARDT, Ph.D., M.D.

Professor and Head, Department of Bacteriology,
College of Medicine, University of Utah

and

DEAN A. ANDERSON, M.S., Ph.D.

Professor of Microbiology, Chairman, Division of Natural Sciences,
Los Angeles State College of Applied Arts and Sciences

WITH 49 ILLUSTRATIONS

ST. LOUIS
THE C. V. MOSBY COMPANY
1954

Press of

THE C. V. MOSBY COMPANY

St. Louis

PREFACE

The general university or college student who pursues a course divorced entirely from some physical or biological science has lost a great deal of the effectiveness of higher education. There has been a trend in higher education toward requiring a series of related courses in broad fields such as the physical and biological sciences, humanities, and other base broadening courses.

Among these courses, a general course in microbiology not only stimulates the student but also broadens his outlook on his daily living, with a better understanding of previously mysterious things, such as souring of milk, spoiling of food, infections and diseases, pure water supplies, etc. This book has, therefore, been directed to this group of university and college students to enable them to gain a general knowledge of a science so vital to their daily activities. The microbic world presents new and challenging ideas that stimulate independent thinking and imagination. It may also have the effect of fertilizing the minds of some, resulting in the growth and development of a career in microbiology.

The basic concepts of microbiology are presented in such a manner that the average college student, with few prerequisites in either chemistry or biology, can digest and fully understand. Ideas will be introduced in simple language; then as the foundation is laid, a more technical terminology will be used. Thus the student will not be thrown into a complex terminology without a foundation. Practical applications of microbiology in the student's daily life are stressed, so that not only the scientific but also the practical point of view is gained.

The first section of the text deals with the basic and fundamental principles of microbiology; the second section deals with sanitary and industrial microbiology; the third part deals with the disease-producing microorganisms—pathogenic microbiology. All three aspects are important for a rounded general knowledge of microbiology.

The inclusion of a moderate amount of material on the pathogenic bacteria has several purposes. First, the student learns that

5

there are many microbial diseases to which he may become heir. Second, the seriousness of some of these disease agents and their frequent mastery of man are interpreted. Third, intelligent thinking is better activated on the part of adults when either children or adults develop a disease, and immediate counsel with their physician is sought. All too frequently lives needlessly end abruptly because of lack of knowledge, delay or deficiency of knowledge of the severity of certain disease processes, resulting in home treatment, the delay frequently ending tragically. A general knowledge of methods that can prevent disease is essential in maintaining a sound, healthy populace.

A general understanding of the problems of sanitary, agricultural, and industrial microbiology also increases an individual's value as a citizen. Recognition of unsanitary practices frequently not only protects the individual but also provides intelligent and effective arguments to be presented to health authorities for proper curative measures. Thus a better understanding of the microbic world in which the individual lives and a fuller appreciation of the impact that microbes have on our daily living can be better judged by students who have completed a course of study in microbiology.

Many individuals have given advice, constructive ideas, and criticism. We especially wish to express our gratitude to Doctors P. S. Nicholes, S. Marcus, J. G. Bachtold, D. L. Larson, L. R. Curtis, S. P. Hayes, D. M. Appleman, J. W. Bartholomew, S. C. Rittenberg, and M. J. Pickett, and to Mr. P. B. Carter, Mr. E. P. Hess, and Miss Betty Kazan. Sincere appreciation is also acknowledged to the many students who have offered the student point of view of presentation, content, and methods of approach to this subject.

The line drawings were generously contributed by Mrs. Rita Bachtold. The general photography was done by Dr. J. G. Bachtold and the microphotography was done by Mr. Howard Tribe, Medical Photography Division, College of Medicine, University of Utah.

Salt Lake City   L. P. G.
Los Angeles   D. A. A.

CONTENTS

CHAPTER 16

CHAPTER 17

CHAPTER 18

CHAPTER 19

CHAPTER 20

MICROBIOLOGY

SECTION I

GENERAL PRINCIPLES OF MICROBIOLOGY

CHAPTER 1

INTRODUCING THE MICROORGANISMS

As a part and parcel of the living world, mankind is surrounded by a great multiplicity of living things. Knowledge of the various living forms has been concentrated and catalogued into distinct fields of science as a heritage of mankind. Botany, as a plant science reeks with antiquity, while zoology assumed its role as a science in the days of Aristotle. Both of these deal with the visible plants and animals. A third science, which deals with the world of plants and animals invisible to the naked eye, is a mere infant, whose birthday coincided with the development of the microscope. While visible plants and animals have a profound influence on mankind, the impact of the microscopic plants and animals is probably even greater. Because of their size, the enormous influence they have or can have on our life and economy is often unrecognized. The harmful effects are often dramatic, sometimes spectacular, and often tragic. On the other hand, the beneficence of these plants and animals is unrecognized by most of us even though continued life on the earth without these forms would be impossible.

The purpose of this book will be to open the door of knowledge and afford a glimpse into this world of unseen plants or animals which are all about us.

A television detective aptly portrayed the everywhereness of these invisible forms with the comment: "The clues in this case are like microbes, all around us, only we can't see 'em."

Impressed with the vast powers of this invisible world, Pasteur exclaimed, "The infinitely small are infinitely great." On another occasion during the course of a fiery scientific meeting, he effectively silenced his critics with the barb, "Gentlemen, the microbes will have the last word."

To be convinced of the power of these infinitely small forms one needs only to consider black pages of history in which disease ruled the world. In that period, disease-bearing microorganisms spread by the blind ignorance of the populace, often changed the course of history by the inglorious defeat of armies unconquerable by man or by completely disrupting life in cities or even nations by wiping out in a single epidemic "the population increase of a century."

It has been aptly said that children were born into the world with the "sound of the clods falling on their coffins." Theirs was a world in which autumn was a melancholy season because typhoid and dysentery could be expected to take a heavy toll. Cholera could strike with such malevolence that the child, laughing and happy in the morning, could be a drawn-faced corpse in the evening. Diphtheria could and sometimes did strangle entire families in their beds. Smallpox struck at the young, strong, and beautiful and maliciously cut short their lives or left them scarred and disfigured, maimed or blind. From the vantage point of today's security, obtained by control of this harmful segment of the world of unseen living forms, we are prone to take much for granted and to fail to appreciate in what abject slavery mankind was held by the deadly microorganisms.

The child of that period had a life expectancy of about nineteen years, and old men were a rarity. Today, life expectancy is fast approaching the biblical three score years and ten and we are becoming concerned with the problem of a rapidly developing population of old people who have survived, largely because they were protected from the killer microorganisms by the wise application of bacteriological knowledge.

In spite of vaccines and antibiotics, chlorinated water, and pasteurized milk, we still have our bouts with the world of microscopic living forms. All of us share the common experience of pimples and boils (which we sometimes give to others in the family), sore throats, sinusitis, appendicitis, ringworm, and food poisoning. The common cold represents an unpleasantry which may strike nearly everyone one or more times a year.

Many of these diseases are acquired unnecessarily because of a lack of knowledge. Through infected nosedrops a whole family may share a cold, and the common towel or wash cloth may represent a convenient means of passing on pimples or ringworms.

Misconceptions such as the common belief that all microbes are dangerous; pimples or boils come from the inside as a result of poison in the blood, or that quickly dunking a needle in alcohol effectively sterilizes it, may make microbe-phobes out of us. Correct knowledge, therefore, affords protection against both the microorganisms and ungrounded fear of them.

It is apparently the development of a particular set of enzymes which causes one kind of organism to be a disease producer while another is harmless. How this occurred is not known, but a number of theories have been advanced to explain it. On the other hand, in our body are found many harmless or sometimes beneficial microorganisms which live with us, producing our vitamins, acting as scavengers, producing antibiotics against harmful organisms or immunizing against them as a sort of natural vaccine.

Other effects of microbes may be less direct. Spoilage organisms force us to resort to canning and refrigeration which increases our food bill. The price of fresh tomatoes in the corner grocery must be high enough not only to pay for those squeezed to death by pernickity housewives, but also those spoiled by the molds.

These harmful effects are spectacular, but vastly less important in the economy of nature than the beneficial effects of bacteria. The harmful microbes represent a mere handful compared with beneficial or harmless forms. Of paramount importance is the innumerable host of soil microorganisms which assure us a continued supply of indispensable nitrogen for protein, and carbon for all the organic matter of the body. Without them, the supply of these irreplaceable and indispensable plant food compounds would soon be dissipated and life as we know it on earth would cease to exist.

A few of the other beneficial effects are the production of sauerkraut (questioned as a benefit by some), dill pickles, vinegar, yoghurt, cheeses, the aroma and bouquet of fine butter, antibiotics, solvents, and vitamins.

The price of beef would be impossibly high if it were not for microscopic animals in the stomach of the cow which digest the cellulose in hay and other inexpensive roughage. Beef from a cow fed exclusively on a refined diet such as we must eat would be worth its weight in gold or platinum in contrast with the present beefsteak, worth only its weight in silver. Other microbial activities include the process of converting sugar to ethyl alcohol and the digestion by one-celled animals of chewed wood in the gut of termites.

Several different terms are used to describe the microscopic plants and animals. *Microorganism* (micro-, small; organism, that which has life) is perhaps the most descriptive and will be largely used in the ensuing chapters. *Microbe* which has essentially the same meaning (micro-, small; bios, life) was coined by the French.

The microorganisms include: (1) Bacteria: (Greek, diminutive of rod or stick). These are simple, one-celled plants without chlorophyl, occurring as rods, spheres, or coils. Many are so small that when laid endwise it would require 25,000 or more to extend one inch. You have often seen them as orange or creamy masses on potatoes or slime on improperly refrigerated meat or the gluey mass termed mother of vinegar. (2) Yeasts: These are also single-celled organisms, usually elliptical forms somewhat larger than bacteria. Tame yeasts are used in making bread or in brewing beer. The common yeast cake represents billions of yeast cells. Wild yeasts cause spontaneous fermentation or spoilage of fruit juices. Cottage cheese when spoiling may show on its surface pink masses of yeast growth. (3) Molds: The familiar cottony or powdery grey, blue, green, or black masses on cheese, oranges and spoiled foods in general are more than one-celled and represent a relatively more complex type of plant growth. (4) Viruses: These forms of disease agents are several times smaller than bacteria and are generally invisible under the light microscope. (5) Protozoa: These are one-celled animals of which amoeba is a familiar example, and they are widely distributed in nature. They are primarily important as causes of malaria and amebic dysentery.

These five groups will each be discussed more extensively in subsequent chapters to provide a clearer picture of their character and importance.

The general field of biology has been divided into several sciences, i.e., botany, zoology, microbiology, genetics, etc.

The science of microbiology has further been divided into a number of distinct branches as illustrated in Table I. Also the relationship of microbiology to the sciences of botany and zoology is presented in Table I.

From the diagram it is apparent that microbiology is useful to a great many individuals. Some of those who use this knowledge directly or indirectly include the physician, the dentist, the nurse, the medical laboratory technician, the sanitarian, the pharmacist, the biochemist, the farmer, the baker (yeast and ropy bread), the brewer (yeast fermentation and spoilage), the physiologist, the geneticist, the botanist, the zoologist, the biology teacher. Likewise, the housewife applies microbiology knowingly or unknowingly whenever she places food in the refrigerator, pickles or cans food, or even washes dishes.

TABLE I

RELATION OF MICROBIOLOGY TO OTHER BIOLOGICAL SCIENCES
AND SUBDIVISIONS OF MICROBIOLOGY

BIOLOGICAL SCIENCES

ZOOLOGY
(macroscopic
animals)

BOTANY
(macroscopic
plants)

PROTOZOOLOGY
(microscopic
animals)

MYCOLOGY
(study of molds
and higher fungi)

MICROBIOLOGY — General microbiology or bacteriology (Deals with "pure science" or theoretical aspects of microbiology)

Medical microbiology (oldest branch of the science; includes pathogenic bacteriology, disease production aspects, etc.)

Immunology (production of antibodies, etc.)

Sanitary bacteriology (sanitary aspects of water, sewage, milk, food, etc.)

Agricultural or soil microbiology (concerned with relation of microorganisms to soil fertility)

Industrial microbiology (production of solvents, vitamins, antibiotics, etc.)

Dairy bacteriology (quality control, cheeses, butter, etc.)

Food microbiology (preservation, food fermentation)

Microbial genetics (inheritance of physiological characteristics, etc.)

Pathogenic protozoology

VIROLOGY — Deals with virus diseases of plants, animals, and bacteria, these agents being invisible under the ordinary light microscope

RICKETTSIOLOGY — Deals with disease forms caused by the rickettsiae, forms barely visible by light microscopes

In fact, it is safe to say that every intelligent man or woman is able to live a safer and saner life if he knows something of this world of microscopic forms of life which surround us and is able to apply that knowledge intelligently in his everyday living.

With continued development of the science of microbiology, many new and intriguing occupational opportunities have developed. These opportunities are: medical microbiology (hospitals, clinics, research), industrial microbiology, food microbiology, agricultural and soil microbiology, dairy microbiology, public health microbiology, marine microbiology and several others.

Evidence is now accumulating that microorganisms, including the minute viruses, possess an inheritance mechanism and may be crossed to produce forms differing from either parent. Microorganisms are, therefore, being used to further study the laws of inheritance, particularly of biochemical or physiological characters. The value of these forms is apparent when one realizes that a population of a billion or more cells can be attained in twenty-four hours, and the physiological characteristics of members of this vast population can be readily studied. On the basis of the concept of comparative physiology it is recognized that the metabolism of bacteria is not far different from that of man. Therefore, much of the physiological information obtained with populations of bacteria may be directly or indirectly applicable to man. Many of these experiments could never be conducted successfully with human subjects because of the long generation time of man and because of man's inherent cussedness.

The intriguing effects of microbes mentioned here will be discussed more completely in subsequent chapters.

CHAPTER 2

HOW MICROBIOLOGY BEGAN

HISTORICAL DEVELOPMENT

At the time of the ape man of Java, dating back some 500,000 years ago, the struggle for existence was of paramount importance. Food, shelter, protection against enemies and fear of unseen and unknown enemies, *disease*, were paramount in the determination of the existence and development of man.

Disease and sickness were unknown mysteries and were feared by and baffling to early man. Unknown evil spirits entered the body and only by driving out the evil spirits and/or preventing their entrance into the body could man stay well. Thus the development of the shaman and the medicine man.

Medicine men practiced their art of curing disease and preventing disease by chants, weird noises, weird and ghastly dress, fearful and gruesome headdresses, all conceived and dedicated to frighten evil spirits of harm and disease.

No written record of early man is wanting in statements regarding pestis or pestilentia, referring to epidemics of disease, with, of course, no knowledge of their cause. Fear and horror struck their hearts and minds, and mass exodus from stricken areas by the populace was not uncommon. Hippocrates magnified seasonal changes of the atmosphere as the cause of disease. This idea influenced writers, mostly philosophic discussions, on the cause of disease.

The risk of contact as a means of spreading disease is suggested in the disease leprosy as set forth in Biblical writings (Leviticus XIV; Numbers XII) with a further idea of segregation or banishment for a given period of time.

Numerous observations have been recorded by lay writers on the probability of the cause of the plague or black death epidemic of Italy. Benedetti (1493—De observatione in Pestilentia, Venetiis) pointedly expressed conviction that to touch a sick person invited disease in a well person, and he further suggested that clothing must be cleaned or purified after wearing by a sick person. No primafacie evidence, however, was forthcoming as to how disease was transmitted by touch. The great Italian physician-poet, Fracastoro (1546) was perhaps the first to conceive the existence of actual disease-producing agents, seeds of disease that were probably capable of being transferred from one person to another. He suggested that direct contact, distance contact (probably sneezing, coughing) and articles touched or worn by an ill person containing the seeds of disease could all spread contagion. Fracastoro also wrote on the contagiousness of other specific diseases, but more particularly, the naming of the disease syphilis. A sheep-herder by the name of Syphilis had contracted the disease we now know as syphilis, and Fracastoro wrote a poem about the man and the disease. However, the actual cause of disease was at best speculative and philosophical.

THE ERA OF THE DISCOVERY OF MICROORGANISMS

Although lenses had been ground successfully and spectacles were used prior to the seventeenth century, the actual use of lens combinations was first established by Lippershey in 1608 when he constructed a simple telescope. Galileo Galilei in 1609 produced a sufficiently powerful telescope to see the satellites of Jupiter. Shortly thereafter Jansen constructed a compound microscope as did others such as Kircher and Hooke. It was not, however, until Anthony van Leeuwenhoek (1676) reported the existence of animalculae and wee beasties (his terms for bacteria) that his simple lenses actually revealed what we know as true bacteria or microorganisms. He found bacteria in water, excreta, and in cultures of pepper water. He saw and described the three important forms we know today, *rods* or *bacilli*, *round forms* or *cocci*, and *curved forms* or *spirillae*. He also described red blood cells, spermatozoa, and other microscopic structures.

The microscope quickly opened new fields for the microbiologist and not only were bacteria studied, but animal, insect, and human cell structures were discovered and studied.

THE MICROSCOPE

Since Leeuwenhoek's simple water immersion lens microscope, great strides have been made in the construction of compound microscopes, raising the magnification from a few hundred to nearly two thousand diameters.

The mechanics of using the microscope are not difficult to master, but the student should handle this instrument with great care, for, although sturdily built, it is a precision instrument and can be damaged by carelessness.

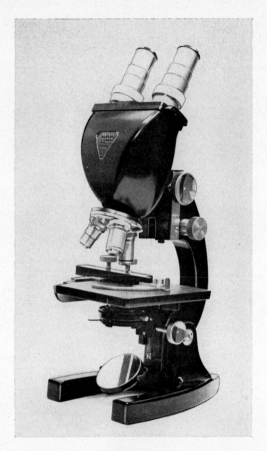

Fig. 1.—Modern binocular microscope with three objective lenses: low, high dry, and oil immersion lenses.

The component parts are the eye piece, the objective lens (the lens closest to the microslide or object you are observing), and the condenser. The monocular has but one eye piece or lens while the binocular has two. Usually microscopes are supplied with two or more eyepiece lenses of different powers of magnification. That is a 3×, 10×, etc., simply means that the image is magnified three times, ten times, or whatever the eye piece is marked. The objective lens may be marked in working distance from the object, that is, 16 mm., means the lens is 16 mm. from the object; 1.9 means the lens is 1.9 mm. from the object, etc. The farther the objective lens from the object (slide or specimen) the lower the magnification. The shorter the working distance, the greater the magnification.

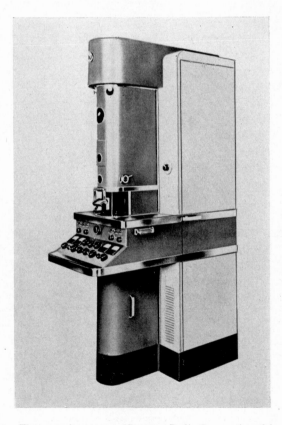

Fig. 2.—Electron microscope.   (Courtesy Radio Corporation of America.)

In general most modern microscopes are equipped with at least three objective lenses mounted on a revolving dual cone nose piece, a low power lens (10-16X) a high power lens (40-45X) and an oil immersion lens (90-95X). When the objective lens (a tiny lens) such as the oil immersion is too close to the object, light rays are slightly distorted or bent by the air between the lens object. Cedarwood oil or high grade mineral oil has approximately the same refractive index as the lens (glass), therefore producing a bright, clear image. Without oil, using this lens (oil immersion), the image is barely discernible and is fuzzy.

Simple magnification is not the determining factor in producing clear images. Many things must be considered, such as the type of lens (quartz, fluorite, etc.), kind of light, etc., and also the resolving power of the lens. The resolving power of modern oil immersion lenses is about 0.25 micron (one micron[1] is 1,1000 of a millimeter). This means the ability of the lens to separate two closely placed points. A simple explanation of this phenomena would be to test our own eyes on the headlights of an automobile traveling a straight piece of road. At first, at a great distance away there is a single blob of light. As the car approaches there is a suggestion of two lights; the nearer the lights get to you the single blob of light becomes partly separated until you are able actually to see two distinctly separated headlights.

As microscopic things approach the limit of visibility of the light microscope a new principle must be applied. This is the use of electrons instead of light waves. Thus the development of the electron microscope using electromagnets instead of lenses and electrons instead of light rays brought into view otherwise previously unseen microbiological entities as well as physical entities. The electron microscope will magnify and resolve from 10,000 to 20,000 or more diameters.

SPONTANEOUS GENERATION

Before the Christian era, it was believed that growth and reproduction of plants, insects, and small animals were brought into existence by the action of putrefaction, sunlight, air, and soil, and spontaneously arose with no conception of growth and reproduction which were known to exist in higher animals. Even Bibilical writings suggested spontaneous generation or abiogenesis (Judges, Ch. XI). Homer, however, knew the origin of maggots or fly larvae (Illiad XIX, 23-37), but not until the middle of the seventeenth century did Redi[2] prove this point. He covered a sample of fish with gauze and a second sample he left uncovered. The uncovered fish became maggoty while

---

[1] 1/25,000 of an inch.

[2] Francisco Redi: Experienze intorns all generazione degl' insetti,, 1688 (english transl. Chicago, 1909).

the covered fish did not. The developing larvae were later seen to develop into flies.

After Leeuwenhoek's discovery of bacteria a great flurry of speculation was presented as to the origin of animalcules. Leeuwenhoek believed they came from the air in the form of seed germs. However, Needham,[3] Buffon,[4] and others believed that matter was composed of organic molecules, each being capable of combinations with others, forming varied patterns of life. Others believed that higher animals, upon death, disintegrated into smaller lower forms, and on the death of these smaller forms, they too disintegrated into still smaller forms.

Experiments, growing microorganisms in meat broths and soups, were conducted by many of the early workers. Boiling, steaming, and other methods were used to kill the microorganisms. Different interpretations were placed on these varied experiments to fit the beliefs of the experimenter. The believers in spontaneous generation argued that the life forces of the molecules were destroyed and that such heating processes were too severe a test. In some of the experiments meat broths that were teeming with microorganisms failed to be sterilized by boiling several hours. This point favored the spontaneous generation group.

Plugging of bottles and tubes with cotton was tried by Schulze[5] and he found that a sterile (boiled) vegetable or meat broth remained sterile. This technique was later improved by Schröder and von Dusch. Schulze suggested that microorganisms were carried into the flask on dust particles in the air, but cotton removed the particles by filtration. Tyndall[6] likewise proved that dust particles carried microorganisms.

A great deal of time was spent on the problem of spontaneous generation and Pasteur (1861-62) undertook the task of establishing the truth. Painstaking, ingenious, laborious work eventually settled the problem. He proved beyond doubt that all life had a similar precursor and that offspring was the result of a parent and not spontaneously derived. He proved that certain microorganisms were highly resistant to heat, these being the spore-bearing rods or bacilli. He found that the spores (round to oval bodies within the bacterial cell) of certain bac-

---

[3] A System of Bacteriology, Paul Filders and J. C. G. Ledingham, editors, for Medical Research Council, London, 1930, His Majesty's Stationery Office, vol. 1.
[4] Ibid.
[5] Schulze, F.: Annals Psycek Chemie. **39**:487, 1836.
[6] Tyndall, J.: Philosophical Trans. 166, 1877.

teria would survive several hours of boiling and would reproduce if placed in a suitable culture media.

Pasteur continued his experiments on microbial life and discovered many microorganisms, saved the French silk industry by his work on pebrine, a silkworm disease; discovered and perfected vaccines against several diseases. He saved the wine and beer industry by his method of partial sterilization, the general idea we now apply in the pasteurization of milk. Pasteur was the father of modern bacteriology and immunology.

Many names will remain immortal in the historical development of microbiology and will forever be model examples for future scientists —such as Koch, Ehrlich, Roux, Yersin, Loeffler, Bordet, Park, Francis, and a host of others.

Selected Reading References

Bayne-Jones, S.: Man and Microbes, New York, 1932, Williams and Wilkins Company.

Bulloch, W.: History of Bacteriology, A System of Bacteriology, London, 1930, His Majesty's Stationery Office, vol. 1, pp. 15-103.

Dubos, R. J.: Louis Pasteur, Free Lance of Science, Boston, 1950, Little, Brown & Company.

Nordenskiold, Erik: The History of Biology, New York, 1935, Tudor Publishing Company.

CHAPTER 3

NATURE AND KINDS OF MICROORGANISMS

In order to learn something about microorganisms we must know their structure, how they grow and reproduce, what types of food they require, how to grow them, and how they behave under different conditions. After we learn some of the above fundamentals about these microorganisms, we can then begin to pigeonhole them or place them in definite phyla, subphyla, orders, families, genera, and species. This latter phase of study will be discussed as we know a little more about the nature and kinds of microscopic life that exists. Thus the study of microscopic plants and animals will be the theme of this book, microbiology.

The microorganisms to be discussed belong to either the plant kingdom or the animal kingdom. Some are parasitic (exist in living hosts), others are saprophytic (nondisease producing). Some are harmful; some, beneficial.

ANIMAL CELLS

Animal tissue is made up of thousands of microscopic sized cells, whether it be man (vertebrate) or mosquito (invertebrate). These multicellular elements combine to form organs or well-defined structures, such as glands, muscles, nervous tissue, skin, etc. Each cell is composed of distinct components and is dependent on the whole organism for its nutrition, multiplication, and behavior, etc. Animal cells, being protein structures, vary in size, shape, and in chemical nature, depending from which organ or tissue they are derived. For instance, the trained histologist or pathologist can tell at a glance, by the use of the microscope, using stained thin slices (5-10 microns thick) of tissue, whether the tissue is a piece of skin, liver, lung, intestine, or other organ.

Single cell animals such as the protozoan parasites, trypanosomes, plasmodia, etc., are self-supporting cells, not requiring the complex whole animal organism to maintain their existence and reproduction.

PLANT CELLS

Higher plants are constructed somewhat like animals in relation to their cells. They are organized multicellular units, the individual cells possessing characteristics similar to animal cells. Likewise the plant pathologist can tell quickly, using his microscope, the type of plant cell he is observing.

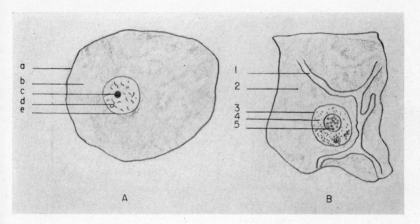

Fig. 3.—*A*, Animal cell; *a*, cell wall; *b*, cytoplasm; *c*, nucleoli; *d*, chromatin of nucleus; *e*, nucleus. *B*, Plant cell; *1*, cytoplasmic canals; *2*, cytoplasm; *3*, nucleus; *4*, chromatin of nucleus; *5*, nucleoli.

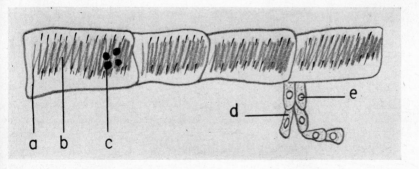

Fig. 4.—*Draparnaldia* sp. of Algae; *a*, cytoplasm; *b*, chloroplast; *c*, centers of starch production; *d*, branching filament; *e*, nucleus.

However, as we go down the scale in the plant kingdom the structure and nature of the cells and organism as a whole become less complex physically, but more complex chemically and as we go still further down the scale we are unable to distinguish or actually see the whole organism without the aid of a microscope. We will therefore deal only with those plants that are microscopic, since this course deals only with microbiological entities.

Algae

These are microscopic forms of plant life that grow primarily in water. They are plants without roots, leaves, or stems, but do possess chlorophyll to aid them in their metabolic activities. Some are unicellular, others multicellular. The greater majority are green and frequently form a green jellylike mass on the surface of water, particularly still or stagnant water. Others will grow in fresh water lakes, giving a greenish cast to the water, thriving best on the surface or at depths to which light penetrates.

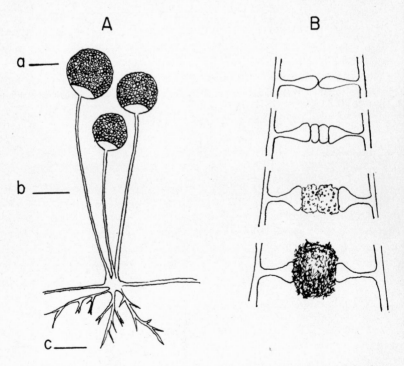

Fig. 5.—*A*, Rhizopus sp. of mold; *a*, sporangium with sporangio-spores; *b*, hypha; *c*, rhizoids. *B*, Mucor sp. showing matings of gametes to produce zygote.

Molds

These rather simple plants are multicellular and when growing in large masses, as on fruit, pieces of bread, or on artificial culture media, they present a fluffy, raised cottony appearance. Many of these are colored: green, brown, black, white, red, pink, or various shades of colors. As we see many thousands of these growing in a small area we might conceive of the viewer on a high mountain looking at a forest several miles away. We cannot see the structure of the individual trees. Therefore, in order to see individual parts of the mold structure, we must use a microscope.

Many thousands of species exist, but only a few illustrations of the microscopic structure will be given. They are threadlike filaments, branching like the roots of plants, some possessing specialized reproductive bodies called spores. The indiscriminate branched, rootlike structures, called mycelia, may be broken into compartments (septate) or smooth, regular, unbroken, flowing protoplasm (nonseptate).

These microscopic agents multiply by sexual and asexual methods. Some are imperfect, however, reproducing only asexually, these being called fungi-imperfecti.

Some of these microscopic agents are capable of producing disease, but the majority are harmless or they may actually be beneficial in producing valuable medicines (antibiotics, such as penicillin), aiding in soil fertilization, industrial production of many chemicals, etc.

Yeasts

Yeasts are round or elliptical unicellular microscopic plants, and reproduce either by budding, by fission, or by forming spore sacs with spores. They are usually smaller in size than are the molds and do not produce hyphae or mycelium. Only a few are disease producing; many are useful. The making of bread, fermented beverages, industrial alcohol, and many industrial chemicals are dependent on the activities of yeasts. The commercial production of certain vitamins, namely, B_1 and riboflavin, is accomplished by the metabolic activities of certain yeasts.

Probably the most frequent use of yeast is in breadmaking, whether in the home or in commercial bakeries. Yeast played an important role in the survival of the early prospectors, in Alaska particularly. The early Alaskan prospector, or trapper who lived months to years out of contact with civilization has been referred to as a "sourdough." This name was derived because these individuals carried in their pack or pocket small bits of bread dough for a bread start. The yeast continued to multiply in these pieces of dough, when the temperature was favorable, frequently producing a great amount of acids, or souring. Bread subsequently made from these "yeast starts" was sour. A hermit who lived on the Salmon River in Idaho for about forty-five years maintained a pure yeast culture in bread dough for over forty-two years.

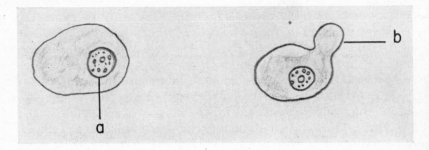

Fig. 6.—Yeast cells; *a*, nucleus; *b*, yeast bud.

TRUE BACTERIA

There are many thousands of species of bacteria all of which are unicellular plants, reproducing by fission. Leeuwenhoek in 1676 described the three important forms: the cocci, the bacilli, and the spiral type microorganisms. These microorganisms are considerably smaller than yeasts varying from 0.5 micron to 20 or more microns long. All are visible under the oil immersion lens of the microscope.

Many of these bacteria produce diseases, such as diphtheria, scarlet fever, boils and abscesses and many others produce serious to mild infections. Many, however, are beneficial in that they aid and assist in soil fertilization, produce useful chemicals by their metabolic activity, and aid in the manufacture of dairy products (e.g., cottage cheese). Even in our gastrointestinal tract certain microorganisms produce certain vitamins or vitamin precursors.

Staining of Bacteria

In order to observe, through the microscope, the size, shape, arrangement, and the presence of special granules, appendages, etc., bacteria must be stained. A number of staining procedures are used by the bacteriologist to make microorganisms visible.

The most useful stains for the study of bacteria are the Gram stain, the acid-fast stain, the capsule stain, and the spore stain.

The Gram stain separates microorganisms into two great groups—the gram-positive which retain the purple gentian violet and the gram-negative ones which take up the counter stain and are generally stained red or pink. This stain likewise allows one to observe, under the oil immersion lens of the microscope, the size, shape, and arrangement of the organisms on the stained smear.

The next important stain is the acid-fast stain. In this stain the original dye (usually red) absorbed by the bacteria is washed with an alcoholic-acid solution. Those that are acid-fast retain the pink to red stain, the nonacid-fast organisms absorb the counter stain, usually blue. This stain is useful in the study of the bacterium causing tuberculosis and related diseases.

The spore stain is used to make visible certain structures contained within the bacteria called spores.

The capsule stain is used to make visible, when a capsule is present, the envelope or capsule surrounding a single bacterium or group of bacteria.

Other special stains will be mentioned as they may be necessary to study the structure of specific genera and species of bacteria.

Size and Shape of Bacteria

Bacteria of the same species frequently are similar in size and shape. However, there are exceptions to this and these exceptions will be briefly discussed as individual species are described. Different species may vary from just visible microorganisms (0.25 micron[1]), when examined under the oil immersion lens, to rather long filamentous microorganisms of the genus *Streptomyces* and *Actinomyces* which may be from 5 to 50 microns long.

1. **The Cocci.**—The coccoid or round microorganisms vary from 0.5 to 1.2 microns in diameter. All are not exactly round, some being flattened against each other. Organisms of some of the genera, particularly the micrococci, form clusters of tiny round cells; other genera may produce chainlike groups of bacteria; others may form groups of

---

[1]A micron is one thousandth of a millimeter. There are 25.4 millimeters per inch.

two (diplococci), four or eight organisms per group. Some are gram-positive and some are gram-negative. None of this group is acid fast.

2. **The Bacilli.**—The bacilli are rod-shaped bacteria varying from 0.3-0.4 micron wide to over 1.0 micron wide and vary in length from 0.5-0.6 micron to several microns long.

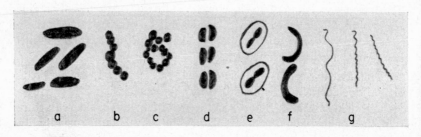

Fig. 7.—*a*, Rod forms of bacteria; *b*, chain forms of cocci (streptococci); *c*, group or bunch form of cocci (micrococci); *d*, diplococci (*Neisseria* sp.); *e*, capsulated diplococci (*Diplococcus* sp.); *f*, curved forms of bacteria (*Vibrio* sp.); *g*, spirochetal forms.

3. **The Spirilliform Bacteria.**—These are rigid, curved, rod-shaped organisms, similar to the bacilli in size.

4. **The Spirochetes.**—These organisms are generally longer than the bacilli, frequently resembling a coiled spring structure or up and down nature of the teeth of a dull hand saw. They are flexible and are able to bend their bodies. They vary from about 0.3-0.5 micron thick to as long as 20 microns.

Bacterial Appendages and Internal Structure

Capsules.—Some bacteria, more frequently seen in the cocci group but also seen on some rod forms, possess an envelope covering or capsule. This capsule helps protect the bacteria from destruction and also frequently aids the bacteria in producing disease.

Spores.—Some bacteria, particularly the rod-shaped organisms, possess an internal structure that may be round or oval and may be terminal, subterminal, or centrally located. These bodies, called endospores, frequently extrude from the bacterial cell and are seen in specially stained smears (spore stain) as tiny round to oval bodies. These spores will regenerate into the vegetative or ordinary bacteria of the specific species, under proper environmental conditions.

Flagella.—Bacteria, especially many genera of the bacilli or rod-shaped microorganisms, possess whiplike locomotion appendages, called flagella. Some species may have a single polar flagella, others a tuft of flagella on one or both ends, while others may have flagella projecting from all parts of the bacterial cell.

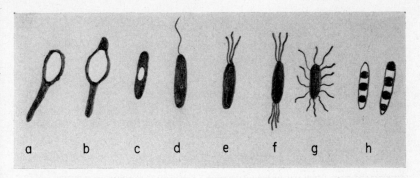

Fig. 8.—*a*, Rod form of bacteria with terminal endospore; *b*, subterminal endospore with bulging of the bacterial cell wall; *c*, central endospore *d*, single flagella (monotrichate); *e*, *f*, tufts of flagella (lophotrichate); *g*, peritrichate; *h*, metachromatic granules inside bacterial cell body.

RICKETTSIAE

The rickettsiae are the smallest microscopically visible microorganisms, many of which can be seen only after special staining and using the best available light microscopes and oil immersion lens. They vary from 0.25 to 0.5 micron, some being coccoid, some bacilli shaped and some rather irregular in shape. Some of these produce disease, such as typhus fever, Rocky Mountain spotted fever, etc.

VIRUSES

These agents are frequently called the filtrable viruses because they pass through bacteria-tight filters. The viruses are the smallest microorganisms known and only a few can be seen using the oil immersion lens of the microscope. Many, however, have been seen and photographed by means of the electron microscope. They vary between 10 millimicrons,[2] and 450 millimicrons. The virus group of microorganisms produce such diseases as influenza, smallpox, mumps, poliomyelitis, etc.

---

[2] A millimicron is 1/1,000 of a micron.

FOOD REQUIREMENTS

Animal and plant cells depend on the organism as a whole for their subsistence. Carbohydrates, fats, protein, vitamins, mineral salts, and oxygen are utilized.

Algae require only simple foods and sunlight. Thus from food sources in water, plus sunlight they are able to synthesize their own food and to proliferate rapidly.

Mold and yeast cells are individual complete entities and each cell must fend for itself for its food supply. The needs of these cells are simple, and are easily grown on simple food supplies. They possess a multiplicity of enzyme systems and are, therefore, capable of synthesizing their own food if supplied with a few basic elemental compounds, along with water. They are, therefore, easily grown in the laboratory on simple foods.

Bacteria vary a great deal in their food requirements, some being able to grow on simple basic foods as do the yeasts. Others may require complex amino acids, blood, or extra sources of vitamins and growth-promoting factors. Bacteria are, however, fairly easily grown in the laboratory if proper foodstuff is furnished. Various extracts of meat, eggs, or other protein sources are frequently used.

Viruses and rickettsiae, on the other hand, fail to reproduce on any type of lifeless media yet developed. They fail to reproduce their kind using any of the various types of media or food on which bacteria, yeast, molds, or algae may be grown. The viruses and rickettsiae require living cells for propagation, such as animal or plant cells, insect cells, incubated fertile eggs, etc.

Selected Reading References

Dubos, R. J.: The Bacterial Cell, Cambridge, Mass., 1945, Harvard University Press.

Hegner, R.: Big Fleas Have Little Fleas, or Who's Who Among the Protozoa, Baltimore, Md., 1938, Williams & Wilkins Company.

Henrici, A. T.: The Biology of Bacteria, ed. 3, New York, 1948, D. C. Heath & Company, pp. 43-165.

Smith, G. M.: Cryptogamic Botany, VI. 1 (Algae and Fungi), New York, 1938, McGraw-Hill Book Company, Inc.

CHAPTER 4

MICROORGANISMS ARE FOUND EVERYWHERE

Distribution of Microorganisms

Since the discovery of the existence of microorganisms by Leeuwenhoek in 1676, investigators have proved that microorganisms may be found practically everywhere. Leeuwenhoek found microorganisms in water, pepper infusion, and scrapings from his teeth as well as from several other sources.

Cultivated soils possess huge numbers of microscopic organisms. Bacteria, yeasts, and molds make up the greatest population of microscopic life. The richer the soil in organic matter the greater the number of microorganisms. Most of the microscopic agents found in soil are beneficial, aiding in the fertilization of the soil; occasionally microorganisms are found in the soil which are able to produce disease. In soil contaminated with human sewage the likelihood of harmful microorganisms increases. In moist soil the microbic population may reach a hundred million or more per gram of soil; in desert dry soil there may be a few thousand per gram of soil.

Water, irrespective of its source, contains microorganisms, many of which are derived from the water running over or through the ground. Snow, during the first part of a snow storm, collects large quantities of dust particles which harbor microorganisms. Later in a snow storm, the snow decreases markedly in microbial content. Deep driven wells and springs have usually the fewer numbers of microorganisms. Lakes frequently have a fairly high population of microscopic organisms, bacteria, yeasts, molds, algae, and other microscopic life. Microorganisms may be found growing in water which is nearly boiling. The colorful hot springs in Yellowstone National Park contain special varieties of microorganisms, the color of the predominating organisms thus imparts its color to the hot spring or pool. These organisms thrive in these hot temperatures. Conversely, certain kinds of microorganisms are found in cold springs and in glacial streams, thus some prefer the cold.

Since microorganisms are carried by dust particles, any article on which dust may settle will show the presence of bacteria and other microscopic life, when cultured on suitable food (i.e., microbial medium).

The surface of one's body harbors many varieties of bacteria, some of which may be harmful if one sustains a wound, thus introducing these microorganisms into the tissue.

The nose, throat, mucous membranes of the mouth, and the digestive tract of humans and animals are always inhabited by many different species of microorganisms. Most of these are classified as the "normal flora" of the individual, but at times pathogenic microorganisms set up housekeeping in one's nose, throat, or digestive tract and as they multiply they produce disease.

Since there are many millions of bacteria in the normal person's digestive tract, any discharge of fecal material will thus deposit these microorganisms on the soil, in the water, or other areas. If the person in question has a microbial intestinal disease or is a carrier of a pathogenic organism in his gastrointestinal tract he may become a hazard to his fellow men. Animals only occasionally carry microorganisms harmful to man in their fecal discharges, thus soil fertilization with manure is relatively safe. However, the use of human fecal material for soil fertilization as is practiced in many parts of Asia is extremely dangerous.

Since most foods that we eat are heated, few microorganisms are likely to be present. However, if cooked foods are allowed to remain in the open air, dust particles containing microorganisms may carry these into the food. Most of these organisms are harmless. Occasionally harmful bacteria may get into food and cause illness in those who eat the contaminated food.

Various insects harbor many kinds of microorganisms. Some microorganisms are found on the feet and bodies of flies and other insects, and some are within their body cavities and organ structures. Flies, mosquitoes, ticks, and other insects may, at times, therefore be responsible for carrying a disease. Fortunately, most of the time the microorganisms carried by insects are the same as those harmless ones found in water, soil, and in the air.

It is therefore a dictum that microscopic life exists practically everywhere and is universally distributed throughout the world. Most of these forms are harmless or may be beneficial; however, some are of the type which may produce disease in man, animals, fishes, insects, and plants. Even bacteria may suffer from disease.

Selected Reading References

McKendrick, A. G.: The Dynamics of Crowd Infection, Edinburgh M. J. **47**:117, 1940.
Meyer, K. F.: The Animal Kingdom, Reservoir of Human Disease, Ann. Int. Med. **29**:326, 1948.
Sonkin, L. S.: The Role of Particle Size in Experimental Air-Borne Infection, Am. J. Hyg. **53**:337, 1951.

CHAPTER 5

HOW MICROORGANISMS ARE GROWN AND STUDIED

CULTIVATION AND IDENTIFICATION OF MICROORGANISMS

The subject of this chapter is a complex one representing the accumulated knowledge of more than a century in which microbiologists have been attempting to subdue the invisible and elusive microorganisms. To secure accurate knowledge regarding any given microorganism it must first be isolated in "pure culture" separate from any other forms. This involves not only the selection of the proper food or medium for its growth but also the providing of proper temperature, acidity, or alkalinity, oxygen tension, or other necessary factors. The knowledge of how microorganisms are grown not only is essential to the professional microbiologist but also has many practical applications in our daily living.

CULTURE MEDIA

A culture medium (plural, media) bears the same relation to microorganisms as soil does to plants. With an acceptable medium, the proper acidity or alkalinity, oxygen tension, incubation, temperature, etc., a majority of the microorganisms can be grown successfully. Some forms, however, have resisted every attempt to grow them outside their habitat on artificial media.

Man has long grown microorganisms on culture media without necessarily recognizing the fact. The Biblical statement, "a little leaven leaveneth the lump" describes the growth of a mixed culture of bacteria and yeasts (the leaven) in an artificial medium (the dough).

Leeuwenhoek, when he added peppercorns to water in 1676, unknowingly supplied a culture medium in which a mixed culture of bacteria grew. Spallanzani, in the study on spontaneous generation which he reported in 1777, was in reality preparing a culture medium (mutton broth) which he sterilized by prolonged boiling.

A compilation of microbial media made by Levine and Schoenlein[1] in 1930 indicates the multiplicity of formulas which have been developed by microbiologists for growing microorganisms. They found in the literature more than 7,000 media listed. Of these, 2,543 which were distinctly different formulas were finally selected for inclusion in their compilation. These media formulas might be looked on as the cookbook recipes of the microbiologist. Since 1930, the discovery of vitamins and related compounds has provided the basis for a host of new media, most of which have been developed during the last decade.

Culture media may conveniently be divided into two categories on the basis of their components: (a) naturally occurring, nonsynthetic or nonreproducible media and, (b) synthetic, chemically defined or reproducible media. Further subdivisions may be made on the basis of the physical states of the media. These divisions are: (a) liquid, (b) solid, and (c) liquefiable solid media.

Nonsynthetic Media

The nonsynthetic media are represented by an almost infinite variety of naturally occurring substances which may be used alone or in combination with other substances for growing microorganisms. Examples of nonsynthetic liquid media are: (a) broths or infusions made from lean meats, liver, brains, hog stomachs, carrots, beans, peas, and many other animal and plant tissues; (b) naturally occurring liquids such as milk, blood serums, ascitic fluid, and coconut milk.

The nonsynthetic solid media include potatoes (cylindrical sections cut diagonally and placed in test tubes), carrots, meat, and similar materials. Solid media have also been made from eggs or blood serum which, while initially liquid, form irreversible solids on heating.

The liquefiable solid media are the most useful of all. They are prepared by adding a substance such as gelatin or agar-agar to a liquid medium. On heating to the desired temperature they become liquid but when cooled form jellylike solids which are reversible, i.e., may become liquid when again heated.

The use of liquefiable solid media grew out of the need, by early microbiologists, of a means of separating individual bacterial cells from a mixture of bacteria so as to yield a "pure culture," the progeny of a single cell.

---

[1]Levine, Max, and Schoenlein, H. W.: A Compilation of Culture Media for the Cultivation of Microorganisms, Baltimore, 1930, Williams and Wilkins Company, Inc.

With liquid media it was almost impossible to secure a culture of only one kind of bacteria from a mixture of bacteria except by preparing a series of dilutions, the higher dilutions of which might yield a pure culture of the organism predominating in the mixture. Sir Joseph Lister used this method in 1873 to secure a pure culture of *Streptococcus lactis*, the common milk-souring organism.

With the idea in mind of medium which would make it possible to secure a pure culture, Robert Koch in 1881 first used gelatin made from beef tendons for growing bacteria. He found that when a suitable dilution of a mixed culture of bacteria was added to liquefied gelatin and it was then solidified, the bacteria which were trapped in the solid gelatin began to grow and in a few hours or days formed a visible mass of growth (a colony). This generally represented the progeny of a single organism hence was a "pure culture." (We now know that a large colony may contain billions of cells.) Such a pure culture is just as fundamental to microbiological research as "chemically pure" chemicals are to chemical research. Koch's simple discovery, therefore, laid the foundation for much of our present knowledge in the field of microbiology.

Koch soon found to his dismay that gelatin possessed two very real disadvantages. It would liquefy at room temperature on a warm day and some bacteria would digest it to form a foul-smelling and often dangerous-to-handle liquid. However, at room temperature or lower and with nonliquefying bacteria it provided a valuable means of securing pure cultures of bacteria. A start was thus made toward solving the mystery of what organism causes a given disease.

An opportune suggestion, made by an American-born woman, provided the answer which revolutionized microbiology. Frau Fannie Hesse met her husband, young Doctor Hesse, while she was on a cycling tour of Europe. He was attached to the office of the board of health in Berlin and was one of Koch's collaborators. On hearing of the difficulty encountered with gelatin, Frau Hesse told her husband that her mother had used agar-agar, an oriental seaweed, to thicken soups. He tried out the suggestion in collaboration with Koch and found agar-agar to be an ideal solidifying agent. It would liquefy just below the boiling of water and formed a firm jelly a little above body temperature. In addition, it was not attacked by any of the bacteria used (although a small handful of agar-liquefying bacteria has since been discovered in the ocean). Koch, when given the information, capitalized on its tremendous possibilities.

Synthetic Media

The synthetic media are made with chemically pure compounds and used primarily in research work. With such a chemically defined medium an investigator in one laboratory can duplicate the medium prepared by another worker in another laboratory. With the naturally occurring substances this cannot be done. For example, one investigator reported that a certain strain of the tuberculosis organism grew well on a medium prepared with potatoes grown in one area while a similar medium prepared with potatoes from a different area failed to support the growth of the organism.

Since agar itself may vary in composition, the need of a chemically pure, inert solidifying agent was recognized by an early bacteriologist, Beijerinck. He discovered that the addition of hydrochloric acid to sodium silicate (water glass) produced a firm, clear jelly known as silica gel. This is widely used for the study of certain bacteria, the autotrophes, which oxidize inorganic compounds and which may be inhibited by the presence of organic materials such as agar or gelatin.

Some microorganisms have never been grown successfully on an artificial medium. These organisms require the nutrients and conditions supplied by living tissue in order to grow. The viruses, rickettsia, and certain of the spirochetes fall in this class. To provide animal tissue, the organisms may be injected into a suitable host animal, a fertile egg in which the embryonic chick is developing, or a culture of growing tissue taken from an animal.

Since it is apparent that many kinds and varieties of media are used, no attempt will be made to give their formula except where it becomes a part of the discussion of the cultivation of a specific organism. Only one medium will be considered at this time along with some of its modifications. This medium is nutrient broth and it is the one most commonly used by the general student in microbiological laboratories.

It is prepared as follows: (a) beef extract 0.3 per cent, (b) peptone 0.5 per cent. Sodium chloride (common salt) 0.5 per cent may or may not be added; however, if whole blood is added to the nutrient agar to make blood agar, the presence of salt is indispensable to prevent the breakdown or hemolysis of the red blood cells. Beef extract is made by evaporating the liquid in which beef has been steeped. This liquid contains the water-soluble constituents of the beef. The final product is a sticky brown mass and is often used commercially in preparing broths, bouillons, soups, etc.   Peptone, a gray powder, represents a dehyrated product of the partial enzymatic breakdown of meat.

If to nutrient broth is added 128 grams of gelatin per liter, the medium then becomes "nutrient gelatin."   If agar, at the rate of 1.5 per cent is added, the medium is known as "nutrient agar."

While nutrient broth, gelatin, or agar will support the growth of most of the common saprophytic organisms, a fair proportion of bacteria, particularly those parasitic on or pathogenic to warm-blooded animals, will not grow on this medium.

NUTRITIONAL CLASSES OF MICROORGANISMS

Microorganisms vary greatly in their nutritional requirements. However, all of them require carbon and nitrogen in some form since these elements represent the primary components (plus H and O of water) of the cell protoplasm.   The character of the carbon and nitrogen source which will satisfy the nutritional requirements of various microorganisms may vary greatly.   Microorganisms may be conveniently grouped as follows on the basis of complexity of their nutritional requirements:

The autotrophes possess amazingly simple nutritional needs.   The chemosynthetic autotrophes have the enzyme systems necessary to build their body structures out of CO_2 and some simple nitrogen compound such as ammonia (NH_3).   Energy is secured by the oxidation of such inorganic compounds (or elements) as ammonia, nitrites, carbon monoxide, hydrogen, or sulfur.   The photosynthetic autotrophes are those bacteria which possess chlorophyll-like pigments which enable them to secure their energy from sunlight.   They can likewise utilize CO_2 and NH_3 or similar compounds directly.

In contrast with the autotrophes, which constitute an important but numerically infrequent group of organisms confined largely to the soil or to water, a vast majority of the microorganisms show much more complex nutritional requirements.

This latter group of microorganisms, the heterotrophes, ranges from the saprophytes (rotting plants) of the soil to the parasites or pathogens found in the body of man.   All of them require organic compounds as sources of carbon.   However, they vary considerably in their additional nutritional needs.

Some forms, particularly many of those found in the soil, can use simple nitrogen sources such as ammonia for the synthesis of their cellular protoplasm. Others are somewhat demanding since they require some particular amino acid. Still others require several amino acids and perhaps certain vitamins particularly those of the B-complex. This latter group represents the so-called "fastidious" microorganisms. They are primarily obligate parasitic and/or pathogenic forms.

Nutritional requirements apparently depend largely on the enzyme systems possessed by given microorganisms. This in turn determines their ability or lack of ability to synthesize needed amino acids, vitamins, or other "growth factors."

Nutrient broth, nutrient gelatin, or nutrient agar lack certain of these "growth factors" and as a result fail to support the development of the more "fastidious" bacteria. Enrichment of nutrient broth, etc., with blood, ascitic fluid, and related substances is necessary before growth of these forms on nutrient agar will occur. Specific vitamins, amino acids, or similar "growth factors" may be added to a synthetic medium and thus supply the nutritional requirements of certain demanding forms.

OTHER FACTORS INFLUENCING GROWTH

Not only must the proper nutrients be supplied but the following factors must also meet the special needs of the organism: (a) acidity or alkalinity of the medium, (b) suitable oxygen tension, (c) proper incubation temperature.

Effect of Hydrogen Ion Concentration

All life processes go on in a water solution whether this solution is largely within the cell or surrounding it. One of the potent factors is the presence in the solution of hydrogen ions which are positively charged hydrogen atoms (H^+). The sour taste of lemonade, for example, is due to the effect of hydrogen ions on our taste buds. We say that such a solution is acid. With acids, such as hydrochloric acid (found in our stomachs) a large proportion of the molecules when in a water solution ionize to form H^+ and Cl^-. Thus $HCl \rightleftarrows H^+ + Cl^-$. An excess of such ions in the stomach produces an "acid stomach" or "heartburn." An alkali such as sodium hydroxide (lye), breaks down into negative hydroxyl ions and the positively charged sodium ion, $NaOH \rightleftarrows Na^+ + OH^-$. Water on the other hand breaks down to form a very minute quantity of both hydrogen and hydroxyl ions. $HOH \rightleftarrows$

H+ + OH⁻. Only about one out of every 10,000,000 water molecules is thus divorced. Pure water is termed neutral because it has an equal number of H+ and OH⁻ ions. Since hydrogen (and hydroxyl) ion concentration, i.e., acidity and alkalinity, has a profound effect on all life processes, whether they be those within our own body or within the bodies of bacteria, any great deviation from the neutral point in either direction might well cause the death of the cells.

Life processes go on primarily in solutions where the concentrations of hydrogen ions are extremely low, generally within the range of 0.0001 to 0.000000001 gram per liter. The highest concentration tolerated by living things is about 0.01 gram per liter and the least is about 0.0000000001 gram per liter, which is accompanied by a hydroxyl (OH) (alkaline) ion concentration of 0.0001 gram per liter.

Manifestly figures of this type are very clumsy to use and a Danish worker, Sϕrenson, in 1909 proposed a simple scale which would change these unwieldy fractions into simple positive numbers. This was termed the pH scale. It is a mathematical device for avoiding decimal fractions. Technically it is the logarithm of the reciprocal of the hydrogen ion concentration expressed in grams per liter. Its range extends from 1 gram (pH = 0) down to 0.00000000000001 gram (pH = 14). In using the pH values one's thinking needs to be thrown in reverse since the larger the value the less acid the solution. Also, each change of one pH represents a tenfold increase or decrease in acidity.

Table II gives a picture of the pH scale and the life process which go on at different pH levels.

Modification of pH may be of practical value to the housewife. For example, the addition of a small amount of vinegar to boiling water before poaching an egg produces an acid reaction which hastens coagulation. Commercial bakeries often make their bread dough slightly acid to prevent the development of ropiness in the bread (due to growth of bacteria). Fruits are much easier to can than vegetables because of their greater acidity.

Most bacteria make their best growth in the neighborhood of pH 6.8 to 7.6. However, *Thiobacillus thiooxidans* which oxidizes sulfur to sulfuric acid grows best at pH 2.0-2.8 and may even grow at a pH slightly more acid than 1.0. *Lactobacillus* found in cheeses, yoghurt, etc., has as its optimum pH 5.4-6.6. The Asiatic cholera organism, *Vibrio comma* is isolated by means of a medium with a pH of 9.6 which effectively inhibits practically all other bacteria which might otherwise grow on this medium.   Yeasts have as their optimum pH 4.0-5.8, but their range is from about 2.5-8.0.   Molds appear to be able to grow on almost everything if given time; however, they prefer a pH of 3.8-6.0 while their range is from 1.5-8.0.

TABLE II

pH Scale—Showing Relative Concentrations of H+ and OH⁻
AND pH OF SOME BIOLOGICAL SYSTEMS

CONCENTRATION GRAMS H+ PER LITER	CONCENTRATION — GRAMS OH- PER LITER	pH		BIOLOGICAL SYSTEMS
1.0	0.00000000000001	0		
0.1	0.0000000000001	1	Increasingly acid	Most acid tolerant of all biological systems *Thiobacillus thiooxidans*
0.01	0.000000000001	2		Highest acidity of the stomach
0.001	0.00000000001	3		pH of tomatoes, pineapple, orange juice—pH 3.0-4.5
0.0001	0.0000000001	4		Botulism organism does not grow in food more acid than 4.5
0.00001	0.000000001	5		
0.000001	0.00000001	6		Milk begins to taste sour
0.0000001	0.0000001	7		Neutral—pH of pure water pH of blood 7.35
0.00000001	0.000001	8	Increasingly alkaline	pH of many drinking waters
0.000000001	0.00001	9		pH of black alkali soils
0.0000000001	0.0001	10		
0.00000000001	0.001	11		
0.000000000001	0.01	12		
0.0000000000001	0.1	13		
0.00000000000001	1.0	14		

A pH range of 2-8 covers the pH requirements of most biological systems.   The pH of solutions may be determined electrically with the

"glass electrode pH meter" or by means of indicators. These indicators are dyes which change color with a change in pH. Litmus is a
well-known indicator. There are many others. The action of an indicator can be demonstrated easily in the home. If vinegar is added to
purple cabbage the water in which purple cabbage has been cooked
turns red due to the acidity. Baking soda, by producing an alkaline
reaction changes it back to purple.

Oxygen Requirements of Microorganisms

Although man and animals in general perish when deprived of
atmospheric oxygen, microorganisms range from those growing only in
atmospheric oxygen to those which will not grow in its presence. They
have been divided into the following classes on the basis of their oxygen
requirements:

(a) **Aerobes (Air-Living).**—These are microorganisms requiring oxygen in order to grow. Man is likewise an aerobe.

(b) **Anaerobes.**—Microorganisms in this class live only in the
absence of oxygen. Obligate anaerobes are apparently unable to grow
when oxygen is present. Why oxygen is so inimical to the growth of
anaerobes is not known, but it is believed that in the presence of
oxygen, hydrogen peroxide is produced which accumulates in toxic
concentrations because the organisms lack the enzyme catalase which
is necessary to break it down as fast as it is formed. Most aerobes
produce catalase while most anaerobes do not.

Many of the microorganisms in our environment are neither strict
aerobes nor anaerobes but are able to live either in the presence or
absence of air. These are spoken of as *facultative* microorganisms or
facultative anaerobes.

(c) **Microaerophilic Organisms.**—Some forms, particularly
some of the parasitic or pathogenic bacteria found in warmblooded
animals, require some oxygen but less than that present in the atmosphere. The organisms which require a reduced oxygen supply are
termed microaerophilic (small-air-loving) forms. Certain forms can be
initially isolated from tissues only under microaerophilic conditions.

Apparently the major factor which determines whether microorganisms are aerobic, facultative, microaerophilic, or anaerobic is the
type of respiratory enzyme system they possess.

Effect of Temperature on Bacteria

Microorganisms also differ in the temperature range in which they can grow. The lowest temperature at which a given microorganism will grow is spoken of as its *minimum growth temperature*; the range of temperatures at which the growth is best is termed the *optimum temperature* range and the highest temperature at which growth occurs is the *maximum growth temperature*. These temperatures are sometimes spoken of as "cardinal temperatures." On the basis of "cardinal temperatures" microorganisms may be divided into three groups: *thermophiles* (heat-loving organisms), *mesophiles* (loving a middle temperature), and *psychrophiles* (cold-loving organisms). However, these groupings are not too clear-cut since there is considerable overlapping between these three groups.

Psychrophiles grow best below 20°C. (room temperature). The optimum temperature as listed by different authors ranges from 10-20°C. The minimum temperature is about 0°C. and the maximum is listed as high as 30°C. The psychrophiles grow in the ocean, deep lakes, cold water streams from glaciers, in refrigerators, etc. Slow spoilage due to the growth of psychrophiles can occur in refrigerators and even near the freezing point.

Mesophiles have their optima in the middle temperature ranges, i.e., between room temperature and body temperature (about 20°C. and 37°C.). Actually the mesophiles tend to fall into two distinct groups: (a) the soil and similar organisms, largely saprophytes, which grow best at 20-25°C., and (b) the parasites and pathogens of warm-blooded animals which grow best at 32-37°C. The minimum growth temperature is listed by various authors as 5-10°C. (Household refrigerators are usually operated at about 40-50°F. or about 5-10°C.) The maximum growth temperature for mesophiles is at or near 45°C.

Thermophilic bacteria are found occurring naturally in hot springs, manure piles, etc. Some will grow only at high temperatures. These are termed *obligate thermophiles* while many will grow either in the thermophilic or mesophilic range. These are termed *facultative thermophiles*. The optimum temperatures given by various authors for the thermophiles range from 45-65°C., with minimum temperatures ranging from 25-40°C. and maximum temperatures from 70-80°C. The incubation temperature used for making thermophilic bacteria counts on foods or milk is 55°C.

Indications are that a relationship may exist between the maximum growth temperature for an organism and the temperature at which the respiratory enzymes of the particular organism are inactivated.

METHODS FOR SECURING PURE CULTURES

Since microorganisms as they grow in nature generally represent a mixture of different forms, it is often highly important to secure bacteria of one kind only or *pure cultures*. Special techniques must generally be employed in order to secure pure cultures. In a few instances, it is possible to secure a pure culture by *direct isolation* or *direct transfer*. This can be done only in those situations where pure cultures occur naturally. As examples, in diseases, such as typhoid, paratyphoid, streptococcus septicemia, etc., the microorganisms may be present as a pure culture in the blood stream of a patient. In such cases a pure culture may be obtained by drawing blood directly from the blood stream using a sterile hypodermic syringe and inoculating the blood into a suitable medium. Pure cultures may also be obtained in some cases from the spinal fluid of a meningococcus patient.

Occasionally masses of growth of a single organism may develop in decomposing material. If this occurs, a small amount of the material picked from the outermost portion of the growth may be a pure culture.

For certain types of special research work it may be necessary to be sure that a given pure culture represents the progeny of a single cell. This may be accomplished with the micromanipulator, a specially equipped microscope stage in which an infinitely small amount of material, containing a single cell may be drawn into a minutely fine pipette while viewing the operation under a microscope. This is a most delicate operation and requires extreme care and skill, plus the use of very finely designed and extremely well-machined equipment. Successful isolation can be made only in a relatively small proportion of the attempts.

Perhaps the most practical and most widely used method is streaking. The technique used consists of pouring a suitable sterile agar medium into a sterile Petri plate and allowing the medium to solidify. By means of a sterile inoculating needle or wire a small amount of growth, preferably from a broth culture or a bacterial suspension is streaked back and forth across the surface of the agar until about one-third of the diameter of the plate has been covered.

The needle is then flamed and streaking resumed at right angles to and across the first streak. This serves to drag bacteria out in a long line from the initial streak. When this streaking is completed the needle is again flamed and the plate is streaked at right angles to the second streak and parallel to the first streak but on the opposite side of the plate. The third streaking drags along the bacteria from the second streak. If the number of bacteria used in the first streak is not excessive, isolated colonies, which are frequently pure cultures, appear along the streak lines in the second or third streakings. These isolated colonies may then be transferred to an agar slant or a broth culture for subsequent use.

Plating

This method consists of diluting out a mixture of bacteria until only a few hundred or thousand bacteria are present in each milliliter of the suspension. A small amount of the dilution is then placed in a sterile Petri dish by means of a sterile pipette or sterile loop. A suitable agar medium which has been melted, then cooled to about 45°C., is poured into the plate and the bacteria and agar well mixed. When the agar is solidified the bacteria will be held in place and will grow into a visible colony. This may then be transferred to an agar slant as a pure culture.

Dilution

This method is now employed only for organisms which cannot be readily isolated by the streaking or plating methods. When several organisms are present in a mixture with one organism predominating, it may be isolated by the dilution method. The mixture of bacteria is suspended in a suitable diluting fluid and a loopful of each dilution is transferred to a suitable broth medium. Several tubes may be inoculated from each dilution. The highest dilutions showing growth may represent a pure culture of the predominating organism. However, it is cumbersome and the results often unsatisfactory so, whenever possible, streaking or plating is employed in preference to dilution.

Enrichment Media

Where a particular organism is present in only very small numbers compared with the total number in the mixture, it is often difficult to isolate it in pure culture. For example, the intestinal discharges of a typhoid carrier may have the organisms present in extremely small numbers when compared with *Esch. coli* and other forms. When feces are streaked directly on a Petri plate it is often almost impossible to isolate the typhoid organisms since they represent only a fraction of a per cent of the total organisms present. Media have, therefore, been devised which will permit the rapid growth of the organism desired while at the same time inhibiting the interfering growth of the other organisms.

Tetrathionate broth is an example of such a medium. It favors the rapid growth of the typhoid organism, *Salmonella typhosa*, and similar forms but inhibits the growth of *Escherichia coli*. With this enrichment medium it is often possible to isolate the typhoid organism from intestinal discharges even though it is initially present only in extremely small numbers. Selenite F broth is another enrichment medium used especially for isolating the paratyphoid and bacillary dysentery organisms from feces.

Selective Media

As a further aid in isolating desired organisms from mixed cultures, such as feces or sputum, *selective media* have been devised. These are media, usually agars, which favor the growth of the desired organism while definitely inhibiting the other microorganisms which are present.

As an example, a medium containing 7.5 per cent sodium chloride, (common salt) has been devised for isolating pus-forming micrococci from pus, throat swabbings, feces, etc. This high salt concentration effectively inhibits most other bacteria. Another example is brilliant green bile agar or broth. The combination of the dye, brilliant green, and bile serves to inhibit practically all bacteria except *Escherichia coli*.

Differential Media

In addition to being *selective*, media have been devised which are *differential*. These differential media, which may or may not be selective in their action, permit the formation of distinctive and easily recognized colonies. This serves to differentiate the organism desired from other organisms in the mixture. For example, on eosin-methylene blue agar, *Escherichia coli* produces distinctive blackish green colonies with a greenish metallic sheen while on the same medium *Aerobacter aerogenes* produces a considerably larger colony with a light violet periphery and a dark violet to brown center and the absence of the metallic sheen.

These selective media become extremely valuable for the rapid diagnosis of bacterial diseases, particularly the intestinal and respiratory diseases.

Variations in incubation temperatures may also be employed in the isolation of particular microorganisms. For example, thermophiles are isolated by incubation at 55°C., psychrophiles at 5-10°C.

Oxygen tension or gas concentrations may be varied for the isolation of various organisms. Thus anaerobic conditions must be employed if obligate anaerobes are to be isolated while *Brucella abortus* and *Neisseria gonorrhoeae* require a concentration of 5-10 per cent carbon dioxide in the atmosphere, at least for primary isolation.

The relative acidity or alkalinity (pH) is also used as an aid in isolating various microorganisms. In a medium with a relatively strong acid reaction, pH 4.0, for example, practically all bacteria are inhibited while the growth of molds is often definitely enhanced. At least, bacterial competition is sufficiently controlled to permit the molds to grow free from bacterial antibiotics or from nutritional competition.

Animal Inoculation

Isolation of certain pathogens is much easier when animal inoculation is employed. For example, when sputum is inoculated, intraperitoneally into a white mouse, it is often possible to secure a practically pure culture of *Diplococcus pneumoniae* whereas it would be much more difficult to isolate this organism by streaking directly on blood agar.

IDENTIFICATION AND DIFFERENTIATION OF BACTERIA

After securing a pure culture of a given organism the microbiologist is faced with the problem of identifying this organism as one previously identified or of differentiating, describing, and naming the organism if it is found to be a completely new and previously unidentified one. With higher plants and animals this identification and differentiation is generally based on physical characters such as shape, size, color, markings, etc., but with microorganisms the problem is not so simple. Many bacteria which have the same shape and staining characteristics might be quite different in their action on the human body.

For example, it would not be possible to differentiate *Escherichia coli* from the intestinal pathogens such as the typhoid-paratyphoid-bacillary dysentery organisms on this basis alone.

Identification, therefore requires a series of rather specialized procedures. It may involve one or a combination of all of the following tests: (a) shape or morphology; (b) staining characteristics; (c) cultural characteristics, i.e., appearance of colonies or growth on various media; (d) physiological (biochemical) reactions; (e) serological reactions; (f) phage typing (i.e., susceptibility to bacterial viruses may sometimes be employed to separate strains of a given species of bacteria).

(a) Morphology

The terms used to describe bacteria are those given in Chapter 7. The mere fact that an organism is a nonsporing rod immediately narrows the search down to several hundred species.

(b) Staining Characteristics

Since the shape of bacteria can be determined only by microscopic examination, staining procedures become extremely valuable to the microbiologist.

However, bacteria may be examined directly without staining. This is not too satisfactory because the bacteria refract the light to about the same extent as the glass of the slide or as the suspending medium. Therefore, direct microscopic examination of unstained bacteria has definite limitations. The so-called *wet mounts* of bacteria, yeasts, etc., are prepared by placing a drop of bacteria, suspended in a suitable liquid, or a loopful of a broth culture of the microorganisms on a microscope slide. A cover glass is then dropped on the preparation and it is examined using a reduced amount of light. Wet mounts are used primarily for determining the motility of bacteria or studying bacterial agglutination tests.

The dark-field microscope represents a very satisfactory method for examining certain microorganisms, notably the spirochetes causing syphilis. It is simply an ordinary microscope equipped with a special condenser which prevents the direct passage of light through the bacteria. This condenser is equipped with a stop or disc in the center which prevents the light from passing through the preparation being examined. However, light passes around the edges of the stop so as to

give side illumination of the suspended bacteria against an unillumi-
nated background. The objects appear as brightly lighted, scintil-
lating forms against a dimly lighted to black background.

A recent and extremely valuable modification of the microscope
is the phase microscope. In this instrument, both the condenser and
objective are equipped with a series of concentric rings which break up
the light. With this device, wet mounts may be examined very successfully and structures not revealed in stained mounts or dark-field
preparations may be clearly visible.

Occasionally "negative staining" has been used. This consists
of providing an opaque background in which bacteria appear as unstained or clear objects. The most common procedure is to mix china
(or India) ink with bacteria and spread the suspension thinly on a
slide. When dry the colorless bacteria are clearly visible against a
black background.

Because bacteria are so difficult to see in the unstained state,
Weigert in 1876 hit upon the idea of using aniline dyes for staining
microorganisms. The dyes commonly used are methylene blue, gentian or crystal violet, basic fuchsin, Bismarck brown, safranin, etc.
These are chiefly the so-called basic dyes.

The method is relatively simple; a small quantity of growth from
a culture or colony is transferred, by means of a loop to a drop of water
on a microscope slide. After suspending the bacteria in the liquid the
suspension is spread out to form a thin smear and allowed to dry in the
air. To "fix" the bacteria to the slide so they will not be washed off
during the staining operation, the slide is passed three times quickly
through the flame of a Bunsen burner, smear side uppermost. This
drives off the excess moisture and attaches the bacteria firmly to the
slide. The smear is then ready for staining.

If only one stain is used, such as methylene blue or carbol fuchsin,
the procedure is spoken of as a *simple stain.*

However, in order to differentiate between bacteria which take
and retain a particular stain from those which fail to retain the stain
when decolorized, the counterstaining or double staining procedure is
used. A number of such stains have been devised. The most important counterstaining procedure is the Gram stain.

The Gram Stain.—The Gram stain as we now designate it, was
developed by Christian Gram, a Dane who was working in the municipal hospital of Berlin. He was attempting to work out some procedure
which would stain bacteria in tissue while the tissue could be decolor-

ized (and stained another color if desired). The procedure was briefly mentioned in 1883 and published in detail in 1884.

Today the method used consists of staining the smear with ammonium oxalate-crystal violet.[2] The slide is then flooded with Gram's iodine, a potassium iodide-iodine solution. This has the effect of fixing the dye firmly in certain bacteria but with other bacteria it becomes easy to remove when a decolorizing agent is used. The slide is then flooded with the decolorizing agent, (most commonly a 50-50 mixture of alcohol and acetone). This part of the operation is critical and the length of time decolorization is allowed to proceed needs to be determined experimentally.

Microorganisms have been found to fall into two classes, those which retain the stain quite tenaciously when decolorized and those which are quickly decolorized. Greatly prolonged decolorization will make even a strongly gram-positive organism appear gram-negative.

To make the decolorized microorganisms (or tissues) visible, they are "counterstained" with a contrasting stain usually aqueous safranine which is red in color. Organisms which retain the first stain and as a result appear purplish or purple black are termed "gram-positive" while those which lose the initial stain and are colored red or pink with the counterstain, safranine, are termed "gram-negative."

The gram-positive organisms apparently are physiologically different from the gram-negative forms. Very few naturally occurring materials other than the bodies of certain bacteria, yeasts, and molds are gram-positive.

For further information the excellent review of Bartholomew and Mittwer[3] may be consulted.

The Gram stain is widely used for the differentiation of bacteria and the diagnosis of such diseases as gonorrhea.

The Acid-Fast Stain.—The tuberculosis and leprosy organisms, together with some related forms, are identified by the "acid-fast stain."

These organisms are difficult to stain by ordinary procedures. The method most frequently used consists of staining the organisms for a few minutes with hot (heated to steaming) carbol fuchsin, decolorizing with acid-alcohol (95 per cent ethyl alcohol plus about 1

---

[2]Gentian violet was formerly used in place of crystal violet. It is a mixture of methylated rosaniline dyes of which crystal violet is the most active agent. We now use the crystal violet because of its greater staining power.

[3]Bact. Rev. **16** (1): 1-30, 1952.

per cent hydrochloric or other acids) then restaining with the counter-stain, methylene blue. The bacteria which do not decolorize will be stained red or pink against a blue background. These are spoken of as "acid-fast" bacteria just as nonfading clothing may be spoken of as "tub-fast."

Members of the genus *Mycobacterium* are acid-fast while some of the actinomycetes are partially acid-fast. The acid-fast stain is, therefore, useful for the diagnosis of tuberculosis and leprosy.

Metachromatic Granule Stain.—The diphtheria organisms are identified by the use of the *metachromatic granule* stain. The stain employed is an alkaline methylene blue known as Loeffler's methylene blue. When cells of *Corynebacterium diphtheriae* are stained by this method they appear as slender blue rods, often club shaped, with blue-black bands or beads (the metachromatic granules). With another metachromatic granule stain, the Albert's stain which involves a counterstaining procedure, the organisms are light green with bluish black granules.

For demonstrating endospores, a number of spore stains involving the use of counterstains have been developed. With these procedures the spores will be one color while the remainder of the cell will be stained a contrasting color.

Flagella, while difficult to stain, can be stained successfully if extreme care is employed. A number of different flagellar stains have been developed. However, they are not as widely used as formerly because of the excellent pictures of flagella which can be secured with the electron microscope.

(c) **Cultural Characteristics**

Since various microorganisms produce types of growth which vary considerably in appearance, these "cultural characteristics" are often helpful in differentiating between various bacteria or as an aid in identifying them. This is often important where a ready recognition of an organism is desired such as may be necessary in a hospital laboratory.

The Society of American Bacteriologists have prepared a chart on which cultural characteristics (with standard terminology) may be recorded. These descriptions include appearance, shape, size, color, etc., of colonies on agar or gelatin plates; appearance of growths on agar slants and type and appearance of growth in broths, milk, etc.

Thioglycollate broth, a recently developed liquid medium which contains a small amount of agar dispersed through it, and which is sufficiently reduced to favor the growth of anaerobes as well as aerobes often produces a distinctive colony-like type of growth dispersed throughout the medium. This may aid in the rapid identification of certain forms of bacteria.

Cultural characters are especially important as a basis for identifying yeasts and molds, particularly the human pathogens.

(d)   Physiological Characteristics

Various microorganisms possess certain distinctive sets of enzymes which appear to be inherited as part of their species pattern just as their morphology, staining characteristics, and cultural characteristics appear to be inherited and peculiar to their particular species.

In instances where morphology, staining and cultural characteristics fail to differentiate or identify the organism being studied, it may be necessary to determine their physiological characteristics. This often provides a ready and relatively sure differentiation between closely related forms.

The major physiological reactions used in differentiating microorganisms are those which detect biochemical changes in the following compounds:

Carbohydrates.—

Sugars, Alcohols, Etc.—Determinations are made of the ability of organisms to produce acid or acid and gas from various sugars and such complex alcohols as glycerol, mannitol, etc. The ability to ferment some specific sugar may provide a ready means of differentiating between two rather similar organisms. For example, *Escherichia coli* and *Salmonella typhosa* and related forms are immediately separated on the basis of the ability of *Escherichia coli* to ferment lactose with the production of acid and gas. In fact, the intestinal pathogens belonging to the typhoid-paratyphoid-bacillary dysentery group are often termed the nonlactose fermenters. In order to differentiate between closely related forms a large number of sugars and similar carbohydrates may be used.

Alcohol and aldehyde production from certain carbohydrates may also be determined.

One interesting reaction, often used for differentiating bacteria belonging to the *Escherichia-Aerobacter* group is the Voges-Proskaeur reaction. Certain bacteria can produce acetyl-methyl-carbinol from carbohydrates. If after a culture has grown for 4 days at 37°C. strong sodium or potassium hydroxide is added to the medium (which must contain peptone) a pink to eosin color will develop at the surface. This test is also used to separate species in the *Escherichia-Aerobacter* group.

Starch digestion is another biochemical reaction which may be used for differentiation. When microorganisms are streaked on a plate of starch agar, a clear zone will develop around the colonies of the starch digesting forms. If the plate is flooded with iodine solution the undigested starch will give the characteristic blue starch-iodine color while the digested zone will be uncolored.

Proteins.—A number of reactions involving proteins or protein-like compounds of various sorts may also be used for identifying bacteria. These reactions include the ability of bacteria to liquefy (digest) gelatin, the power to coagulate milk casein and to digest the coagulated casein, ammonia production from peptones, indole production from peptone media rich in tryptophane and similar reactions.

Nitrates.—The ability of bacteria to reduce nitrates to nitrites or sulfates to hydrogen sulfide may also be used as differential tests.

Enzymes.—Some tests are set up to determine possession by the microorganisms of specific enzymes. An example is the catalase test. Catalase splits hydrogen peroxide to water and oxygen. If a plate culture of bacteria is flooded with hydrogen peroxide solution, a vigorous production of bubbles of oxygen gas will take place if the organism is a catalase producer. Quick tests for the production of other enzymes have also been devised.

(e) Serological Tests

Widal in 1896 discovered that the blood serum of a patient suffering from typhoid fever would clump a suspension of the typhoid fever bacteria even when the serum was greatly diluted. The clumping which occurred was called *agglutination* and the antibody which caused the cells to stick together in clumps was termed an *agglutinin*.

It was found by subsequent investigators that a good many pathogenic bacteria, especially the gram-negative rods, caused the production of agglutinins in the patient.

Other workers reasoned that if the diseases could be diagnosed by using known bacteria, then unknown bacteria could be identified by

using known antisera.   These known antisera are prepared by injecting killed bacteria such as the typhoid organism into a rabbit or other
suitable animal.   After a series of such vaccinations, spaced a few days
apart, the blood is drawn from the animal, allowed to clot and the
serum separated from the clot as a straw-colored liquid.   This serum
thus contains a known antibody, in other words an antibody against
a known organism.   Diagnostic antisera have been prepared for the
identification of a good many species and strains of bacteria.

In identifying bacteria, either the slide or test tube method is
used.   The slide method is relatively simple.   It consists of placing a
drop of physiological salt solution on a microscope slide and suspending
a small amount of the unknown bacteria in it.   A drop of 1:10 solution
of the known antiserum is then added and the two mixed, usually by
rocking the slide.   In a relatively short time clumping will occur if the
unknown organism is the one for which the antiserum is effective.   If
not, no clumping will be noticed.   Thus a typhoid organism will not be
agglutinated by an antiserum which has been prepared against the
paratyphoid A bacterium.

This method has its limitations because some organisms are so
closely related antigenically that they can be separated only by quantitative tests.   These are made by diluting out the serum in a series of
test tubes using physiological salt solution as the suspending agent.
The dilutions are usually made in steps of 2 times, i.e., 1:2, 1:4,
1:8, 1:16, etc., sometimes going as high as 1:1024 or even 1:2048.   The
highest dilution in which agglutination occurs is spoken of as the
"agglutination titer."   Thus, if agglutination occurs in the slide test
with the unknown organism and both antiserum *A* and antiserum *B*
but in the test tube procedure the titer was only 1:20 for *A* and 1:256
for *B* it would justify the conclusion that the unknown organism was
the same as organism *B*.

The diagnostic antisera are widely used in hospital and public
health laboratories as an aid to the rapid identification of unknown
bacteria.   With this test a diagnosis can often be made in 18-24 hours
which might take as many days if made by the use of staining, morphological, cultural, and physiological characteristics alone.

A large number of bacterial strains have been characterized serologically, particularly members of the genus Salmonella, which includes the typhoid and paratyphoid bacteria.   More than 200 "antigenic varieties" have been described in this genus.

Several Salmonella and Shigella "typing centers" have been set up in various countries to aid in the identification of these strains as they are isolated from human or animal diseases. Strains within a given species may also be separated by means of Vi agglutinins, a special type developed by using freshly isolated, virulent forms.

In addition to agglutination, a number of other serological tests have been devised such as precipitin tests (the Kahn test for syphilis and the Lancefield grouping for streptococci are examples); complement fixation tests (the Wassermann test is of this type) and opsonic index. These tests are used primarily for the diagnosis of disease but may be utilized occasionally as an aid in identifying an unknown organism.

"Phage typing" may be used to separate or identify strains of bacteria within a given species. This is based on the observation that not all strains of a given species of bacteria are equally susceptible to attack by the phages (viruses which attack bacteria). Thus by using a number of different strains of phage which will attack some species of bacteria, the various bacterial strains can be identified. This may be useful for tracking to its lair the organism causing a particular epidemic. Phage typing is especially useful for separating strains of the typhoid organism or the pus-former, *Micrococcus pyogenes* var. *aureus*. One investigator, by means of this method, found that a left-handed patient suffering from sinusitis caused by a certain strain of *M. pyogenes* var. *aureus* had this organism on the back of his left hand but not on the back of his right hand. (Men wear buttons on the sleeves of their coats for the same reason.)

With these many procedures available, the ready identification of microorganisms is still far from being an accomplished fact. New and ingenious aids are and need to be developed in order to make the task of labeling microbes an easier and surer one.

Selected Reading References

Buchanan, R. E., and Buchanan, E. D.: Bacteriology, ed. 5, New York, 1951, The Macmillan Company.

Burrows, William, et al.: Jordan-Burrows Textbook of Bacteriology, ed. 15, Philadelphia, 1949, W. B. Saunders Company.

Salle, A. J.: Fundamental Principles of Bacteriology, ed. 3, New York, 1948, McGraw-Hill Book Company, Inc.

CHAPTER 6

MICROBIAL DIGESTION OF FOOD

PHYSIOLOGY AND BIOCHEMISTRY OF MICROORGANISMS

Without realizing the feverish physiological activity which is involved, all of us have observed the metabolic processes of the microorganisms at work as they are busily engaged in digesting and assimilating their food.

The souring of milk, the fermentation of beer, the rising of bread, the production of dill pickles and sauerkraut, the retting of flax, the rotting of wood, mildewing of cotton goods damped for ironing, the stench of a dead carcass, the suppurative weeping of an infected wound all are visible evidences of the biochemical activities of microorganisms.

This chapter will attempt to present a simplified and necessarily incomplete picture of the mechanism of these changes and indicate practical applications of the metabolic processes of microorganisms. Because their physiology has been studied extensively the metabolism of bacteria will be given primary consideration.

METABOLISM OF MICROORGANISMS

Metabolism as here discussed is looked upon as the total of all the chemical changes involved in the growth of microorganisms. It includes digestion, which is a breakdown of complex molecules into simpler ones, and respiration which represents the further molecular changes by which the energy of the molecules is made available to the organism. These two processes are spoken of as dissimilative processes. In addition, metabolism includes assimilation or the series of synthetic processes by which the proteins, carbohydrates, etc., which make up the cell structure are synthesized.

Bacterial metabolism is intense and vigorous. The cells when growing actively display metabolic rates many times greater than the metabolic rates of animal cells. Considering their small size, bacterial cells usually metabolize prodigious amounts of foodstuff, often amounting to many times their own weight, in a short period of time.

To compete in metabolic activity with bacteria we would probably have to consume half a ton or more of food a day. The enormous activity of bacteria is due to their small size which gives them manyfold greater relative surface area and their extremely short generation time which brings about a tremendous increase in cell numbers in a few hours. A single cell can literally have millions of progeny in a single day.

Much of the food necessary for the nutrition of microorganisms is in the form of complex molecules such as the protein, carbohydrate, and fat molecules which are far too large to pass through the semipermeable membrane of the bacterial cell. These substances are useless to bacteria until they are broken down into diffusible compounds.

For example, the protein molecule, while it contains the basic stuff out of which bacteria and all other cells are made, is useless to the bacteria until it has undergone digestion.

Protein molecules are built primarily out of amino acids. Since there are some 25 different amino acids and part or all of these may enter into the protein molecule in varying proportions it can easily be seen that proteins can be almost infinite in their nature and variety. The relative complexity of the protein molecule is apparent when one recognizes that the molecular weight may range from about 10,000 to several million. In comparison, water (HOH) has a molecular weight of 18; carbon dioxide (CO_2), 44; and ammonia (NH_3), 17.

The shape of the protein molecule is not definitely known but folded, platelike, and spherical structures have all been suggested. The important point is that the molecule can be split into components (amino acids) which are small enough to pass through the semipermeable membrane of the bacterial cell.

Since protein makes up about 50 per cent of the dry weight of bacteria, substantial amounts of protein constituents are required for the building of the millions of bacterial cells which can develop in a medium in the matter of a few hours.

However, this does not apply to all bacteria since certain ones can synthesize their own proteins out of such simple compounds as ammonia, sulfates, phosphates, and sugars.

Cellulose, which is the major constituent of woody tissue, cotton, paper, etc., is an example of another extremely complex molecule. It is built out of simple sugar (glucose) molecules. Microorganisms, such as certain bacteria, wood-rotting fungi and some protozoa, possess the enzyme systems necessary to break this resistant cellulose molecule down to glucose. In this form it is immediately assimilated. The rotting of wood or of paper when in contact with the moist earth is a manifestation of microbial digestion of the cellulose molecule.

Other similar carbohydrates such as starch may be readily attacked by a much larger variety of microorganisms. Starch digesting ability of microorganisms may be easily demonstrated by streaking bacteria on a Petri plate containing starch agar. As these bacteria grow and form colonies, a clear zone will develop around the starch digesting colonies. When tested with iodine this clear zone will not give the blue starch reaction indicating that the starch has been broken down into simple sugars.

The digestive process which takes place outside the cell is due to extracellular enzymes (synthesized in the bacterial cell and excreted into the medium). The protein, cellulose or starch digesting processes are hydrolytic reactions. In its simplest form hydrolysis consists of the splitting of these complex molecules by the process of introducing a water molecule at the proper point. This water molecule (HOH) divides the other molecule or group of molecules. A simple example, the splitting of a disaccharide or double sugar is shown below.

$$C_{12} H_{22} O_{11} + HOH \rightarrow C_6 H_{12} O_6 + C_6 H_{12} O_6.$$

In the hydrolytic process the H^+ of the water molecule goes to one simple sugar molecule while the OH^- is attached to the other.

Assimilation

Assimilation consists of the intake of foods through the semipermeable membrane and the building of these food substances into cytoplasm or other components of the cell. The reaction is essentially endothermic in that energy is stored in the process. Energy must, therefore, be supplied for the operation and for all the work connected with it. Energy is needed not only for the synthesis of protoplasm from amino acids and other related components and the building of capsules, cell walls, semipermeable membranes, etc., but also for the synthesis of the enzymes responsible for these actions and for movement and rearrangement of molecules, etc.

This energy is therefore provided by the respiratory process which is an oxidative process resulting in the production of energy. In this connection it is well to point out that energy is chemically bound within the sugar, amino acid, or other molecules which may be oxidized. However, the ultimate source of all energy is the sun and the energy within the sugar molecule represents that energy trapped by photosynthetic action in some plant. For example, the glucose used in a bacterial medium in a laboratory in California or Maine may be providing the bacteria with energy trapped by photosynthesis from the sunlight shining on a cornfield in Iowa.

Respiration

This is a general term applied to oxidative processes going on in the presence or absence of air which result in the release of bound energy. Some of this energy may be used by the bacterial cell but much of it is dissipated as heat. This heat loss may amount to as much at 9/10 of the energy produced.

If respiration takes place aerobically (in the presence of atmospheric oxygen) the end products are likely to be those which have lost a large part of their bound energy such as carbon dioxide, water, ammonia, nitrates, sulfates, etc.

On the other hand, if the process is carried on anaerobically the end products will represent substances which still possess a sizeable part of their bound energy. These will be represented by such compounds as lactic, butyric, proprionic, and other acids; ethyl, propyl, butyl, and other alcohols; aldehydes of various sorts and a variety of other compounds. Aerobic respiration results in a high yield of energy while anaerobic respiration produces only a fraction as much.

As an example, sugar oxidized by yeasts in the presence of atmospheric oxygen yields carbon dioxide and water and 673,000 calories of energy per mole. Under anaerobic conditions the end products will be ethyl alcohol, carbon dioxide, and only 26,000 calories of energy.

Because of the relative inefficiency of the anaerobic respiratory process, microorganisms must work over very substantial amounts of raw material in order to supply the energy needed for life processes. As a result, large amounts of these partly oxidized compounds tend to accumulate. We take advantage of this in such ways as lactic acid formation in the dill pickle fermentation or alcohol and carbon dioxide accumulation in the production of beer. Equipping of the home cider barrel with a gas trap is partly dictated by a desire to maintain anaerobic conditions and obtain a higher yield of alcohol.

THE CHEMISTRY OF MICROBIAL METABOLIC CHANGES

The metabolic changes produced by microorganisms are legion. Many of these are still not understood or perhaps not even recognized. No attempt will, therefore, be made to discuss more than a mere handful of these innumerable processes. It is recognized that this oversimplification may result in some loss of accuracy, but it is hoped that this discussion will afford a practical mental picture of what is taking place.

Most changes produced by microorganisms take place through the intermediation of enzymes. By definition, enzymes are organic catalysts which speed up reactions without being themselves changed in the process except as "friction" occurs in the system resulting in a "wearing out" of the enzymes. A slow but constant replacement of enzymes is, therefore, necessary in order for reactions to continue. These enzymes are extremely reactive since the proportion of enzyme molecules to substrate molecules is of the magnitude of about 1 enzyme molecule to 1,000,000 molecules of the substrate. The rapidity with which these enzymes react is equally amazing.

Enzymes appear to be protein in nature. They have many proteinlike properties being colloid, chemically complex, and unstable. They may be "poisoned" by such substances as hydrocyanic acid (HCN), hydrogen sulfide (H_2S), etc. Many enzymes appear to be made up of both a protein and nonprotein portion. Most of the enzymes are rather highly specific in the substances they attack.

Naming of Enzymes

Certain types of enzymes and enzyme actions have been known for a considerable period of time. These earlier known enzymes have been named without regard to any system. They include ptyalin in saliva which converts starch to sugar, pepsin and trypsin, protein-digesting enzymes found in the gastrointestinal system, and rennin which curdles milk proteins as part of the digestive process.

To obviate the confusion caused by such unrelated or unidentifiable names the terminology of enzymes has been systematized. The suffix -ase (or occasionally -ese) is now used to designate an enzyme. This name may describe the substance attacked as carbohydrases, proteinases, gelatinases, etc., or activities such as oxidases, hydrolases, dehydrogenases, etc.

The enzymes which are extracellular are largely hydrolases and serve primarily to digest the substrate prior to passage of its simplified components through the semipermeable membrane of the bacterial cell.

The intracellular enzymes are primarily those which release energy within the cell, that is, oxidases, dehydrogenases, etc., and those which carry on a synthesis of or building up cytoplasmic proteins, capsular carbohydrates, etc.

The subject of enzymes and enzyme reactions is complex and difficult to study; however, new methods of approach such as the use of "labelled" (radioactive) atoms, and such highly sensitive laboratory procedures as electrophoresis, paper chromatography, ultra centrifugation, etc., are doing much to reveal the mysteries of enzymology.

There are probably many kinds of enzymes and each seems to be a definite compound. They are generally relatively large molecules. Cytochrome C of plants, one of the small enzymes, has a molecular weight of about 13,000; while urease of plants, one of the large ones, has a molecular weight approaching 500,000. Most of the enzymes appear to be about one-half this larger value or less.

As mentioned before, many of the enzymes, particularly the respiratory enzymes, are made up of two parts: (a) the protein (nondialyzable) portion, and (b) a prosthetic group (nonprotein and dialyzable). Many of these prosthetic groups are members of the vitamin B complex.

Several hundred different enzymes are known. How many of these may represent duplicates of previously recognized forms has not yet been fully established. In general, any given enzymatic process appears to be carried on by several different enzymes working as a series rather than being carried on in one step by a single enzyme. The advantage of such a series of orderly steps, each releasing some energy or producing some intermediate product is apparent particularly with the respiratory enzymes. The sudden release of energy by one single reaction might easily prove fatal to the cell whereas a steady flow of energy keeps it functioning smoothly and effectively.

Likewise, the synthetic process, representing as it does the building of simple molecules into a complex molecule analogous to building a brick wall, must proceed in an orderly fashion. Even so, enzymatic reaction occurs at an amazingly rapid rate.

Hydrolases

As previously mentioned, the hydrolases, or digestive enzymes, are largely extracellular. Their nature depends on the nature of the compound which is attacked. The mode of operation is essentially one of introducing a water molecule into a vulnerable spot or spots in a complex molecule and thus reducing it into two or more simpler molecules. The splitting of a complex molecule often proceeds stepwise until the ultimate building blocks are reached.

Thus the protein molecule may first be split into proteose molecules, then these are progressively split into peptones, peptides, and finally amino acids.

Starch may first be split into dextrin, then by degrees into maltose and finally glucose. The following simplified reaction illustrates this change:

$$\underset{\text{(starch)}}{(C_6H_{10}O_5)_n} + n(H_2O) \text{ yields } \underset{\text{(maltose)}}{n(C_{12}H_{22}O_{11})} \text{ which } + n(H_2O) \text{ yields } \underset{\text{(glucose)}}{n(C_6H_{12}O_6)}$$

Protein, the stuff out of which living things are made, is a huge molecule. Only a relatively small number of bacteria have the enzyme systems necessary to make a direct attack on the protein molecule itself. As a result, in bacterial media we use peptones which many more bacteria are equipped to handle successfully. Whether the molecule is folded, in the form of stacked plates, or spherical is not too important. The important point is that it is a mosaic made up of amino acids. These have an amino (NH_2) group and a carboxyl (COOH) group plus a radical. This radical may be simply one or more CH_3 groups or it may be a complex ring compound. There are some 25 or more amino acids.

The following is the general molecular structure for the amino acids:

$$(R)-\overset{\displaystyle NH_2}{\underset{\displaystyle H}{\overset{\|}{\underset{|}{C}}}}-\overset{\displaystyle O}{\underset{\displaystyle O-H}{\overset{\|}{\underset{|}{C}}}}$$

The components present in the complete amino acid may include not only carbon, hydrogen, oxygen, and nitrogen, but also sulfur.

The protein molecule may well be looked on as a large wall made of assorted sizes and colors of bricks. When pushed over on the ground as the first step in wrecking it (digestion) it breaks along lines of weakness into several smaller chunks, these in turn are broken into progressively smaller clusters of bricks and finally into the single bricks (amino acids). When these "bricks" pass through the semipermeable membrane of the cell they may be built up once more into another "brick wall," the protein molecule.

Fat-Splitting Reactions

Fats (and many of the esters) are fairly large and complex molecules. The fat moleucle is essentially a glycerol (glycerine) molecule to which fatty acids are attached.

The following molecular picture illustrates a typical fat structure:

glycerol	*fat*	
H	H	
H—C—OH	H—C—O—fatty acid	Note that the fatty acid has replaced the H on the glycerol and the O H on the fatty acid
H—C—OH	H—C—O—fatty acid	
H—C—OH	H—C—O—fatty acid	
H	H	

These fatty acids are long-chained molecules, some of them with as many as 15 to 17 or more carbons in the chain.

$$
\begin{array}{c}
\text{HO} \\
\diagdown \\
\text{C} \\
\diagup\diagup \\
\text{O}
\end{array}
\begin{array}{c}
\text{H H H H H H H H H H H H H H H H H H} \\
| \ | \ | \ | \ | \ | \ | \ | \ | \ | \ | \ | \ | \ | \ | \ | \ | \ | \\
\text{C—C—C—C—C—C—C—C—C—C—C—C—C—C—C—C—C—C—H} \\
| \ | \ | \ | \ | \ | \ | \ | \ | \ | \ | \ | \ | \ | \ | \ | \ | \ | \\
\text{H H H H H H H H H H H H H H H H H H}
\end{array}
$$

Splitting of the fat molecule by hydrolysis consists of the introduction of a water molecule between the glycerol and fatty acid molecules.

In addition to hydrolysis, molecules may be broken down through the intermediation of the phosphoric acid molecule. This process is known as phosphorylation.

Action of Synthetic Enzymes

After digested constituents, such as amino acids and sugars, have been absorbed through the cell membrane, as mentioned before, these constituents may be built up again into proteins, carbohydrates, capsular material, enzymes, etc. This synthetic process, like other enzymatic reaction, proceeds through a series of steps but the final result is a restoration of the bonds which originally held together the proteins, carbohydrates, etc., before digestion took place. The operation is the opposite of hydrolysis since the various molecules are bonded together by the removal of water molecules at the point of union. The enzymes responsible for this reaction have been termed anhydrases.

Synthesis is an endothermic reaction, i.e., energy is stored in the synthesized molecule. These synthetic operations which are the processes by which bacterial growth takes place require energy. The process of cell proliferation or growth is an amazingly rapid one. Under ideal conditions a single bacterial cell will have grown and divided into two cells in a matter of 20 to 30 minutes.

The energy necessary for growth is supplied through the action of the respiratory enzymes. Respiration by definition represents those oxidative processes which yield energy. While such oxidation might consist of the direct addition of oxygen to a molecule it is more commonly accomplished by the removal of hydrogen. The molecule or portion of the molecule which gives up the hydrogen is termed the hydrogen donor while the molecule or portion of the molecule which takes up the hydrogen is termed the hydrogen acceptor.

Hydrogen acceptors may be

(a)   Molecular oxygen
(b)   Easily reduced organic compounds
(c)   Another portion of the same molecule.

For example, in the laboratory, methylene blue may be used as a hydrogen acceptor. When reduced (by the taking up of hydrogen) methylene blue becomes colorless.

This process of hydrogen transfer is not accomplished by one single enzyme but proceeds in a stepwise fashion involving the action of a considerable number of enzymes. These enzymes release a small portion of energy from each atom of hydrogen which they carry along. The result is that a steady flow of energy is produced rather than an explosive release of energy which might be fatal to the organism.

The respiratory enzymes are given the general name oxidases. In addition, as a further subdivision, the name oxidase has been applied specifically to those respiratory enzymes which bring about a transfer of oxygen and dehydrogenase, those involved in hydrogen transport.

Among the oxidases is *catalase* which changes hydrogen peroxide to water and oxygen.

$$2 \; H_2O_2 \rightarrow 2 \; H_2O + O_2$$

This enzyme protects bacteria against the toxic effects of hydrogen peroxide which accumulates during the growth of aerobes.

Catalase is an amazingly active enzyme. A single molecule of the enzyme has been found capable of breaking down more than two million five hundred thousand hydrogen peroxide molecules in one minute. The bubbles of gas which arise when an open wound is treated with hydrogen peroxide are bubbles of oxygen. Their rapid release is due to the catalase present in the tissues of the open cut. Peroxidase is another oxidase. It causes the transfer of oxygen from hydrogen peroxide directly to some reduced organic compound. The result is the oxidation of the organic compounds and the release of a molecule of water as the residue from the hydrogen peroxide molecule.

In general, the oxidases are complex compounds containing iron or less frequently copper in their molecular structure. Hematin of the blood is one of these compounds.

Mutases

Among the dehydrogenases is an interesting group of enzymes known as the mutases. These enzymes oxidize a molecule by transferring hydrogen from one part of the molecule to another.

Carboxylases

This group of enzymes yield energy by removing carbon dioxide from the carboxyl group of an organic acid. They are important also in the breakdown of sugars such as glucose.

The above-described enzymes are only a few of the many respiratory enzymes. However, they are representative of the various respiratory enzyme groups. Most of these respiratory enzymes are made up of a protein (apoenzyme) plus a coenzyme. The coenzymes are non-protein dialyzable molecules or prosthetic groups. As previously mentioned, they are primarily members of the vitamin B complex.

Descriptions of the coenzymes read like the label on a bottle of multiple vitamins. Included in the group are nicotinic acid, biotin, thiamin, riboflavin, pyridoxin, folic acid, pantothenic acid, vitamin B_{12}, etc. These vitamins are also found in our own bodies where they likewise yield energy. Hence, if we are deficient in the B complex vitamins, we also suffer a lack of the energy-producing vitamin. The result is the tired feeling so aptly described by the television salesmen. When some single enzyme-vitamin complex is lacking, energy reactions can proceed only to a certain point and then stop. This brings about the physical sensations of vitamin deficiency and also may result in the production of lesions in the tissues of the body. With microorganisms it simply results in a cessation of growth or fermentative reactions.

ANTIBIOTIC EFFECTS

Recent experiments have indicated that the effect of antibiotics is probably one of blocking the action of one of the respiratory enzymes. In some instances, the antibiotic agent may merely interfere with the vitamin-enzyme complex or in other instances may enter directly into the enzyme complex in the place normally occupied by the coenzyme.

PICTURE OF VITAMIN ACTIVITY

To recapitulate, if we could look into a bacterial culture and watch the reactions taking place immediately adjacent to the bacterial cell and inside the cell proper, we would observe a veritable beehive of industry. Immediately outside of the cell, the large nondiffusible protein and carbohydrate molecules would be breaking down under the impact of the enzyme hammer driving in water wedges. As their molecular components became water soluble, they would be diffused through the semipermeable membrane of the cell. Immediately these molecules would be drawn into a maze of enzyme machinery. A glucose molecule, for example, might be pulled into the respiratory enzyme machine and forced to give up its energy. If the machine was operating under aerobic conditions the end products in addition to energy would be carbon dioxide and water. Under anaerobic conditions, the end product might be lactic acid, butyric acid, ethyl alcohol, propyl alcohol or some other intermediate compound. Another glucose molecule might enter the carbohydrate synthetizing enzyme system and be made into capsular material or it might be drawn into the protein synthetizers and be built into the cytoplasm of a growing cell. Amino acids might likewise either be robbed of their amino group and converted into energy or taken bodily by the protein synthetizers and become cell protoplasm.

Each one of these reactions would represent a whole series of changes moving along at an amazingly rapid rate. This concept becomes particularly remarkable when one realizes that a single microbial cell in a matter of a few hours might develop millions of progeny.

During the course of this rapid enzymatic action, waste products tend to accumulate rapidly. The result is that the number of cells increases up to a certain point and further growth then ceases. A slow or rapid death of the microorganisms may follow. This is a fortuitous circumstance since bacterial growth if it were allowed to go on unchecked would soon overrun the entire earth. It has been estimated that a single bacterial cell dividing at 30-minute intervals would in five days produce an amount of growth greater than the entire volume of the earth, if proper growth conditions could be maintained.

GROWTH PATTERN OF BACTERIA

When foodstuff is available in adequate quantities, adequate moisture, temperature, pH, and the necessary co-factors, enzymes and vitamins are properly adjusted, bacterial populations grow in a more or less constant pattern. These optimal conditions will produce the greater number of bacteria per unit time. A variation of any one of the above items will change the bacterial population. Their growth may be so rapid that reproduction or *generation time* (G.T.) may be only 18 to 21 minutes. Thus there is a doubling of the population of microorganisms in as many minutes as it takes days for mice or fertile eggs to reproduce their kind.

When bacteria are seeded to another media, it takes from two to four hours for them to become adjusted to their new surroundings. There is therefore a *"lag"* period before they begin to reproduce to any great extent. As they begin to reproduce they gather momentum and there is then an accelerated growth pattern. This acceleration of growth may require only an hour or two, then they begin to multiply in earnest, reproducing to several million or even billions in only a few hours. This rapid growth is termed the *logarithmic period* because if the logarithm of the numbers of bacteria is plotted against time of growth, a straight line results. Growth then begins to slow down. During the period of maximum growth, comparatively few microorganisms die. During the slowdown period of growth, many of the organisms begin to die and the generation time increases. Growth then levels off as a plateau, with reproduction and death of the microorganisms being about equal. This period may last several hours, then there is a general decline in the total number of living bacteria, more dying than reproducing. A growth curve is presented in Fig. 9.

The increasing death rate and decreasing reproduction rate are due to many factors. Age of the microorganisms, depletion of readily available food, decreasing oxygen supply, crowding, increased amount of waste products, some of which are toxic to the bacteria, change in pH and a variety of other conditions.

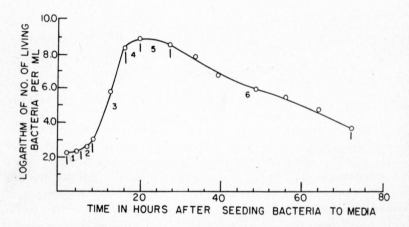

Fig. 9.—Bacterial growth curve of *Escherichia coli* in synthetic media at 21° C. *1*, Lag period; *2*, acceleration growth period; *3*, logarithmic growth period; *4*, decreasing growth period; *5*, stationary period; *6*, logarithmic death period.

MANIFESTATIONS IN NATURE OF MICROBIAL ENZYME ACTION

Over the years, man has learned to apply in a practical way many of the physiological activities of microorganisms. These uses include fermentative changes involved in the production of cheeses, butter, and various fermented foods such as sauerkraut and dill pickles and the various industrial fermentations. These will be discussed in some detail in subsequent chapters.

On the other hand, microbial enzymes are responsible for many other interesting and sometimes spectacular activities in the world about us. A few of these will be briefly discussed as examples.

LIGHT PRODUCTION BY MICROORGANISMS

In walking through the woods on a dark night one may sometimes encounter a dead stump which glows with an eerie light, or a stroll along the beach may be punctuated by the sight of a dead fish which shines weirdly in the dark. The phosphorescent wake which follows ships as they ply through tropical waters is another manifestation of the same type of microbial activity. The stump glows because luminescent fungi are growing actively on the constituents of the wood, and bacteria, likewise growing vigorously, impart the phosphorescence to the dead fish or the crest of the wave rolling out in the wake of the ship.

Microorganisms, of course, have no monopoly on this phenomenon since they share it with the glow worm, the firefly, and a number of other representatives of the animal kingdom. However, among the plants, fungi and bacteria are probably the only forms capable of producing bioluminescence.

Certain tropical fish termed "lantern fish" appear to have taken advantage of the light-producing power of the luminescent bacteria. Some of them have special structures, near the eye, on which a glowing mass of luminescent bacteria is found. Some of these fish have developed eyelidlike structures which permit the "headlight" to be turned on and off at will. Since many of these forms live at rather great depths in the ocean, the light may be either used merely for illumination or as a means of attracting prey.

It appears that some of the luminous bacteria may also be pathogens since some of them have been found to be capable of producing infections, sometimes fatal, in sand fleas, shrimps, caterpillars, etc. Most of the luminescent bacteria are gram-negative nonsporing rods and they are found most commonly in sea water.

Why light is produced is not known but it is thought to be merely an incidental product of a respiratory enzyme system in which a pigment gives off light when oxidation is taking place. That oxygen is necessary is witnessed by the brightness of the freshly churned water in the wake of the ship and the relative rapidity of its subsitsence after the ship has passed.

It is postulated that a pigment, luciferin, is oxidized by the enzyme, luciferinase, to oxyluciferin, water and light. The amount of light produced by an actively growing culture is often considerable. Photographs have been taken with the aid of light produced by a culture and a well-aerated culture may produce several thousand times as much light as a luminous watch dial.

HEAT PRODUCTION BY MICROORGANISMS

Heat is produced whenever microorganisms carry on respiratory activities or other energy-yielding processes. The growth of molds, yeasts, and bacteria is always an exothermic reaction, that is, a reaction in which heat is produced. Part of this energy may be used directly to carry on certain vital activities or be stored endothermically in the synthesized cytoplasm, capsular material, etc. However, most of it is dissipated as heat into the area adjacent to the growing microorganism. Therefore, if growth is taking place in a small container, such as a test tube or a flask, the heat may be lost into the atmosphere almost as fast as it is formed. On the other hand, if microorganisms are growing in organic matter which serves as an insulator or holds the heat, such as the conditions found in a large vat of fermenting beer or a pile of hay or manure, enough heat may accumulate to cause a marked increase in temperature. This may produce a "self-pasteurization" of the mesophilic microorganisms. As a result, the fermentation may be stopped. In large commercial fermentation vats holding thousands of gallons of the medium, it is often necessary to cool these vats artificially to keep the growing organisms from "burning" themselves to death.

When penicillin was first being produced commercially, the process consisted of setting up the cultures in large rectangular flasks. Hundreds or even thousands of such bottles were often used. It was found that in the center of a large bank of such flasks, penicillin production practically ceased due to the "self-pasteurizing" effect of the mold-produced heat.

On the other hand, in a heating manure pile or a stack of hay where the temperature may often rise to 70-80°C. due to the combined effect of microbial and plant enzymes, heat-loving (thermophilic) bacteria may continue to grow.

Microorganisms which produce such high temperatures are sometimes termed thermogenic forms. This term is probably a relative one because all microorganisms are to a degree "thermogenic" when they are growing vigorously. The only difference is that these forms while producing large amounts of heat can tolerate their own accumulated heat. In fact, certain methane-producing, anaerobic spore-forming bacteria which have been used experimentally to produce methane from materials such as corn stalks, do not function actively unless the temperature is artifically raised at the beginning of the growing period.

In general, these thermophilic forms have certain rather unusual characteristics. They reproduce very rapidly, with generation times as short as 5 to 17 minutes having been reported. In addition, the amount of any fermentation product, such as lactic acid, which may be formed may be several times that produced by mesophiles in the same period of time. The result is that the thermophils often lead a "short life but a merry one."

While the experimental evidence is far from complete, "the few enzymes which have been studied in growing cultures and resting cell preparations apparently have slightly higher optimum and maximum temperatures than the mesophilic forms."[1]

This may explain why the thermophiles can tolerate their own heat while the mesophiles succumb.

RELATION OF MICROORGANISMS TO PETROLEUM

In recent years evidence is accumulating that bacteria through their enzyme systems may have played an important role in the formation of petroleum and even today may be changing the properties of the petroleum which has accumulated under the surface of the earth.

How the long-chain hydrocarbon molecules (made up of carbon and hydrogen) which make up the main compounds of petroleum may have been formed is not known. However, it appears that two factors may have played a part: (a) The hydrogen formed by bacteria may have completely hydrogenated certain fatty acids and/or, (b) the hydrogen sulfide produced by the sulfate-reducing bacteria may have had a part in reducing fatty acids and similar oxygen-bearing compounds.

In addition to producing petroleum, bacteria may also have the power to break it down, since specific bacteria have been found which can utilize as food such compounds as natural gas, gasoline, kerosene, crude oils, mineral oils, lubricating oils, etc.

The character of the mixture of which crude oil is comprised may be changed by bacteria which "preferentially attack" some hydrocarbons while leaving others.[2]

---

[1]Gaughren, E. R. L.: Bact. Rev. **2**(3) 212, 1947.

[2]ZoBell, Claude E.: Functions of Bacteria in the Formation and Accumulation of Petroleum, Oil Weekly, Feb. 18, 1946.

OTHER MICROBIAL ACTIONS

In marshes, swamps, on lake shores, etc., where masses of organic matter are undergoing decomposition under anaerobic conditions, bacteria may be actively producing "marsh gas" which is mostly methane (CH_4), with some ethane (CH_3CH_3) and sometimes a small amount of propane, etc. Sometimes on a dark night a small puff of gas may be "burped" out of the swamp and as the compounds in the ball of gas are oxidized by the atmospheric oxygen a weird glow may be visible to the dark-adapted eyes of a spectator. The eerie light given off by this ball has gained it the name of "will o' the wisp."

During the early days of commercial penicillin production, it was discovered that some flasks of mold which were accidentally contaminated with certain bacteria from the air would have lost most of their penicillin even though the uncontaminated flasks showed a high yield of antibiotic. It was found that certain bacterial contaminants possessed an enzyme which broke down the penicillin molecule and as a result was given the name, penicillinase. Later the discovery was made that many strains of *Escherichia coli*, the common intestinal form, were producers of penicillinase. This offered a partial explanation of why the administration of penicillin by mouth is not particularly effective.

The varied and exotic tastes of bacteria assumed a new significance when it was found that World War II airfields in which runways had been surfaced by mixing plastic with the soil, suddenly gave way and planes were nosed over as their wheels broke through. It was discovered that certain soil bacteria had developed a strange appetite for this new and foreign food.

The examples which have been given represent only a very small part of the ways in which the microbial enzyme systems operate in the world about us and exert a constant and ever-present influence on our daily lives.

Selected Reading References

Foster, J. W.: Chemical Activities of Fungi, Academic Press, New York, 1949.
Porter, J. R.: Bacterial Chemistry and Physiology, John Wiley and Sons, New York, 1946.
Prescott, S. C., and Dunn, C. G.: Industrial Microbiology, ed. 2, New York, 1949, McGraw-Hill Book Company, Inc.

CHAPTER 7

FAMILY TREES OF MICROORGANISMS

Microbial Classification

The microorganisms represent a vast and motley crowd ranging from relatively complex mold bodies to the simplest single bacterial cell. Thousands of these forms have been described and have been assigned scientific names, usually descriptive, which separate them from their fellows. Manifestly any attempt to assimilate this vast array of knowledge at one time would lead only to profound mental indigestion. However, properly selected bits of this information can form a digestible, mental fare which will leave the partaker nourished and satisfied. The result will be a bit of the extra intellectual polish which comes from knowing the right microbes well enough to call them by their first names or at least their surnames. This is of some importance when they insist on causing your pimples, sore throats, pneumonias, and kindred discomforts and add insult to injury by spoiling your foods and thus robbing your pocket book.

About two centuries ago Linnaeus recognized the impossibility of gaining a mental picture of the vast array of living forms then known and developed a scheme for giving them names and placing them in neat packages which in turn he filed into neatly labelled pigeonholes. He made his names and groupings permanent by using Greek and Latin forms since these, being dead languages, were fixed and not subject to change. To a biologist in Sweden or England a plant or animal name in Greek or Latin would, therefore, convey the same meaning. The scheme of biological classification developed by Linnaeus can be a very useful tool. However, some teachers who insist on classification for classification's sake have largely lost sight of the usefulness of this valuable tool.

Linnaeus knew virtually nothing about the microorganisms. In fact, he revealed this general lack of knowledge by assigning microorganisms to the order, *Chaos*. Microbiologists soon applied the princi-

81

ples laid down by Linnaeus as the world of microorganisms began to be discovered.

Classification is the means by which our finite mind assembles like forms together in small groups so as to allow us to understand, recognize and assimilate mentally the infinite variations in form and function which characterize nature.   Alexander, the colloidal chemist, suggests that while nature is continuous, "it is we who are simple, not nature."   In our simplicity, we need to set up the pigeonholes into which we place like bits of information.   This practice extends into all our learning.   Any young male college student classifies all young female college students into categories for which he often has terms as unintelligible to the oldster as Linnaeus' Latin is to the youngster.

This method of learning is most economical, since once we have established such pigeonholes of likeness it is often unnecessary to learn the characteristics of each individual; instead, a knowledge of the group characteristics suffices.

Under Linnaeus' system of *Biological Nomenclature* all living things are divided first into the plant and animal kingdoms and the kingdoms are then divided into phyla (plural for phylum Latinized from the Greek phylon meaning a race).   Four phyla have been assigned to the plant kingdom.   In one of these, the molds, yeasts and bacteria are placed.   The plant phyla are described below:

Phylum I   Spermatophytes
              Seed plants.   The major plant group.
Phylum II   Pteridophytes—Ferns.
Phylum III   Bryophytes
              Liverworts and mosses are included in this phylum.
Phylum IV   Thallophytes
              The thallus plants, i.e., plants without distinct leaves, stems, or
              roots belong to this group which includes both the algae and fungi.
              Some of the algae forms are extremely large, notably certain kelp
              plants which are often hundreds of feet in length.   On the other hand
              the tiniest of the Thallophytes are the single-celled, simple bacteria
              which are often less than one micron (1/25,000 of an inch) in length.

In Table III the relationship of the microorganisms to other plants is shown.   It will be noted that the bacteria, yeasts, and molds are indicated as the chief members of the group.   The mushrooms, puff-balls, and other similar fungi are omitted because it is not felt that they fit into the group of microorganisms which we propose to study. However, the blue-green algae and certain other one-celled algae may be properly considered with these microorganisms.

However, because of their minor economic importance, the algae will not be given more than brief mention. The classification of the yeasts and molds will be considered in a later chapter.

TABLE III

RELATION OF BACTERIA, YEASTS, AND MOLDS TO OTHER PLANTS

KINGDOM	PHYLUM	SUBPHYLUM	CLASS
Plant	Spermatophyta (seed plants)		
	Pteridophyta (ferns)		
	Bryophyta (liverworts)		
		SCHIZOPHYTA	SCHIZOMYCETES (splitting fungi, the BACTERIA) blue green algae
	THALLOPHYTA (Phylum of the thallus plants)	Fungi	YEASTS, MOLDS, etc.
		Algae	Green, Brown, Red, etc.

The following descriptions are confined to the Schizomycetes (splitting fungi) or as they are more commonly designated, the bacteria. The term "splitting fungi" is applied because these one-celled forms reproduce by transverse fission or splitting of the cell. They are classed as fungi because of their lack of chlorophyll.

The classification which follows is not a complete one. Only the bacteria specifically discussed in the text are included to avoid burdening the student with the rare or relatively unimportant forms, which are of interest only to the specialist. The scientific names used will be made functional by giving the meaning of each. Suggestions will be given as to the distribution or nature of the activities of each of the groups. In a number of instances the species names may be omitted.

The terminology and classification used in bacteriology has developed over nearly two centuries. Muller in 1773-1786 made a simple classification of bacteria and introduced the term *Vibrio* which we still use. Ehrenberg in 1828, gave the name *Bacterium* to a rod-shaped form and in 1838 added *Spirillum* and *Spirochaete*, terms which also are still in use. *Bacillus* was first used by Cohn in 1872-1875 for the spore-forming rods. During this developmental period a number of classifications were issued and many new terms added. Many of these new names have since been found invalid and have been discarded. However, some of them still rear their heads to confuse the student.

In 1890-1894 Migula published classifications which served as bases for later classifications. The Society of American Bacteriologists, recognizing the variability which existed in classifications and terminology, appointed a committee in 1917 under the chairmanship of David H. Bergey to prepare a descriptive classification of the bacteria. The first *Bergey's Manual of Determinative Bacteriology* was published in 1923. The manual has been subject to constant revision and enlargement. Several new editions have appeared, the last of these, the Sixth Edition, appeared in 1948. It will be used as the basis for the descriptive matter which follows.

It is recognized that no classification can be complete, final, or unequivocally accurate with such a varied and probably changing group of forms as the bacteria. However, this does not lessen essentially the usefulness of a classification in affording a picture of the common bacteria and their relationships.

With the development of our knowledge of bacterial genetics a further complicating factor is added. However, our major concern is still with the "wild types" of bacteria which occur in our environment and still fit fairly well into the artificial scheme of the descriptions used rather than with the one-in-a-million mutant forms (the albino blackbirds of bacteriology), which are being so vigorously studied in the test tube. No doubt our future descriptions and classifications will be profoundly influenced by this newer knowledge. As a further complication, strains of bacteria which show resistance to antibiotics are often genetically different from the original antibiotic-sensitive "wild strain."

Beginning with the phylum, the microorganisms are subdivided into smaller groups which have like characteristics (subphyla); these in turn are divided into classes; these in turn into orders and stepwise down to the species. This series of subdivisions is shown as follows:

Phylum
 Subphylum
  Class
   Order
    Family
     Tribe
      Genus
       Species

Bacteria belong then to the kingdom *Plants*; phylum *Thallophytes*, subphylum *Schizophyta* (splitting plants) and class *Schizomycetes*.

It may be noted here that order names end in *-ales*, family names in *-aceae*, and tribe names in *-eae*. The scientific name, so-called, is made by combining the genus and species name. For example the scientific name for mankind is *Homo sapiens* (the thinking man) although man's habits of making war on his fellows belies this optimistic estimate. *Homo* is the genus name; *sapiens*, the species name.

A right and a wrong way exists for writing the scientific names of bacteria. Properly written, the genus name is capitalized, the species name not capitalized. In print the scientific name is set off in italics, in typewritten or longhand material it should be underlined.

The class, Schizomycetes is divided into five orders. These orders are described in Table IV.

The True Bacteria

Of the five orders, the Eubacteriales contain the lion's share of the important bacteria. These bacteria are all simple, single-cell forms. They may be spheres, rods, or spirals or characteristic groupings of these.

A graphic picture of the characteristic shapes of the true bacteria is shown in Table V.

Shape or morphology alone is not considered a sufficient basis for separating bacteria into families since bacteria which are identical in shape might be completely different in physiological characteristics or pathogenicity. The present classification divides the true bacteria into thirteen different families. These, together with the most important genera and some of the important species are described in Table VI.

TABLE IV
BRIEF DESCRIPTION OF THE ORDERS OF THE SCHIZOMYCETES

CLASS	ORDER

THE TRUE BACTERIA

Eubacteriales
(true-rod-order)

Schizomycetes
(splitting-fungi)
Simple, single-celled
forms without
chlorophyll, repro-
duce by splitting.

The major group of bacteria. Contains a wide variety of forms including (a) such disease-producers as those responsible for pimples, boils, septic sore throat, gonorrhea, meningitis, typhoid, bacillary dysentery, plague, tularemia, brucellosis; (b) beneficial forms which carry out processes such as the production of cheeses, sauerkraut, dill pickles; ripening of tobacco, retting of flax, and a multitude of other helpful activities; (c) soil forms which fix nitrogen, produce nitrates, decompose organic matter. These activities are essential to continued life on the earth. (d) certain plant pathogens; (e) spoilage bacteria such as those causing the rotting of meats, carrots, celery, and other vegetables.

HIGHER BACTERIA

Actinomycetales
(ray-fungi-order)

These bacteria have some characteristics of both true bacteria and molds. Vary from simple long rods to fine branching filaments. Found in great numbers in the soil. Also found in our mouths associated with mucin plaques. In soil, actinomycetes are largely beneficial, however, potato scab and other plant diseases may be caused by actinomycetes. Human and animal diseases including leprosy, tuberculosis, and lumpy jaw in cattle are caused by members of this order. Such antibiotics as streptomycin, aureomycin, and terramycin are produced by actinomycetes.

Chlamydobacteriales
(sheathed-rod-order)

Essentially water forms which are filamentous, colorless and apparently intermediate between the true bacteria and the filamentous blue-green algae. No diseases or fermentations are produced by this group. They may have played an important role in the early geological developmental period of the earth, particularly in relation to iron and sulfur. Now primarily of academic interest.

Myxobacteriales
(slime-rod-order)

The slime bacteria apparently have characteristics of both the true bacteria and the slime molds. Not common forms but seen occasionally as slimy masses on rotting wood. Decomposers of organic matter, especially woody tissue. No disease-producers or fermenters.

Spirochaetales
(coiled-hair-order)

These organisms are long, thin, coiled, flexuous spirals. They resemble certain protozoan forms. May be considered as intermediate between the protozoa and the true bacteria. Unlike most of the other bacteria these forms can rarely be grown on artificial media. Many water forms are in this group. Nonpathogenic, parasitic spirochetes are found in our intestines, mouths, external genitals. The disease producing spirochetes include those which cause syphilis, yaws, trench mouth, and recurrent fever.

TABLE V

BASIC SHAPES AND GROUPINGS OF THE TRUE BACTERIA (EUBACTERIALES)

MAJOR SHAPE	GROUPINGS	NAMES OF GROUPINGS
Coccus (from Gr. kokkus, a seed)		Micrococcus (small seed) (staphylococcus)* (cluster seeds)
		Streptococcus (curved or flexuous seeds)
		Diplococcus (double seed)
		Sarcina (packet)
		Gaffkya (after Gaffkya)
		Neisseria (after Neisser)
Bacterium (Gr. bakterion, a small staff)		Usual isolated forms— diplobacillus-streptobacillus†
Bacillus (N.L.—small staff or rod)		Usual isolated forms— diplobacillus-streptobacillus
Spirillum (N.L.—small coil)		Spirillum (a coil)
		Vibrio (to vibrate) (curved rod, part of a spiral)

*Formerly Micrococcus was the genus name of the saprophytic clustered cocci, while staphylococcus was genus name applied to pathogenic forms of the same grouping. Now both are placed under the genus, Micrococcus. Apparently staphylococcus has assumed the status of a common name.

†Same descriptive terms apply to both regardless of presence or absence of spore formation, i.e., diplobacillus, streptobacillus (essentially common names).

Table VI

Summarized Description of the Families of the True Bacteria
(Eubacteriales)

FAMILY	DESCRIPTION	SHAPES	GENERA AND IMPORTANCE
1. Nitrobacteriaceae (soda*) (rod) (family) 9 genera, 22 species	Autotrophic (self-nourishing) bacteria, found in soil and water; use CO_2 as a source of carbon in place of sugars and other carbohydrates; various genera oxidize ammonia, hydrogen, or sulfur, respectively, to secure energy	Rods and cocci (nonsporing)	*Nitrosomonas, Nitrosococcus, Nitrobacter.* Produce from ammonia, the nitrates so necessary for the growth of plants. Earth would probably be barren of higher plants without these bacteria. *Thiobacillus, Hydrogenomonas.* Oxidize sulfur to sulfuric acid, hydrogen to water
2. Pseudomonadaceae (false) (unit) (family) 12 genera, 257 species	Mostly soil or water bacteria, some plant pathogens; includes organisms producing green pus infections in man and animals, cholera and ratbite fever; "mother of vinegar" bacteria which oxidize alcohol in wine or cider to produce acetic acid fall in this group	Motile straight or curved rods. Some rigid spirals (nonsporing)	*Pseudomonas aeruginosa* (green pus) *Spirillum minus* (ratbite fever) *Vibrio comma* (cholera) *Acetobacter* (vinegar producer)
3. Azotobacteriaceae (nitrogen) (rod) (family) 2 genera, 4 species	Widely distributed in soil; have ability to take nitrogen gas from air and transform it into their own protoplasm; one of the most important of the "nitrogen fixers"; lives independent of higher plants, as a result are classed as nonsymbiotic nitrogen fixers	Large rods or cocci; often resemble yeast cells but do not bud; probably nonsporing	*Azotobacter* assists in keeping earth life from starving by replenishing supply of nitrogen compounds lost from "nitrogen cycle"

4. *Rhizobiaceae* (root) (living) (family) 3 genera, 13 species	Major members of the family are the bacteria which form nodules on the roots of legumes such as peas, beans, alfalfa, clover; live symbiotically (together for mutual benefit) with plants; (symbiotic nitrogen fixers) utilize nitrogen gas from air and transform it into protein in the plant, greatly increasing their protein content	Small gram-negative rods (nonsporing)	*Rhizobium* (root dweller) the major genus, profoundly influences the nutrition of mankind by giving the legumes preeminence as a major source of protein for man and animal
5. *Micrococcaceae* (small) (seed) (family) 3 genera, 33 species	Spherical bacteria found in soil, water, food, dairy products, on the skin, etc.; the few parasites include the pimple or boil organisms; mostly harmless or spoilage forms; the pathogens were formerly placed in the genus *Staphylococcus*	Grapelike clusters, packets, tetrads (multiples of four); essentially (nonsporing)	*Micrococcus* (mostly harmless but including the pus formers commonly responsible for pimples, boils, osteomyelitis, etc.) *Sarcina* (harmless) *Gaffkya* (parasites of mucous membranes); may be moderately pathogenic
6. *Neisseriaceae* (Family named after Neisser, the bacteriologist who discovered the gonorrhea organism)	All of the bacteria in the family are parasitic on the human mucous membranes; most of them are harmless, a few pathogenic	Primarily gram-negative diplococci resembling double coffee beans (nonsporing)	Most important forms are: *Neisseria gonorrhoeae, N. meningitidis,* cerebrospinal meningitis organism; and *N. catarrhalis,* common inhabitant of nose or throat which may be associated with mild respiratory infections

*From nature soda (sodium nitrate).

TABLE VI—CONTINUED

FAMILY	DESCRIPTION	SHAPES	GENERA AND IMPORTANCE
7. *Lactobacteriaceae* (milk) (rod) (family) 7 genera, 62 species	A varied group of forms placed together because they are relatively active producers of lactic acid (sour milk acid) from carbohydrates; also have other characteristics in common; some pathogens, many harmless and some beneficial forms	Diplococci, streptococci, rods, (all nonsporing)	*Diplococcus* (one genus causes lobar pneumonia) *Streptococcus*, (a) pathogens causing such infections as "Strep throat," scarlet fever, (b) beneficial streptococci found in buttermilk, butter cultures, sauerkraut, etc. *Lactobacillus*, (a) species may play a part in tooth decay; (b) another species is the yogurt organism; (c) some species are active agents in dill pickle production
8. *Corynebacteriaceae* (clubbed) (rod) (family) 3 genera, 29 species	Irregularly shaped, generally gram-positive rods, found as parasites on animals and plants; some produce disease; may be found in soil, water and dairy products	Rods, nonmotile, vary from small rods to rather large forms resembling baseball bats; banded or beaded, (nonsporing)	Most important member is *Corynebacterium diphtheriae*, the cause of diphtheria
9. *Achromobacteriaceae* (without color) (rod) (family) 3 genera, 45 species	Mostly soil and water forms, some are nonpathogenic inhabitants of the intestines, some are plant pathogens	Small to medium rods, nonsporing	No important producers of human diseases or fermented products

10.	*Enterobacteriaceae* (intestinal) (rod) (family) 5 tribes, 8 genera, 62 species plus 220 serological species	A large group which have, as characteristics in common, the ability to grow well on laboratory media and vigorously attack sugars with the production of acid or acid and gas; many are inhabitants of the intestine, some are plant pathogens, some are widely distributed in nature	Gram-negative rods (nonsporing)	Most important are: *Escherichia coli*, common intestinal form *Salmonella* spp. including typhoid, para-typhoid, food poisoning organisms *Shigella* spp. causes bacillary dysentery *Erwina* spp. plant pathogens
11.	*Parvobacteriaceae* (small) (family) 4 tribes, 10 genera, 56 species	All members of this family are parasitic in or pathogenic to warm-blooded animals; cause a number of important diseases; often difficult to grow on ordinary media	Small gram-negative rods (nonsporing)	Important forms: *Brucella* spp. cause Brucellosis (undulant fever); *Pasteurella* spp. cause plague and tularemia; *Hemophilus* spp. cause whooping cough, soft chancre, pink eye
12.	*Bacteriaceae* (rod) (family) 1 genus, 56 species plus some miscellaneous forms	A varied, unrelated group of bacteria; a sort of catch-all for those forms not clearly related to established families	Nonsporing rods; gram-positive or negative	No important pathogens or fermenters; some cellulose and agar decomposers included
13.	*Bacillaceae* (small) (staff) (family) 2 genera, 94 species	All rods which produce endospores are placed in this family; two genera, *Bacillus*, aerobic (air living) and *Clostridium* (small spindle), anaerobic (not air living) forms; primarily soil forms, fortunately only a few pathogenic forms; some inhabit the intestine	Sporeforming rods	*Bacillus anthracis* (anthrax) only important pathogen in genus *Bacillus*; *Clostridium* spp. include those causing tetanus, gas gangrene, botulism

HIGHER BACTERIA

The so-called "Higher Bacteria" are more complex either in shape or growth habit or both than the simple one-celled forms making up the True Bacteria. Of the four orders in the Higher Bacteria group only two, the *Actinomycetales* and the *Spirochaetales*, are of sufficient practical importance to merit further discussion.

Actinomycetales

Most microbiologists look upon the *Actinomycetales* as occupying a position in the phylogenetic tree somewhere between the true bacteria and the molds. The simplest of the *Actinomycetales*, *Mycobacterium*, is generally found as a long, thin, and often curved acid-fast rod. Occasionally these forms may develop filaments. On the other hand, the more complex filamentous members of this order may show

TABLE VII

FAMILIES, GENERA, AND SPECIES OF THE ACTINOMYCETALES

FAMILY	BRIEF DESCRIPTION, HABITAT, ETC.	MOST IMPORTANT GENERA AND SPECIES
Mycobacteriaceae (fungus-rod-family) slender, often curved rods, occasionally forming meager filaments; one genus, 13 species described	Five species pathogenic to man or other warm-blooded animals; 5 species infecting cold-blooded animals; 3 species are saprophytes found in dust, soil, or hay, etc.	*Mycobacterium tuberculosis* var. *hominis* causes tuberculosis in man; *M. tuberculosis* var. *bovis* cause of tuberculosis in cattle; *M. leprae*, believed to be the cause of leprosy
Actinomycetaceae (ray-fungus-family) form mycelial filaments which may break up into rodlike or coccoidal segments; two genera, 35 species described	Of the genus *Nocardia*, 18 species are apparently from the soil, 14 were isolated from human infections, one from a reduvid bug; a few of the soil forms attack paraffin, benzine, petroleum or phenol; *Actinomyces*, 2 species	*Nocardia madurae* causes madura foot; *Actinomyces bovis* produces lumpy jaw in cattle and a similar disease in man termed actinomycosis
Streptomycetaceae (pliant-fungus-family); form mycelial threads which do not break up; two genera, 78 species	Mostly found in the soil; a few cause rare types of human infections or plant infections	*Streptomyces griseus* is the streptomycin producer; many of the new and valuable antibiotics are derived from members of this genus

a remarkable resemblance to certain mold forms, except that the filaments are extremely fine, having a diameter about equivalent to that of the cells of many of the true bacteria. The families of the *Actinomycetales*, together with the most important genera and species, are briefly described in Table VII.

Spirochaetales

The *Spirochaetales* are slender, undulating, or spiral forms. They differ from *Spirillum* in that they are definitely flexuous while *Spirillum* by comparison is relatively rigid. Very few of the spirochetes have been cultivated successfully in artificial media. Certain of their characteristics suggest that the spirochetes are intermediate between the true bacteria and the protozoa. A brief outline of the order *Spirochaetales* is given in Table VIII.

TABLE VIII

FAMILIES AND IMPORTANT GENERA AND SPECIES OF THE SPIROCHAETALES

FAMILY	BRIEF DESCRIPTION, HABITAT, ETC.	MOST IMPORTANT GENERA AND SPECIES
Spirochaetaceae (coiled hair family)	Flexous, spiral forms found in fresh or salt water and in certain molluscs	No forms of practical importance
Treponemataceae (turning-thread family)	Slender spirals, protoplasm shows no structure; mostly parasitic on man and animals; found in the mouth, intestines, external genitalia	Several species of the genus, *Borrelia* cause relapsing fevers; *Borrelia vincenti* is associated with *Fusobacterium plauti-vincenti* in trench mouth; *Treponema pallidum* causes syphilis; *Treponema pertenue* is the cause of yaws or tropical syphilis

Since this chapter lists only a few species of bacteria it may serve to create a false impression as to the magnitude of the variety of forms in the bacterial world. The 1630 species listed in *Bergey's Manual* represent only those forms which have been rather adequately described.

In many instances, subcultures of the original cultures are available for comparison. The repository for such cultures in the United States is the American Type Culture Collection.

Many additional forms have been isolated and named. However, with most of these the descriptions have not been sufficiently complete or accurate to enable later workers to identify new cultures as being identical with those previously named.

While microbiology as a science is scarcely a century old, during that period of time we have explored our immediate environment fairly well in terms of the microorganisms present. As a result we now know a good bit about most of the bacteria-producing disease in man and a little less about the pathogens found in plants and higher or lower animals. We have a fairly accurate picture of many of the fermenters, the food spoilers, and certain of the soil forms. On the other hand there are still many areas within or outside our environment where microbial explorations have never been made or at best these have been only of a cursory nature. In some of these areas, we have found certain microorganisms to be consistently present but we have not been able to isolate or grow them successfully. As we develop new techniques, many new horizons may loom up in the microbial world and beckon workers on to an exploration of the many unexplored microbic hinterlands. In addition much still remains to be done in clarifying the relationships between the organisms we already know.

Selected Reading References

Breed, R. S., Murray, E. G. D., and Hitchens, A. P.: Bergey's Manual of Determinative Bacteriology, ed. 6, Baltimore, 1948, The Williams & Wilkins Company.

CHAPTER 8

HEREDITY OF MICROORGANISMS

Mutations and Variations of Microorganisms

Applications of the fundamental principles of the Mendelian laws of heredity (1866) have proved successful in studies of man, animals, higher plants, insects, etc., for over fifty years. Mutants and variants from parents have been observed occurring naturally and in many cases have been artificially induced.

It was not until the late 1920's that consideration was given to the genetics of microorganisms, with variants and mutants being observed.[1] Sporadic interest was the rule until after 1940 when an increasing interest developed in this subject.[2, 3, 4]

Microorganisms, in general, reproduce themselves unchanged. For millions and billions of progeny, they remain true to the parent. However, not unlike man and animals, within the same family group, certain changes may be evident in microorganisms due to some mutant change or variant. Not only are these changes observed in light microscope-visible microorganisms, but also may be determined as taking place in submicroscopic microorganisms such as *viruses.*

In any given microbial population there are always a few bacteria out of the many millions in the culture that are different from their parents. Such spontaneous changes occur not only in the test tube, but also naturally in the human and animal body and in disease processes.

PHYSICAL CHANGES

Often the first macroscopically visible change seen in a bacterial population is, under certain conditions, reversible change from so-

[1]Burnet, F. M.: J. Path. and Bact. **56:**423, 1929.
[2]Beadle, G. W., and Tatum, E. L.: Proc. Nat. Acad. Sci. **27:**499, 1941.
[3]Beadle, G. W., and Tatum, E. L.: Am. J. Bot. **32:**678, 1945.
[4]The Genetics of Pathogenic Organisms, Pub. Am. Assoc. Advancement Sci., No. 12, 1940.

called smooth to rough type colonies. The *smooth* colony will be entire, homogeneous, smooth, and regular in size. Its neighbor, originating from the same parent may be a *rough* colony, exhibiting an irregular margin, rather granular appearance of the center, with a dull surface appearance. This change is frequently associated with virulence, the organisms of the smooth colony being the more virulent.

Microscopically other physical changes are apparent. For example, a normally motile (flagellar) species of bacteria may become nonmotile. Whether this is strictly a genetic mutant is not clear, but, nevertheless, it is a variation from the parent. Another physical change that is frequently seen is the loss of capsular substance of those organisms whose parents are capsule producing.

In cultures of rod forms of bacteria, particularly the gram-negative group, one sees in practically every microscopic field of stained microorganisms a variety of sizes. In the genus *Shigella*, one frequently observes a few long, slender bacteria on the slide. Likewise with the coliform group, many involution forms may be seen. In the genus *Hemophilus* there are many organisms in a given microscopic field which may be twice as long as their neighbors. In fact, the variability in length of some of the organisms may suggest contamination, but repeated subcultures may always result in pure cultures.

IMMUNOLOGICAL CHANGES AND VIRULENCE CHANGES

A highly virulent microorganism such as the pneumococcus becomes avirulent on loss of its capsule. Growth on a nutritionally poor medium may bring about this change within a few transfers. The naked cell (somatic antigen) is then immunologically identical to any of the seventy-five types when noncapsulated. On regaining the capsule, most easily accomplished by repeated animal passages, the organisms become capsulated and type specific as well as disease-producing. Transformation from one "specific type" to another specific type has been accomplished. The use of a desoxyribonucleic acid fraction from pneumococcus type III in the medium can convert noncapsulated type II pneumococci to type III.

The use of specific immune serum in culture media may cause certain mutations in microorganisms.

Repeated culture of *Sal. typhosa* on plain nutrient agar frequently produces a loss of Vi antigen, thereby reducing virulence and thus reducing the organism to a nonprotecting antigen when used as a vaccine.

Hemophilus pertussis, when cultured on a minimal growth-promoting medium, may change from Phase I to Phase II or even III, thus losing its immunizing qualities when used as a vaccine.

Many viruses are capable of mutations, from a highly virulent to a less virulent strain, or from one antigenic pattern to another. Reversions to its parent are extremely difficult. The virus of smallpox, in the natural disease, is a highly virulent disease-producing agent. If this virus is passed for several generations on rabbits and skin of the calf it loses its virulence but maintains its immunizing qualities. The virus, now called vaccina virus, is used in the active state for vaccination against smallpox. Rabies virus also appears to be less virulent for its normal host (the canine family) after repeated passages in rabbits. The furious, rabid state is no longer produced in dogs after the virus is "fixed" in rabbits. Antigenic (immunizing) qualities are maintained. When influenza virus (Type A) is repeatedly passed by allantoic inoculation in embryonated eggs many bizzare virus mutants are produced which are antigenically distinct from the parent.

CHEMICAL CHANGES

The sudden change of a gene may produce a lasting mutant with the altered characteristic being relatively stable. Such genic control of biochemical activities are mediated through certain enzyme systems of the microorganism. Such changes may be naturally occurring or they may be produced by x-ray treatment, nitrogen mustard, and other chemicals as well as by ultraviolet light rays. Single gene mutants are the rule, the nutritional requirement of the mutant being frequently a growth-promoting factor, which has become a completely dependent biochemical entity. Examples of this change are the dedependence of the mutant on a single amino acid, such as tryptophane, lysine, tyrosine, etc., whereas the parent could either synthesize the lacking food or exist without its presence. In the dependent mutants the specific substrate requirements must be furnished.

Likewise a certain lactose-fermenting gene may be changed, resulting in an *Esch. coli* mutant failing to ferment lactose or failing to produce gas from lactose, thus differing from its parent which produces both acid and gas from this sugar.

Drug resistance (drug fastness), whether spontaneous or induced, is a common occurrence with certain microorganisms. Streptomycin readily produces mutants that are resistant to this antibiotic while penicillin has less tendency to produce penicillin-resistant mutants.

Some strains of organisms, such as gonococcus and meningococcus, may mutate to a complete reversion dependence, and will propagate only in the presence of, for example, streptomycin. Aureomycin resistant strains of *Micrococcus pyogenes* var. *aureus* may show changed characteristics—such as loss of pigment production, increased oxygen requirements, etc.

There is increasing evidence that bacteria possess nuclei and undoubtedly chromatin and genes.[5,6] Sexual reproduction of higher microorganisms such as yeasts and molds have long been known. Sexuality in bacteria has been suggested, but only recently successful genetic recombinations have been made.[7,8] Single deficient mutants of *Esch. coli* and multiple deficient mutants of the same parent have been mated in cultures and strains reisolated. Such strains have then been found to cross genetically, the recombined mutant being genetically different from either parent.

IMPLICATIONS OF MICROBIAL GENETICS

The theoretical concepts of a better understanding of the behavior of microorganisms have been brought about by genetic studies. Practical applications of these genetic studies of microorganisms are biological assay of amino acids, vitamins, and other biological products. Since the generation time is so short, billions of progeny may be studied in a matter of weeks.

A better understanding of infection and resistance to disease-producing microorganisms is brought about by these studies. Also a more secure understanding of epidemics and epidemic processes in relation to high and low microbial virulence has been gained by such genetic studies. Also a more secure base for antibiotic therapy is attained by genetic studies of mutants produced by antibiotics and the understanding of how antibiotic resistant mutants develop and how they may be eliminated or guarded against in the therapy of disease.

---

[5]Stille, B.:   Arch. Microbiol. **8:**125, 1937.

[6]Robinow, C. F.:   J. Hyg. **43:**413, 1944.

[7]Ledeberg, J., and Tatum, E. L.:   Cold Spring Harb. Symp. Quant. Biol. **11:**113, 1946.

[8]Newcombe, H. B., and Nyholm, M. H.:   Genetics **35:**126, 1950.

Selected Reading References

Braun, W.: Bacterial Dissociation, Bact. Rev. **11**:75, 1947.
Catcheside, D. G.: The Genetics of Micro-Organisms, London, 1951, Sir Isaac Pitman and Sons, Ltd.
Gordon, F. B.: Genetics of Viruses, Ann. Rev. Microbiol. **4**:151, 1950.
Hadley, P.: Microbic Dissociation, J. Infect. Dis. **40**:1, 1927.
Ledeberg, J.: Microbial Genetics, Selected papers, Madison, 1951, University of Wisconsin Press.
Tatum, E. L., and Perkins, David D.: Genetics of Microorganisms, Ann. Rev. Microbiol. **4**:129, 1950.

CHAPTER 9

AGENTS DEALING DEATH TO MICROORGANISMS

Sterilization and Disinfection

Surrounded as we are by an invisible world of microorganisms, we depend to a great extent for our safety and well-being on the use of procedures which kill or otherwise control these forms. These processes may be utilized (a) to prevent the transfer of disease-producing microorganisms, (b) to prevent and cure infections, (c) to permit the isolation and study of microbes in pure cultures, (d) to prevent food spoilage or the microbial deterioration of cloth, leather, lumber, or other materials, (e) to maintain desirable fermentations such as those involved in butter production, the brewing industry, etc., (f) to produce valuable biological products such as antibiotics and certain vitamins and (g) to make possible aseptic surgery.

Sterilization procedures provide the foundation for all bacteriological or sterile techniques and have become commonplace in our everyday life. In a hospital, for example, virtually everything must be sterilized. However, the theory behind many of these procedures is often not well understood and many people because of ignorance rob themselves of the protection thus afforded.

DEFINITION OF TERMS

Sterile is the state of being free from all living microorganisms. Used in this sense, the presence of a single microorganism renders an object nonsterile. The process of achieving this state is termed *sterilization*.

An object, such as a properly cleaned eating utensil, while safe, is not necessarily sterile. The term sanitary is a better one for describing a state in which an object is freed from all dangerous bacteria without necessarily being sterile.

Sanitization is the term describing the process by which an object is rendered sanitary. In the course of the sanitization of an object, not only are the pathogens killed but most of the nonpathogens will also be destroyed. However, a sanitary object can seldom be described as "sterile."

Sterilization of inanimate objects is generally achieved by the use of some physical methods such as flaming or the use of the pressure cooker or autoclave. Although sterilization can be achieved through the medium of chemicals (the so-called germicides), chemicals are slower, less sure, and not so practical to use as the physical methods. Chemicals have their greatest use in the sanitization of inanimate objects or in the treatment of disease.

Disinfection, antisepsis, etc., are terms most generally used to describe the relative freedom from living microorganisms achieved through the use of chemicals.

STERILIZATION BY PHYSICAL MEANS

The most useful of the sterilizing procedures are:

1. Flaming or Incineration

Any object, such as a needle which can be heated to red heat, will be rendered sterile. It is likely that a much shorter exposure to flame might accomplish this purpose. However, red heat is positive assurance of sterility. Of course such sterility will remain only so long as the cooled, sterile object is not contaminated. Dipping a scalpel in alcohol then burning off the alcohol cannot be depended on to produce sterility because the temperature thus obtained may be far below that necessary to kill certain of the bacteria.

2. Dry Heat

Any dry object not damaged by high temperatures may be sterilized by heating in an oven. However, dry heat requires a relatively long period of time to bring about the loss of viability in microorganisms. Bacteria, particularly those in the spore state, are extremely resistant to dry heat. Apparently the initial heating dries the bacteria and the dry proteins do not coagulate. It is known that dry enzymes frequently will remain active even after heating to 100°C.

(the boiling point of water).   According to Rahn[1], death of the microorganisms appears to be due to an oxidation of the cell contents rather
than enzyme inactivation or protein coagulation.

The usual laboratory practice is to heat glassware, etc., at
150°-180°C. for 1 to 3 hours.   The temperature of 150°C. is apparently
the lowest practical dry heat temperature since Koch long ago found
that at 140°C. three hours were required to kill anthrax spores.   At the
higher limit (180°C.) cotton used to plug pipettes or paper wrappings
of glassware are charred excessively.

In the home, dry glassware, etc., can be sterilized effectively by
heating at 300°-350°F. for 1 to 3 hours in the oven of the cooking stove.

Higher temperatures than the above greatly shorten the length
of time required to sterilize; however, the greater thermal shock will
probably cause a more rapid deterioration of glassware.

3.   Steam Under Pressure

The pressure cooker or autoclave provides the most generally
useful method of sterilization.   In the autoclave, sterilization is accomplished in an atmosphere of water vapor (steam) heated above the
boiling point of water and because of the moisture a much shorter
period of exposure will produce effective destruction of microorganisms.

Generally a pressure of 15 pounds for 15 to 20 minutes is used.
This gives a temperature of 121°C. at sea level.   Bacterial spores, the
most resistant of all the microbial forms, can rarely withstand more
than 5 to 10 minutes at 121°C. in an autoclave.

In operating the autoclave or pressure cooker certain precautions
must be observed in order to obtain effective killing of bacteria.   Some
of these are:

(a)   The petcock of the autoclave or pressure cooker should remain open until a plume of steam pours out, indicating that all of the
air has been replaced with water vapor and that sterilization is actually
taking place in an atmosphere devoid of air.   If air is entrapped the
temperature will be reduced materially.   This may produce tragic
results if surgical dressings or instruments are being sterilized or if
nonacid foods such as fish or beans (in which *Cl. botulinum* can readily
develop) are being canned.

(b)   Wherever possible, the temperature, rather than the pressure,
should be read, since pressure gauges are notoriously less reliable than
thermometers.

---

[1]Rahn, Otto: Bact. Rev. **9**(1):1-47, 1945.

(c) Autoclaves or pressure cookers should not be tightly loaded with surgical dressings, etc., because of the danger of producing pockets of entrapped air with a resultant lowering of the sterilization temperature.

(d) Large objects, such as large flasks of broth, gallon cans of food, etc., require much longer heating due to the time necessary for heat to penetrate to the center of large liquid masses.

Fig. 10.—Autoclaves. On the left is a small-sized steam autoclave; the center one is a medium-sized electric autoclave; on the right is a large-sized electric autoclave. The wheels on the doors are the locking mechanism to hold the doors tight when the steam pressure is up to 15 pounds per square inch. This equipment is in the media preparation room of the University of Utah, Department of Bacteriology.

(e) Oils, petrolatum, zinc oxide, and similar materials cannot be effectively sterilized in the autoclave. With these materials dry heat should be used.

(f) Acid fruits or acid media are sterilized easily, at a lower temperature; however, media, canned foods, etc., which have a neutral reaction require the longer period mentioned.

(g) At higher altitudes the pressure or length of time should be increased since the temperature increase due to pressure is super-imposed on the boiling point. The boiling point decreases proportionately with increase in altitude.

4. Boiling

Boiling, while effective in killing nonsporing bacteria in a short time, is often relatively ineffective against bacterial spores. While most vegetative bacteria are probably killed even before the boiling point has been reached, spores generally resist boiling temperatures for several minutes or even for an hour or more. A false sense of the effectiveness of boiling as a sterilization procedure is gained from the so-called "instrument sterilizers" used in the offices of most physicians and dentists. As suggested by McCulloch,[2] these should be called instrument boilers rather than sterilizers. Boiling in this case probably rarely achieves the absolute absence of viable microorganisms necessary for sterility. However, it is generally safe because the instruments are clean before boiling and spores are rare under such circumstances. Also the instruments are used for minor rather than major surgical procedures.

5. Flowing Steam

What has been said regarding boiling is equally true with regard to flowing steam, since the temperatures will be about the same. Although flowing steam may kill vegetative bacteria rather quickly, spores may remain unscathed even after exposure for some time to flowing steam. The device used in the laboratory for sterilization by flowing steam is called the Arnold Sterilizer. While it is not now commonly used it was at one time an important piece of laboratory equipment.

6. Intermittent Sterilization

Many years ago Tyndall hit upon a rather clever dodge for circumventing the resistant spores which refused to die in boiling water. He reasoned that in a liquid medium or broth, the spores which escaped the initial boiling would subsequently germinate, thus becoming vulnerable vegetative cells. The procedure he followed was to boil the

---

[2]McCulloch, E. C.: Sterilization and Disinfection, ed. 2, Philadelphia, 1945, Lea and Febiger.

medium being sterilized for thirty minutes; incubate for 24 hours at a temperature favorable for the germination of spores; boil again to kill the spores which had become vegetative cells and repeat the same procedure on the third day. This process has been termed "tyndallization," fractional sterilization, or more commonly intermittent sterilization. It is highly effective against those spores which oblige by germinating during the first or second incubation period. Occasionally uncooperative spores may refuse to germinate until after this period thus invalidating the process. Of course, boiling for 30 minutes is of itself generally sufficient to kill all but the most resistant spores. These are the ones which must be tricked into germinating before boiling can effect their demise.

Intermittent sterilization is occasionally used for the sterilization of media which can withstand boiling temperatures but would be broken down at the higher temperature used in the autoclave.

7. Filtration

In the bacteriological laboratory it is occasionally necessary to sterilize media or solutions which would be broken down by heating. This is accomplished by the use of specially constructed filters which have pores small enough to hold back bacteria while allowing most or all of the other components of the solution to pass through. These filters are made of unglazed porcelain, asbestos, plaster of Paris, diatomaceous earth, or a combination of the last three, sintered glass, or collodion.

A complicating factor is the presence of an electric charge on some of the filters which may repel and serve to prevent the passage of certain components even though their molecules are far smaller than the pores themselves.

Certain of the filters are used to separate a mixture of viruses and bacteria, since the viruses pass through while the bacteria are retained. The solution so filtered thus becomes bacteriologically sterile but non-sterile so far as viruses are concerned.

By varying the composition of the solvents used to dissolve collodion the size of pores in a collodion membrane can be varied. These filters of different porosity are designated as Gradocol filters and have been used to separate viruses of different sizes or to estimate the size of a given virus particle.

On the other hand, the filters used in water purification or filtering procedures remove most of the bacteria but do not in the true sense produce sterile water. With the removal of perhaps 90-99 per cent of the bacteria, the water may be made safe without necessarily being sterile. Filtration often serves merely to remove enough of the foreign matter to improve greatly the quality of the water and enhance the effectiveness of chlorination.

Air and other gases can be sterilized by passage through dry cotton since the intertwined fibers effectively enmesh the dust particles and bacteria present in the air. This is the basis for the use of the cotton plug in test tube and flask cultures of bacteria.

8. Ultraviolet Light

While the microbe-killing power of ultraviolet light has been known for a good many years, the development during the past two decades of the cold cathode mercury vapor lamp has given us a practical means of using these rays as a kind of invisible machine gun spraying bullets which crash into the bodies of bacteria floating in the air or reposing on surfaces.

These lamps were developed concurrently with fluorescent lighting. About 80 per cent of the rays are ultraviolet with a wave length of 2537 Å[3] which is essentially the most germicidal of all wave lengths. These lamps have proved to be effective in killing microorganisms in the air and on surfaces. As a result, they have been used extensively in hospital operating rooms, hospital wards, and bacteriological laboratories to prevent the transfer of air-borne pathogens. In bakeries, meat packing plants, and other food processing plants the ultraviolet lamps are used to reduce surface contamination of equipment or food products.

These germicidal lamps differ from the so-called sun lamps. Their wave lengths do not produce tanning. Rather the skin is irritated superficially and reddening occurs. Eyes exposed to germicidal lamps without being protected by proper glasses will become red and "bloodshot."

The germicidal rays do not penetrate to any extent through ordinary glass or any appreciable depth of solution so their effectiveness is confined largely to air or surface sterilization. On the other hand, quartz glass test tubes are easily penetrated and are therefore used when studying the effects of ultraviolet light on cultures of microorganisms.

---

[3]Ångstrom unit equals 0.0001 mm.

Sun lamps have far less germicidal effect than the germicidal lamps. Sunlight, itself, has a degree of microbe-killing power due to the wave lengths shorter than 3,000 Å which are found in its spectrum. In addition, the heat and drying effect plus the mild germicidal action of certain visible rays all combine to make sunlight relatively effective as a sanitizing agent.

9. Supersonic Waves

Vibrations more rapid than audible sound waves have been found to be rather highly destructive to bacteria suspended in solutions. The vibration rates used experimentally have generally ranged from about 20,000 to 1,000,000 per second. The procedure used has consisted of placing a vibrating rod in the solution being tested such as milk, broths, blood, and many other solutions or suspensions. The rapid vibrations set up in the liquid apparently tear the bacteria apart and, in addition, the molecular friction induced results in a marked rise in temperature which of itself may be germicidal.

Viruses, because of their small size, tend to resist destruction by these vibrations. As yet few practical applications have been made of supersonic waves other than the breakdown of bacterial cells for certain vaccines or for the chemical analysis of cell contents. Yeast extracts have also been so prepared. The marked variation of various kinds of microbes and spores to supersonic vibrations complicates the practical utilization of this method.

10. Electronic Bombardment

Experiments are being conducted in which electronic bombardment has been used in an attempt to sterilize various bacterial media, food in cans, etc. Technical difficulties and high cost still prevent the utilization of this method although it may become practicable in the future.

11. Other Physical Factors Adversely Affecting Bacteria

Freezing cannot be classed as a sterilizing procedure. However, bacteria do not grow in frozen materials, and some forms die off fairly rapidly. Others merely become quiescent and their numbers in a given medium will decrease only slightly when the material is frozen. This is especially true of bacteria in the spore state. However, when some relatively sensitive bacteria are frozen quickly with a dry ice-acetone or alcohol mixture, then dried under a vacuum they may remain viable for months or even years. This process is known as *lyophilization* and is an effective means of preserving some bacteria which would otherwise be short-lived.

Drying usually causes the death of most vegetative forms of bacteria in a matter of a few hours or days. The cholera and gonorrhea organisms are extremely sensitive. On the other hand the tuberculosis organism is notoriously resistant. It has been shown to retain its infectivity, when dried in sputum, over a period of many months. Without the protection of a sputum coating, however, it is much less resistant to drying.

Most bacterial spores are extremely refractory to drying. Anthrax spores, for example, remain viable and infective for many years. Some experiments have suggested that spores are able to survive in the dried state for centuries.

12. Mechanical Pressure

When subjected to extremely high hydrostatic pressure in liquid media most common nonsporing bacteria are killed. Spores, on the other hand, are relatively resistant to the high pressures.

Bacteria have been found living and functioning actively at great depths in the ocean where the water pressure reaches thousands of pounds per square inch. Certain of these bacteria can be cultivated successfully only when grown under the relatively high hydrostatic pressures analogous to those found in their environment.

13. Osmotic Pressure

High osmotic pressures serve to prevent the growth of many bacteria or to kill them. The high concentrations of common salt (sodium chloride) used in brines for pickling meat apparently exert their preservative effect largely in this manner. Water is drawn from the bacterial cells because the osmotic pressure of their cytoplasm is less than that of the surrounding salt solution. As a result their cytoplasm shrinks greatly and the cell dies or at least loses its power to reproduce.

However, in Nature's Pickling Vats such as the Great Salt Lake and the Dead Sea, where the salt content is in the neighborhood of 25 per cent or sometimes more, a race of hardy microbes has developed which grow only in the presence of salt. These are known as *halophilic* or salt-loving bacteria. Such bacteria may be present in these waters in concentrations as high as several hundred per milliliter. They can be isolated successfully only in bacterial media containing at least 12-15 per cent salt.

When present in large numbers in the salt used for pickling of meats, they may produce spoilage of the product even though the salt concentration is too high to permit the growth of nonhalophilic bacteria.

In preserving jams and jellies the high osmotic pressure of these fruit-sugar mixtures seems to be the factor which keeps them from being readily spoiled by microorganisms. The usual proportion of sugar to fruit is about 50 per cent of each. At this concentration bacterial growth seems to be almost completely inhibited and yeast growth is generally inhibited. On the other hand molds will often grow but at a very slow rate. A concentration of at least 70 per cent sugar seems to be necessary to keep the molds in check. Even at this concentration a very slow development of some molds may occur.

DISINFECTANTS AND RELATED COMPOUNDS

Chemicals have long been used for the purpose of killing microorganisms or keeping them in check. Frequently their use has been based, not on scientific experimentation, but on some happenstance observation. Manifestly many of the physical methods of destroying microbes, such as flaming or dry heat, would be completely unsuitable for sterilizing or sanitizing living tissue or for controlling infections after they start. Chemicals have, therefore, come to occupy an important position in the control of undesirable or dangerous microorganisms. Such chemical agents may be placed conveniently into three categories: (a) the disinfectant-germicide group, (b) the specific chemotherapeutic agents, and (c) the antibiotics.

In the first group are placed the common disinfectants. These are, for the most part, protoplasmic poisons, toxic to most living tissue and not especially selective in their action against different kinds of microorganisms. The toxicity which they manifest is often against the cytoplasm of the cell or against the cell membrane. In addition, enzyme systems may be poisoned and death may result from a respiratory or other metabolic shock.

The chemotherapeutic agents, which include the arsenical compounds such as the famed 606 and the arsphenamines in general, the sulfa compounds and certain of the dyes, are rather highly selective in their action against microorganisms. They are generally bacterio-static rather than bactericidal and appear to exert their influence by blocking certain enzyme systems essential to the life of the micro-organisms.

The antibiotics are chemical compounds which are produced by living cells and which serve to inhibit or kill certain microorganisms. They may be produced by molds, bacteria, actinomycetes, higher plants or even by certain animal tissues. They resemble the sulfa compounds in that they are often highly selective in their action against particular pathogens and appear to exert their influence primarily by interferring with certain enzymatic processes, or blocking certain amino acids from being metabolized.

The term disinfectant means, as the name implies, "ridding of infection." It appeared in the English language some three centuries ago. To be effective, a disinfectant need not kill all microorganisms so long as it kills pathogens.

On the other hand, the term germicide is a much more positive one, since a germicide should properly be able to kill germs, i.e., micro-organisms, without regard to their pathogenicity. Likewise, bacteri-cides, fungicides, viricides, and algacides are, respectively, killers of bacteria, fungi, viruses, and algae.

The term bacteriostat or fungistat implies a mere inhibition of growth of bacteria or fungi.

The term antiseptic had its inception about the same time as disinfectant. Originally it meant "against sepsis or putrefaction." The current usage as defined by the Pure Food and Drug Act considers the term to indicate that an antiseptic is a germicide, "except in the case of a drug purporting to be, or represented as, an antiseptic for inhibitory use as a wet dressing, ointment, dusting powder, or such other use as involves prolonged contact with the body." Chemicals which have prolonged body contact could not be as strong, i.e., con-centrated as those having a shorter contact period.

Preservatives are compounds which preserve foods, etc., by in-hibiting the growth of microorganisms. Frequently they are not espe-cially germicidal. Active germicidal agents in lower concentrations may be "preservative" in their action.

Contrary to the belief that disinfectants work instantaneously (as evidenced by the layman when he plunges a needle into a bottle of rubbing alcohol and removes it immediately), disinfectants do require time in which to operate. The length of time necessary depends on a number of factors but it may range from a minute or less in the case of some highly sensitive organisms to many hours in the case of some bacterial spores.

Several factors must be considered when the effectiveness of a given disinfectant is being discussed. Most important of these are:

(a) **Concentration**

A very high concentration may be highly germicidal, a lower concentration merely inhibitory and a very dilute concentration may actually stimulate the growth of certain microorganisms.

(b) **Temperature**

The germicidal action of chemicals generally increases two or more times for every increase of 10°C. unless a temperature is reached at which the disinfectant undergoes chemical change.

(c) **Time**

As indicated previously, the time required for a disinfectant or germicide to act will depend on the factors listed above (and below) and on the nature of the microorganism involved.

(d) **Other Factors**

These include pH (degree of alkalinity or acidity); toxicity to tissue which may limit to a point where it will no longer be effective the concentration which may be employed; inactivating action of blood, serum, tissue, or other organic matter on the disinfectant being employed and surface tension or wetting power. This last factor determines the penetration which takes place and the ability of the disinfectant to be brought into contact with the organism. Tinctures (alcohol-water mixtures) are often used instead of water solutions because of their ability to spread over the skin and thus penetrate into cracks or cuts in the tissue. Hexylresorcinal (ST37) has as one of its claims for effectiveness as a disinfectant, its penetrating ability. Soaps added to certain cresolic mixtures increase their miscibility in water and reduce their surface tension.

An ideal disinfectant is being and has long been sought. This is one which would be highly toxic to all microorganisms, relatively non-toxic to tissue, free from objectionable taste, color, or odor, stable and inexpensive, and having a low surface tension. While sulfa compounds and antibiotics meet several of these specifications, they lack completely the first requisite—the ability to attack all bacteria. Such an ideal germicide has not been found so the layman is faced with the problem of selecting a disinfectant which will best satisfy his particular need.

These needs include disinfection of skin surfaces and minor injuries; of mucous membranes such as the nose, throat, or eyes; of body discharges as in intestinal diseases, tuberculosis, and diseases in which a profuse purulent discharge may occur; of inert objects such as dishes and other eating utensils, floors, stables, etc., and of water supplies, swimming pools, etc.

Various methods have been employed for comparing the strength of disinfectants. The "phenol coefficient," which is a ratio expressing the strength of a disinfectant as compared with phenol (carbolic acid), has been widely used. Commercial disinfectants frequently have their phenol coefficient given on the label.

The weakness of the method lies in the fact that it employs only two organisms, *Salmonella typhosa* and *Micrococcus pyogenes* var. *aureus* and one set of conditions. With some other set of conditions and other organisms the results obtained might be quite different.

By the use of mice, the embryonated chick, and tissue cultures, the toxicity of a germicide for living tissue has been compared with its bactericidal effect. The ratio is expressed as the "toxicity index." Certain disinfectants which have been found to possess great bactericidal power may be far too toxic to tissue to be useful as skin or wound disinfectants.

NATURE AND USES OF COMMON DISINFECTANTS

Many centuries ago Egyptians used salt and spices to embalm bodies; however, the specific use of chemicals to attain disinfection is of relatively recent origin.

As early as 1827 Alcock suggested the use of calcium hypochlorite (chloride of lime) for its disinfecting and deodorizing properties on board certain British Naval vessels. The compound was first produced commercially in 1799. In 1832 *Lancet* recommended the use of silver nitrate for cauterizing autopsy wounds. Doctors Oliver Wendell Holmes in America and Semmelweiss in Austria both recognized in the 1840's the value of calcium hypochlorite for preventing the transfer of the childbed fever organism by means of the infected hands of attending physicians. Sir Joseph Lister (1870-1887) developed the technique of antiseptic surgery. This involved the spraying of the operative area (and the operator) with phenol (carbolic acid) solution. This procedure was soon supplanted by the technique of aseptic surgery.

The common disinfectants now employed can be placed conveniently into the following groups:

1. The Chlorine-Iodine Group

With the introduction of the chlorination of drinking water at the Chicago stock yards and Jersey City, New Jersey, in 1908, chlorine gas, the most vicious member of the family, gained a position of international importance. It was later found that chlorine in water was relatively unstable so ammonia gas was added at the time chlorination occurred. This transformed the unstable chlorine into chloramine, a steady, dependable compound. As a result, the active germicidal effect of the chlorine persists for a few days instead of merely for several hours. The amount of free chlorine in properly chlorinated water is small, usually considerably less than 1 ppm (part per million). However, it apparently controls the pathogens in water rather effectively as evidenced by the amazing decrease in water-borne diseases which has occurred in the last half century.

Calcium (or sodium) hypochlorite, with or without the presence of ammonium sulfate to give it more stability, is also used for successfully disinfecting water supplies.

Proper chlorination rarely sterilizes a water supply. However, the numbers of bacteria are greatly reduced and the only bacteria remaining are spore formers or other resistant but harmless forms. Cysts of the amebic dysentery organism are also resistant to chlorination. These are removed, however, by filtration.

Hypochlorites have been found extremely useful for sanitizing eating utensils, dairy equipment, and similar objects. A concentration which will give 50-200 ppm of free chlorine is used and the recommended time of exposure is five minutes.

Tincture of iodine is another time-honored member of the family. Its use in surgery was first suggested in 1884. It contains about 3-7 per cent metallic iodine dissolved in 70-83 per cent alcohol. Experiments indicate that it is an effective skin disinfectant with a relatively low toxicity index except in the case of certain sensitive individuals or where the concentration has been unduly increased by evaporation of the alcohol. A starch suspension provides an effective means of neutralizing the iodine in cases of iodine burns.

A mild tincture, 2 per cent iodine, 2.4 per cent sodium iodide in 46 per cent ethyl alcohol, may also be used.[4] This is much less irritating but still actively germicidal. Less coagulation of tissue protein is produced than with the stronger tincture.

Lugol's solution (a water solution of iodide and iodine) is also a rather effective skin disinfectant. However, it has little spreading power because of its inability to wet the skin as effectively as do the tinctures.

McCulloch[5] states that the concentration of iodine necessary to kill microorganisms "does not vary greatly with different organisms" and that "the organisms are killed rather than held in a state of bacteriostasis."

2. The Carbolic Acid Family

Members of this group of disinfectants have in common a benzene ring with one or more OH groups.

Phenol or carbolic acid is fairly effective against vegetative bacteria but spores may survive for several hours even in a 5 per cent solution. For applications to the skin 0.5-1.0 per cent is as strong as can be safely tolerated. It tends to anesthetize the skin and may necrotize (kill) the tissue. It is poisonous and the odor may be objectionable. The cresols are similar to phenol although being somewhat more germicidal and somewhat less toxic in proportion since they can be employed in a weaker concentration. They are active against the tuberculosis organism and are relatively effective in the presence of organic matter. These properties make them desirable for disinfecting tuberculous sputum. However, they share with phenol a lack of effectiveness against spores.

Hexylresorcinol, another of the phenolics, is characterized by its low surface tension and low toxicity. It is active against vegetative bacteria and its low surface tension allows it to penetrate cuts or superficial abrasions. It appears to be a rather satisfactory general household disinfectant.

---

[4]McCulloch: Op. cit.
[5]Ibid.

3. The Mercury-Silver Group

Koch, as early as 1881, found that very low concentrations of mercuric chloride (bichloride of mercury or corrosive sublimate) 1:500 to 1:2000 were apparently active against most bacteria. The effect of this compound has since been shown to be primarily an inhibition of growth rather than actual killing of the organism since mercury-poisoned bacteria have been revived and made to reproduce. The most objectionable feature of mercuric chloride is its extreme toxicity and its tendency to coagulate the outside of a tissue mass without penetrating and inhibiting bacterial growth within. Also, if used repeatedly, it may collect in the tissues and produce mercury poisoning in the patient.

To overcome the extreme toxicity of the mercury, various mercury-containing organic compounds have been synthesized. In many of these, the inhibitory effect against microorganisms has been retained while the toxicity has been greatly reduced. Merthiolate and phenyl-mercuri-nitrate appear to be effective against most bacteria and to be relatively low in toxicity. This gives them considerable value as household disinfectants. Mercurochrome is a related but somewhat different mercurial, being a mercury containing dye.

Silver compounds appear to be very active bacteriostats and the silver ions appear to act as protoplasmic poisons; however, when applied to the skin or tissues a black discoloration often develops due to the precipitation of metallic silver under the action of light. In addition, certain compounds such as silver nitrate may be caustic in their action. Prolonged use of the silver compounds in the eyes is contraindicated because a relatively permanent discoloration of the white of the eye may follow.

4. Alcohols and Aldehydes

Ethyl alcohol appears to be at best only a weak germicide on which too much reliance is generally placed. Isopropyl alcohol appears to be equal to if not somewhat better than ethyl alcohol.

Formaldehyde,[6] on the other hand, is an active germicide and fungicide. For example, a small pledget of cotton wet with formalin and placed in shoes of a patient with athlete's foot apparently rids the shoes of the fungus if they are held in a closed box for at least 24 hours.

---

[6]This compound in the pure state is a gas. However, it is sold in the form of a water solution containing (37-40 per cent) formaldehyde. Thus, it is called formalin.

It is useful for disinfecting tuberculous sputum. However, it has two extremely undesirable properties, high toxicity (except in very low concentrations) resulting in a necrotic effect on tissue and a highly penetrating and irritating odor. Vaccines may be prepared in dilute concentrations of formaldehyde.

5. The Quaternary Ammonium Compounds

These recently developed compounds, which are classed as cationic detergents, appear to show considerable promise as disinfectants, particularly for sanitizing eating utensils, dairy equipment, and similar uses. They are also being used as skin disinfectants. Their effectiveness, however, has not yet been completely evaluated. The type compound of this group is alkyl-methyl-phenyl ammonium chloride; however, numerous modifications of this basic structure have been synthesized. These compounds seem to vary considerably in their effectiveness. Their action appears to be bacteriostatic rather than bactericidal. The quaternaries are available on the market under a variety of trade names such as Phemerol, Roccal, Zepharin, etc.

6. The Detergents

Certain of the new cleaning agents, classed as anionic detergents, of which Drene is a representative compound, appear to have the power to inhibit the growth of certain bacteria especially the gram-positive forms; however, they appear to exert little effect on the gram-negatives. A large number of these compounds are on the market and they vary considerably in their molecular structure. They are classed chemically in the alkyl sulfate and the alkyl sulfonate groups. Their disinfecting power is being studied experimentally; however, their value or lack of value as disinfectants has not yet been established.

7. Soaps

While soaps appear to possess some inhibitory or even germicidal effect against certain sensitive bacteria their greatest value appears to be mechanical rather than chemical. Because soaps reduce the surface tension this serves to pick up the bacterial cells and float them away. The detergents also function in the same manner.

8. Alkalies

<u>Lye (sodium hydroxide) and quicklime (calcium oxide) are both rather effective germicides</u>, but their caustic nature limits their use to disinfecting inert objects not injured by this action. <u>These compounds are useful for the disinfecting of body discharges, especially feces, and the sanitizing of inert objects such as milk bottles, beverage bottles, etc.</u>

SULFA COMPOUNDS AND ANTIBIOTICS

Since these compounds differ from the disinfectants both in the purpose for which they are used and their effect they will be discussed in a subsequent chapter.

Selected Reading References

McCulloch, E. C.: Disinfection and Sterilization, Philadelphia, 1946, Lea and Febiger.
Rahn, Otto: Physical Methods of Sterilization of Microorganisms, Bact. Rev. **9** (1)-1:47, 1945.
Smith, D. T., et al.: Zinsser's Textbook of Bacteriology, ed. 10, New York, 1952, Appleton-Century-Crofts, Inc.

CHAPTER 10

ANTIBIOTICS AND SULFONAMIDES

The economic and social life of microbes somewhat parallels that of plant, insect, and animal life, including man.

Microbes are continually struggling for their existence, and frequently foodstuffs become so scarce that bacteria literally "starve to death." This applies to the helpful as well as the harmful types of bacteria. Nature, in many ways, attempts to strike a balance between living things of one kind or another.

Like people who help others, microorganisms actually help one another to survive and multiply by furnishing food and proper environment. Conversely, many people antagonize each other, and some to the point of murder. Microbes, likewise, antagonize many of their fellow members, some to the point of killing each other. This microbial killing has been termed "antibiosis."[1,2]

The actual use of fungi or products of the growth of yeasts and molds dates back to 1500 B.C. The Chinese used moldy materials for treating boils and abscesses. Both Hippocrates and Pliny recommended mold products in treatment of disease.

Although early in the beginning of the development of the science of microbiology, experimental antagonism was known (Pasteur, 1877 and others), very little consideration was given to this phenomenon as being practically applied in treatment of disease. Pyocyanin, an antibotic substance produced by *Ps. pyocyanea*, was marketed in Germany in 1890. Nearly a decade elapsed between the discovery by Fleming in 1929[3] of penicillin and the application of this antibiotic to the treatment of disease.

---

[1]Behrens, J.: (Symbiose, Metabiose, Antagonismus), Lafars Hanab. Tech. Mykol. Jena, **1**:501, 1940.

[2]Ward, H. M.: Symbiosis, Ann. Bot. **13**:549, 1899.

[3]Fleming, A.: On the Antibacterial Action of Cultures of a Penicillium, With Special Reference to Their Use in Isolation of B. Influenzae, Brit. J. Exper. Path. **10**:226 1929.

Since Fleming's discovery of penicillin, numerous antibiotic agents have been found, but only a few are suitable agents for the treatment of disease.

Certain criteria are necessary for the usefulness of an antibiotic substance to be used universally. First, it must be active against more than one genera and species of bacteria—a moderately wide range of activity. Second, it must not be toxic to the patient in doses effective against the pathogenic microorganisms. Third, it must not produce too many side reactions such as allergy, dizziness, neuritis, nausea, etc., in the patient.

ANTIBIOTICS

Penicillin

This antibiotic is produced by the metabolic activity of a mold, *Penicillium notatum* and other closely related Penicillia, particularly *P. chrysogenum*. In the production of penicillin the mold spores are seeded in large tanks containing thousands of gallons of special media and grown at a temperature about 26-30°C, and aerated during growth. After six to nine days the fluid is filtered, freeing the liquid from the mass of mycelia and spores. The penicillin is then extracted with organic solvents, such as ether, amylacetate, etc., after acidifying the solution. The penicillin is then removed from the organic solvent with an alkali and then purified by chromatography. The sodium or potassium salt of penicillin is then prepared, dried, standardized and used for treatment. The powder must be diluted before it can be injected, since it is highly alkaline. Application of penicillin salts (powder form) directly to the skin or wounds will cause severe damage due to the caustic action.

Various penicillin preparations are available for the treatment of disease and dispensed in 100,000, 300,000, 600,000 and one million units or more. Some preparations are put up in a waxy or oily base, to which is added procaine (an anesthetic). These preparations when injected are less painful than the salt and act as a reservoir in the tissue so that the drug absorbs into the blood stream quite slowly, thus one injection will give a sufficiently high blood concentration for 12 to 18 hours or more. A solution of the salt of penicillin without oil must be injected every 4 to 6 hours in order to maintain a suitable blood level. A higher concentration in the blood is obtained by the injection of the salt, but must be repeated several times during a twenty-four hour period because the penicillin is rapidly eliminated in the urine.   This antibiotic is more effective when given by injections, but may be given by mouth; however, blood levels are not as high as when injected.   Lozenges have been used for local mouth and throat infections, with some curative effects.

Since the antibiotic is eliminated primarily by the kidney, certain drugs, given before penicillin injections, can alter the kidney tubules so that the penicillin will be reabsorbed into the blood stream, thus allowing higher blood concentrations of the drug with less frequent injections, even with the salt of penicillin.

Penicillin is standardized by its ability to inhibit bacterial growth. The standard test involves the use of a strain of micrococcus.   Under standard conditions, 0.6 microgram of penicillin will produce a 22 mm. zone of inhibition of growth of the test organism.   This quantity is termed the International Unit.

At least four penicillin fractions occur naturally in media in which penicillin mold is grown.   These are termed penicillin F, G, K, and X.   Only G is marketed, since the others are less active in vivo.   All types are fairly active in the test tube (in vitro).   In the production of penicillin G, substances are added to the culture media, such as phenylacetamide which probably acts as a precursor or coenzyme in the manufacture of the penicillin G molecule, thus increasing penicillin G production.

Penicillin is used more frequently than any other chemotherapeutic agent, although there are many useful antibiotics and chemicals available.   Its action on organisms is probably through its interference with the glutamic acid metabolism of microorganisms.   Resistant forms synthesize their own required glutamic acid.

Streptomycin

In 1944 a new antibiotic substance, streptomycin,[4] was discovered, a metabolic product of *Streptomyces griseus*.

The preparation of this antibiotic is somewhat similar to the technique used in the production and purification of other antibiotic substances.   The unit of this antibiotic is designated as one microgram of the pure antibiotic salt; thus dosages are given in terms of milligrams or grams, instead of units.

---

[4]Schatz, A., Bugie, E., and Waksman, S. A.:   Proc. Soc. Exper. Biol. & Med. **55**:66, 1944.

This antibiotic is active against many organisms on which penicillin has little or no effect. Two important bacteria that streptomycin destroys are the *Mycobacterium tuberculosis* and *Hemophilus influenzae* bacterium. Neither of these are affected by penicillin. This antibiotic has a tendency to produce streptomycin-resistant strains particularly if sublethal doses are used. When drug resistance develops, in many cases, further use as a treating agent for a given disease is inadvisable. This antibiotic is frequently used in conjunction with the sulfonamides and also as an adjunct with penicillin therapy in certain diseases. Streptomycin is effective by mouth as well as by injections.

The exact chemical structure of this antibiotic is unknown, but it is probably a streptidine plus streptobiosamine. This antibiotic agent has not been synthesized. A newer product from streptomycin, "dihydrostreptomycin" is produced in the laboratory by substituting a hydroxyl group for the carboxyl group in the streptobiosamine molecule. This compound is now used more frequently than the original streptomycin.

Chloromycetin (choloramphenicol) the metabolic product of *Streptomyces venezuelae* was discovered in 1948.[5] This chemical has since been synthesized and it is now produced for commercial use by chemical synthesis. Its structural formula is:

$$NO_2 - \underset{\underset{\displaystyle OH}{|}}{\overset{\overset{\displaystyle H}{|}}{C}} - \underset{\underset{\displaystyle H}{|}}{\overset{\overset{\displaystyle NHCOCHCL_2}{|}}{C}} - CH_2OH$$

This antibiotic is effective by mouth and has a fairly wide range of activity against gram-negative organisms, particularly against *Salmonella typhosa*. It is therefore recognized as the treatment of choice in typhoid fever. Chloromycetin has a somewhat wider range of activity in its antibiotic effects, in that it has a curative effect on certain viral and rickettsial diseases.

This drug has an extremely bitter taste and unless taken in capsules (Kapseals) it is difficult to mask the bitterness. It has a few side reactions but is tolerated well by all age groups. However, in some cases it has been suggested as the cause of serious blood disorders.

---

[5]Ehrlich, J., Gottlieb, D., Burkholder, P. R., Anderson, L. E., and Predham, T. G.: Streptomyces Venezuelae, N. Sp., the source of Chloromycetin, J. Bact. **56**:467, 1948.

The drug is standardized so that one unit is equivalent to one microgram.

Aureomycin, the metabolic product of *Streptomyces aureofaciens* was discovered in 1948. The range of activity against microbes is similar to Chloromycetin and Terramycin. It is fairly effective against gram-positive cocci and rods as well as against many of the gram-negative group of microorganisms. This antibiotic can be taken by mouth and is also effective against certain rickettsiae and viruses. It has gained quite wide usage in local infections as well as being used in ointments for eye infections.

Terramycin, the metabolic product of *Streptomyces rimosus*, was discovered in 1948. The range of antimicrobial activity is similar to that of aureomycin. This antibiotic is also effective against certain rickettsiae and viruses.

Many other antibiotics have been discovered and some of these have a limited use, others are, however, too toxic against tissues to be usable in routine treatment of disease.

Erythromycin (erythrocin) is the metabolic product of *Streptomyces erythreus*, discovered in 1952. It is highly active against gram-positive cocci and also inhibits some gram-negative rods. There is moderate activity on certain strains of microorganisms which have become resistant to penicillin and streptomycin. The antibiotic activity is lost below pH 6 or above pH 8.0.

How Antibiotics Are Discovered

After the discovery of penicillin and establishment of the fact that it is an extremely valuable chemical in the treatment of disease, many research workers have undertaken the search for new antibiotics. The principal method used has been the testing of thousands of soil samples containing a multitude of microbes, using one or more kinds of bacteria as the test organisms. An agar plate is seeded with a known microorganism, for which an antibiotic is desired. Several soil samples or microorganisms grown from soil samples, infected wounds, water, etc., are then streaked across the above plate. If an organism is present that produces an antibiotic against the test bacterium a clear area will surround the test organisms. This prevention of growth signifies an antibiotic effect. Many thousands of soil samples and isolated microbes have thus been tested for antibiotic activity. This is frequently referred to as screening for antibiotic activity. Even though a given microorganism will readily inhibit or destroy other microbes, it does not mean that a new usable antibiotic has been discovered. Many tedious experiments are required to determine if the "new antibiotic" will inactivate a rather large group of different microbes. If the new product prevents or cures disease in experimental animals, its toxic effects on tissue and its relative cost of production will determine its value as a marketable antibiotic.

Testing Antibiotics

The methods used for testing the potency of antibiotics and testing the sensitivity of an organism against an antibiotic are similar. Several methods are used, but two such methods are more commonly employed.

The more accurate method is the serial dilution method. The antibiotic is diluted in a special broth, usually in a logarithmic progression, to one million or more. If an antibiotic is to be used for therapy only very dilute solutions of antibiotics are tested, e.g., fifty, twenty-five, ten, five, one unit etc., (penicillin) or from fifty to one or so micrograms (aureomycin, etc.). The test organism is added and the mixture incubated overnight. The highest dilution of antibiotic which shows no growth is the end point, i.e., the least amount of antibiotic preventing bacterial growth.

The other method is called the disc method. Discs of absorbent paper about the size of a paper punch disc are soaked in different amounts of antibiotic and dried or drained of excess fluid. The antibiotic-containing discs are then placed on agar plates previously seeded with bacteria to be tested. After incubation overnight there will be a clear area (no bacterial growth) surrounding the affective discs.

Besides being useful in the treatment of disease, recent investigations have shown that streptomycin, aureomycin, and Terramycin are dietary supplements, and some of these antibiotics are being added to chicken feeds and certain stock feeds. These "antibiotic supplements" increase growth for reasons not yet understood.

THE SULFONAMIDES

Salvarsan was the first practical chemical to be used successfully against microorganisms when given internally. This chemical was developed by Ehrlich for the treatment of the disease syphilis.   This latter drug has since been replaced by penicillin, for all practical purposes.

In 1933, Domagk published a report that an azo dye, prontosil, had a protective effect in mice infected with streptococci.   In the years that followed, many hundreds of sulfonamide compounds were synthesized.   Only a few, however, are now in general use for the treatment of disease or prevention of infection.

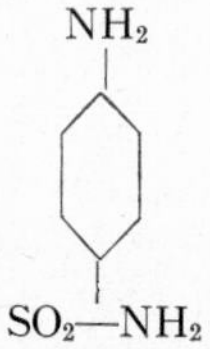

Sulfanilamide

The mother structure, sulfanilamide, forms the basis for many of the used sulfonamide compounds.   One of the most frequently used compounds of this type, sulfadiazine is used either by itself or in combination with antibiotics.

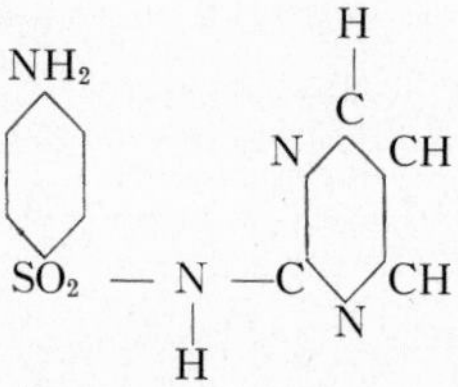

Sulfadiazine

All of the sulfonamide compounds are effective against certain infections when taken by mouth.   Due to their possible harmful effects on the kidneys, if taken in too large doses, these drugs are given only on prescription of a physician.

DRUG RESISTANCE AND MUTATION

Many of the antibiotics as well as chemicals (sulfonamide drugs, bichloride of mercury, iodine, phenol, etc.) will, in sublethal doses for bacteria, gradually build up a resistance in microorganisms, so that increasingly larger quantities of antibiotics will be required to inhibit bacterial growth.   Sometimes this "drug resistance" or "drug fastness" will increase ten- to twentyfold.   Thus if an organism initially susceptible to streptomycin in two microgram amounts, is continually exposed to one-half microgram amounts, it will eventually resist the effect of ten or even twenty micrograms.

Some microorganisms change in their normal characteristics when they become resistant to antibiotics.   Some lose their ability to produce pigment, lose some of their enzyme activity, but continue to be able to produce disease.

This "drug resistance" is an important consideration in the treatment of disease.   If insufficient amounts of antibiotic or sulfa drug are used for a given disease caused by a microorganism, there is a possibility of "drug resistance" developing by the microorganism. In a recent survey of micrococcus infections in a certain area, there appears to be an increasing resistance of these microorganisms to penicillin.   Previously nearly all strains were susceptible.   Now, nearly 50 per cent of these microorganisms are resistant to penicillin in sufficiently large quantities that treatment with this antibiotic is ineffective.

Selected Reading References

Herrell, W. E.:   Newer Antibiotics, Ann. Rev. Microbiol. 4:101, 1950.
Miller, C. P., and Bohnhoff, M.:   The Development of Bacterial Resistance to Chemotheraputic Agents, Ann. Rev. Microbiol. 4:201, 1950.
Waksman, S. A.:   Microbial Antagonism and Antibiotic Substances, ed. 2, New York, 1947, Commonwealth Fund.
Wooley, D. W.:   A Study of Antimetabolites, New York, 1952, John Wiley and Sons, Inc.

CHAPTER 11

INFECTIONS AND HOW THEY ARE RESISTED

Virulence, Resistance, and Immunity

It is important to differentiate the various types of diseases in order to understand infection. A disease is any deviation from the normal, thus gall stones impacted in the neck of a gall bladder produce a disease but it is not an infectious disease. Human cancer is a disease process, but to date no infectious agent has actually been discovered as the cause of this disease. An infectious disease, on the other hand, is caused by pathogenic microorganisms. There are infectious diseases of man, lower animals, insects, and plants, even including bacteria. Not infrequently we even speak of "diseases" of milk, wine, beer, food, etc., which may be caused by microorganisms and spoil the product for human consumption.

Pasteur devoted a great part of his life (1850-1890) studying diseases, how they were spread and the processes by which microorganisms were capable of producing infection. Koch (1882) not only discovered the microorganism of tuberculosis, but studied the mechanisms of this disease process. The mechanism of the virulence of the diphtheria organism was extensively studied by Roux and Yersin (1888) and a potent exotoxin was found that was responsible for the killing effects of this microorganism.

Not only did the early microbiologists study the mechanism of infection in relation to the virulence of microorganisms, but they also devoted much time to the host (patient with a disease) and his resistant state or ability to resist the inroads of these microorganisms. It was found in 1886 by Van Fodor and also Nuttall that normal blood had the capacity to destroy pathogenic microorganisms Bordet (1891) found that on immunization the animal blood showed potent powers of destroying the injected material (sheep red blood cells). Gruber and Durham (1896) found that immune blood serum

126

had the capacity of clumping bacteria with which the animal had been injected. Ehrlich (1899) suggested a theory (side chain theory) accounting for the production of these immune bodies or protecting bodies found in the blood serum.

HOW DISEASES ARE SPREAD

Many of our human diseases are spread by direct contact or are communicated from man to man directly by coughing, sneezing, dirty hands, etc. Such diseases, called *communicable* or *contagious* diseases, are whooping cough, measles, mumps, pneumonia, etc. Other diseases may be spread by carrier agents such as insects. Infected mosquitoes frequently cause disease, when they inject into your body a pathogenic microorganism, in the process of biting you and sucking your blood. Such diseases as yellow fever, malaria, various types of encephalitides, etc., are transmitted in this manner. Other insects such as ticks transmit Rocky Mountain spotted fever, tick fever, etc., if these insects carry these agents. Body lice may transmit epidemic typhus fever. The leaf hopper transmits the agent of curly top disease of beets. Insects capable of transmitting diseases directly are called vectors.

On the other hand, some insects may transmit disease agents indirectly and are called carrier insects. For example, the legs of flies or their bodies may become contaminated with pathogenic microorganisms from sewage, garbage, or fecal material, and as they crawl on exposed food they may leave microbes for your eating pleasure.

Rodents also frequently cause disease in man indirectly by contaminating food with their body discharges, which may contain pathogenic microorganisms.

Other indirect methods of spread of disease are by water which has been contaminated with sewage or fecal material, food contaminated by a human disease carrier and numerous other ways. Things that you put into your mouth that have been contaminated and not properly cleaned may transmit disease. Pencils, paper clips, dirty table ware, dirty drinking glasses, etc., or even your own fingers are all possible indirect methods by which diseases may be spread. Inert objects which transmit disease are called "fomites."

Another frequent method of contracting an infection is by means of a break in the tissue. Wood or metal slivers, scratches, animal bites, rose thorn pricks, nail puncture wounds, deep cuts, gunshot wounds, etc., are but a few of the methods by which we may introduce microbes into our bodies.

HOW MICROBES PRODUCE INFECTIONS

Microorganisms are continually struggling for their existence, and by being able to survive under many diverse and often adverse conditions they are constantly in our presence, greedy for a host to furnish food. Different microorganisms are endowed with a variety of weapons to enable them to overcome the body defense mechanisms. The combined microbic weapons may be referred to as the *virulence* mechanism. When we speak of microbial virulence we must also consider the body defense mechanism which tends to ward off infections by destroying the infectious agent or neutralizing its viciousness.

The factors that go into the make-up of virulence are not all known. Some organisms are at times highly virulent and invade the body with ease and at other times under identical conditions are avirulent or noninvasive. Some are virulent for one host, but not for other hosts of the same species. Some disease-producing agents, such as *P. tularensis*, are capable of producing disease when only a single microorganism of its kind is introduced into the body while other microorganisms such as *Streptococcus pyogenes* may require several dozen or even several hundred or more to produce an infection.

Toxins

Toxins are poisonous substances produced by certain microorganisms during their growth processes. These types of toxins, called exotoxins, are frequently harmful to the host, and may produce a poisonous effect on the tissues, thus allowing the microbe to gain a foothold. These toxins vary in their effects on different tissues. Some of these toxins produce a local killing of tissues; some destroy the red blood cells (hemolysins); some affect nervous tissue (neurotoxins); some destroy the white blood cells (leucocidin) or at least destroy their engulfing power; some destroy the cement substance that binds or cements the tissue cells together (hyaluronidase) thus producing little tunnels through which microorganisms can travel; some toxins destroy glandular tissue and some produce heart damage (viscerotoxins). Some microorganisms produce only one type of toxin while others may produce several different types. Some of these dissolve clotted blood plasmas (fibrinolysin), others cause blood plasma to form clots (coagulase).

Still other types of microorganisms produce no exotoxins, but the microbic cell (cytoplasm) is itself toxic to tissue. These types of toxic cytoplasmic substances are referred to as *endotoxins.*

Capsules

Some microorganisms possess a so-called envelope composed of protein or mucoid and polysaccharide mixtures, this representing the capsule surrounding the microorganism. These capsules may possess virulent properties, thus only a few organisms of the capsular type are necessary to produce an infection. It also serves as a protective mechanism for the microbe against the killing effects of leucocytes and antibodies of the host. When this capsule is lost, the virulence of the particular organism decreases markedly and may require, now, several hundred thousands or several millions to produce an infection. An example of this is the *Diplococcus pneumoniae* organism which, when capsulated, is capable in small doses of killing mice, but when it loses its capsule it is relatively avirulent, and many hundreds of thousands of these decapsulated organisms fail to produce an infection.

Other substances which are integral parts of the microbial cell membrane play a role in infection and virulence. A virulence component, frequently referred to as Vi antigen, and being present in the typhoid bacterium (*Salmonella typhosa*) does, in some way, aid the organism in invading the tissue. When the organism loses this particular component its virulence becomes greatly reduced.

BODY DEFENSE MECHANISMS

Perhaps the first defense line against infection, whether it be in man, insects, plants, or animals is the intact outer covering of the body (skin) as well as the lining of the respiratory and gastrointestinal tract (mucous membranes). The outside skin or epithelium harbors many types of microorganisms, some are harmless (*B. subtilis, Ser. marcescens*) and some are pathogenic (micrococci, streptococci, etc.). Thus a scratch, break, or other injury of the epithelium or mucous membranes allows the entry of bacteria which may be able to overcome the local defense mechanism of the body and produce either a local infection or, if they spread into the blood stream, produce a generalized infection. Normal skin contains about 1,500 to 3,000 microorganisms per square centimeter. A few hours after a flash burn causing only slight redness, this microbial population increases from 30,000 to 3,000,000 organisms per unit area above.[1]

When microorganisms gain entrance into the body tissues, the numerous defense mechanisms of the body begin a counterattack.

Antibodies are immune substances in the blood stream and on the surface of certain tissue cells and can be detected by various means. They are formed when one recovers from a disease or also after successful vaccination. These antibodies are produced by the injection of most all types of so-called complete heterologous proteins, including bacteria, viruses, heterologous serum, egg white, etc. Certain carbohydrates such as polysaccharides or other organic chemicals will also produce antibodies. There are several functions as well as different types of of antibodies which will be discussed later.

Antigens are therefore heterologous proteins which when introduced into the body cause certain body cells to elaborate antibodies. In disease, the microorganism is the antigen and produces specific antibodies against the specific disease agent. Specific antigens will cause the body to produce homologous antibodies, these in turn reacting only with the specific antigen. Some antigens are highly effective antibody producers, others are poor. Killed bacteria, such as the typhoid organism, is called a vaccine. It is an example of an antigen.

ANTIGEN-ANTIBODY REACTIONS

Antibodies are recovered from animals or man by obtaining blood and allowing it to clot, thus allowing serum to be separated from the clotted cells. This fluid part of the blood (serum) contains antibodies. We can therefore test the serum for the presence of antibodies by several means.

Agglutination

When insoluble proteins, such as microorganisms, are added in the test tube to serum containing specific antibodies, and in the presence of an electrolyte (0.9 per cent sodium chloride) the antibody coats the antigen, forming a thin film around the bacteria. If the correct proportions of antigen and antibody are present, the antigen

---

[1]Price, P. B.: Personal communication, Department of Surgery, University of Utah College of Medicine.

(bacteria) is effectively coated, and by physical-chemical attraction and cohesive forces the individual bacteria begin to clump together. When sufficient bacteria adhere to each other they settle to the bottom of the tube, and this clumping and settling out is called agglutination. The antibody causing the agglutination is called an *agglutinin* and the antigen is called an *agglutinogen*.

Precipitation

When a soluble antigen (egg white) is added to specific antibodies (anti-egg white antibodies) in correct proportions, the attachment of the antibody and antigen causes the resultant floc to aggregate into larger flocs and finally settles to the bottom of the tube. This is called a precipitate. If the antigen is carefully layered over the surface of a solution of antibody in the test tube a fine whitish ring or precipitate is formed at the juncture of the two fluids, this being designated the "ring test."

COMPLEMENT FIXATION

Complement is an enzymelike (cofactor) substance found in small quantities in nearly all blood serum. It does not increase during a disease process or during artificial immunization. It is easily destroyed by oxidizing chemicals and by heat, being inactivated at 56° C., in one-half hour. Complement, unlike antibody, is a nonspecific substance and it will combine with antigen-antibody complexes only when the antigen and its specific antibody unite.

In a complement fixation test to determine the presence of a specific antibody or antigen, and either one or the other can be unknown, we mix in test tubes some serum containing the antibody and add to this some antigen. To this we add a specific quantity of diluted and titrated normal guinea pig serum as a source of complement. The serum containing the antibody to be tested has previously been heated to destroy any complement that may be present, and also many of the so-called anticomplementary substances are destroyed by heat.

Let us consider At the antigen, Ab the antibody, and C the complement. The reaction of antibody-antigen-complement, in the presence of isotonic sodium chloride solution, is specific and will react even when the antibody is diluted to such an extent that agglutination or precipitation cannot be observed.

A. **Positive Complement fixation.**

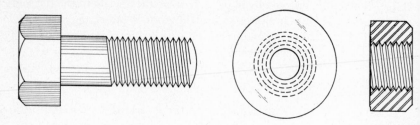

1. Specific antigen (*At*) *Sal. typhosa* + Complement (*C*) + Specific antibody (*Ab*)
against *Sal. typhosa*

Elastic washer *C* (complement) stretches to fit over antigen bolt (At) and specific antibody nut (Ab) screws on to bolt to fix complement (C) firmly.   After incubating 37° C. about two hours, add indicator system.

2. Indicator System: Specific antigen (sheep red blood cells) and specific antibody (amboceptor) or antibody against sheep red blood cells.

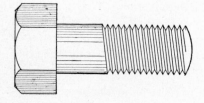 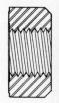

Sheep red blood cells antigen (*At*) + Antisheep red cell antibody (*Ab*)

Since all the complement *C* was used in *A-1* there is none remaining in *A-2* to combine with the indicator system of specific antigen-antibody reaction; therefore, the sheep red cells will not hemolyze.   Therefore, the test is a positive complement fixation, *A-1*.

Fig. 11.—Schematic representation of the complement fixation test.

B. **Negative complement fixation.**

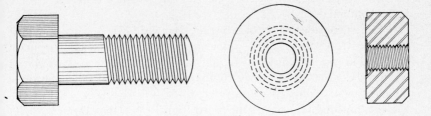

1. Antigen *Sal. typhosa* (*At*) + Complement (C) + Nonspecific antibody anti
Esch. coli (*Ab*)

The elastic washer *C* (complement stretches to fit the antigen (bolt) but the antibody (nut) is too small in this case; therefore, it is not specific for the antigen (At). The complement (*C*) will therefore fall off and thus be available for another antigen-antibody reaction. This test is incubated about two hours at 37° C.

2. Indicator System. Specific indicator antigen (sheep red blood cells) and specific indicator antibody (antisheep red cell antibody) are now added.

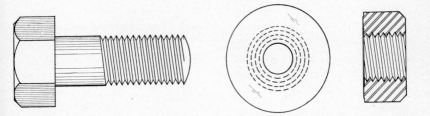

Sheep red blood cell antigen (*At*) + Complement C + Specific antibody for sheep red
         Complement   blood cells (antisheep red
       from *B-1* unused) cells) (*Ab*)

Since in *B-1* the nut (*Ab*) failed to fit the bolt (*At*) the complement (*C*) was free to act in the indicator antigen-antibody combination. The sheep red blood cells are now lysed and the solution is pink in color. The test *B-1* is, therefore, a negative complement fixation.

Complement-fixing antibodies are possibly different from the other previously mentioned types of antibodies.

Fig. 11.—Schematic representation of the complement fixation test—Continued.

$At + Ab + C =$ Fixation of complement to the At-Ab combination if the Ab has been formed in the animal body by the specific At. Complement will attach to the specific At+Ab complex even after these have combined.

Let us consider a bolt of a certain size, say ¼ inch with standard threads as the antigen (At), and a nut to fit this bolt as the antibody (Ab). This specific nut will fit the specific bolt and can be screwed on tightly. Let us consider complement (C) as an elastic washer which is nonspecific, or can stretch to fit any size bolt. When the specific antibody (Nut) is fastened tightly on the antigen (bolt) which contains the complement (elastic washer) the complement cannot be removed. If the antigen (bolt) and antibody (nut) are not specific they will not combine tightly to hold the complement, thus the complement is free to combine with another specific antigen-antibody combination, if added.

An indicator system, containing a specific antigen and antibody, is added to the above to determine if any complement is still available, or unused. The indicator antibody is an *hemolysin* or an hemolytic antibody produced by injecting rabbits repeatedly with washed sheep red blood cells. This hemolytic antibody, frequently referred to as *amboceptor* will cause, in the presence of complement, a lysis or dissolution of the sheep red blood cells (hemolysis), thus producing a pink color to the fluid. If no complement is available the red blood cells will settle to the bottom of the tube leaving a clear supernatant fluid.

The complement fixation test represented schematically would then be as shown in Fig. 11.

Neutralizing Antibodies

One type of antibody develops in the animal body in response to exotoxins, these being called antitoxic antibodies (antitoxins). These naturalize the toxic effects of the toxins. Another type of neutralizing antibody is that kind which neutralizes or destroys viruses. This type of antibody is produced in response to a virus in the body.

Opsonins are antibody-like substances that are present in nearly all normal blood serum, but may also be increased during a disease or by artificially immunizing an individual. The opsonins in some way "flavor" the bacteria so that leucocytes either find the bacteria easier or make them digestible. When the opsonin content of blood serum is low, leucocytes are able to engulf only a few microorganisms, say ten or twelve per leucocyte; when opsonins are high (immune state) in the serum, a leucocyte may engulf fifty of more bacteria. An opsono-phagocytic index, which is derived from the average number of bacteria engulfed per leucocyte, when bacteria are exposed to immune serum, divided by the average number of bacteria per leucocyte when bacteria are treated with normal serum is useful in determining resistant states.

DEVELOPMENT OF ANTIBODIES IN THE ANIMAL BODY

Many theories have been advanced since the side-chain theory of Ehrlich to explain the development and mechanism of action of antibodies, but none explain completely this protective phenomena. More recently Pauling[2] has suggested an infolding of protein molecules to form antibodies and an unfolding of the antibody molecule

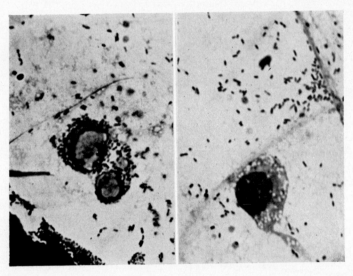

Fig. 12.   Fig. 13

Fig. 12.—Antibody-producing cells. Macrophages from a mouse treated six days previously with *Sal. typhosa* contained in a fine glass capillary which was inserted under the skin (local immunity). A tissue smear was then prepared and to this was added killed *Sal. typhosa*. The preparation was stained with May-Grunwald-Giemsa stain. Note the great number of bacteria attached to the cell.

Fig. 13.—Similar tissue smear from a normal "nonimmune" mouse, the smear being treated with *Sal. typhosa* as in Fig. 12. Note the number of organisms near the macrophage, but there are none fastened to this cell.

---

[2]Pauling, L.: A Theory of the Structure and Process of Formation of Antibodies, J. Am. Chem. Soc. **62**:2643, 1940.

to match the incoming antigen when reintroduced. Heidelberger[3] has done much to further our understanding of the mechanism of antigen-antibody combination.

Antibodies are formed by some mechanism, the exact chemico-biological activity that takes place being not completely understood. The antigen apparently attaches itself to certain reticular cells and/or lymph cells, these in response to this foreign substance, produce antibodies. The globulins thus formed, at the expense of the antigen and some metabolic changes in the cell, are the immune substances or antibodies.

Endothelial cells may also play a role in antibody production. Antibodies may be formed locally in a restricted area or may be formed in many areas throughout the body by such cells and then released into the blood stream. The lymph nodes, spleen, bone marrow, and connective tissue probably play a role. Brain tissue cells may also produce their own antibodies against heterologous antigens. *Immunity* and *resistance* to infection are undoubtedly due not only to circulating antibodies in the blood stream, and antibodies on the surface of certain tissue cells, but also to the mechanical resistance of cells of the body to invasion by microorganisms. We thus have cellular immunity and humoral (antibody) immunity.

ACTIVE IMMUNIZATION

Artificial active immunity is produced by the injection of an antigen into the body. The body responds by producing specific antibodies as well as specific tissue cell resistance; thus the individual develops an immunity. Some bacteria, toxins, and viruses may be treated with chemicals or heat and are still capable of producing antibodies when injected into the body. Not all diseases, however, can be prevented by vaccination because some microorganisms are poor antigens when killed with either heat or chemicals. Some disease-producing agents are used as vaccines that are able to produce a mild infection (e.g., vaccination against smallpox), and this type of artificial immunity lasts longer than that produced by dead antigens. Artificial active immunity varies in its lasting qualities, some types lasting several years, while others persist for only a year and such vaccinations must be repeated, "booster injections" being necessary every year or two.

---

[3]Heidelberger, M.:   Quantitative Absolute Methods in the Study of Antigen-antibody Reactions, Bact. Rev. **3**:49, 1939.

Natural Immunization

Recovery from certain infections (diphtheria, typhoid fever, measles, etc.) leaves most individuals immune for life to those specific infectious agents. Other microorganisms may produce only temporary immunity, lasting only a few months to a year (micrococcus infections—boils, abscesses; streptococcus sore throats, etc.). Many individuals may have a very mild attack of a disease, so mild, in fact, that the individual may not know he has had such an infection. Solid, lasting immunity frequently develops, which is equally as good as when an individual becomes very ill with the same disease and recovers. Poliomyelitis, diphtheria as well as other diseases may produce, in some individuals, such a mild infection and the resulting immunity prevents a second attack throughout the life of the individual.

Passive Immunization

When antibodies or antitoxins are produced in one person or animal and injected into another for the prevention or cure of a disease, this is called *passive immunization*. This type of immunity (borrowed immunity) lasts only a short period of time, three to six weeks at most. Many diseases were formerly treated with antibody (immune serum) but the use of sulfonamides and antibiotics has largely eliminated serum therapy, except for certain toxic and viral diseases.

When a baby is born to an immune mother, antibodies from the mother pass through the placenta and the baby is endowed with a temporary immunity to whatever extent the mother may have had such antibodies. Just before birth the mother produces, by unknown activities of hormonal combinations, extra amounts of antibodies, presumably for the benefit of the child. This temporary antibody immunity lasts only a couple of months, and the infant becomes completely susceptible to a disease for which he had a temporary immunity.

OTHER TYPES OF RESISTANCE TO INFECTION

Plants do not develop diseases of animals and vice versa. Some animals are highly resistant to certain human diseases. For example, poliomyelitis virus will not infect rabbits or guinea pigs. Certain disease agents may be carried by insects but have no effect on the insect, but when the insect bites a susceptible host (man or primates) he will develop the disease (e.g., yellow fever).

Age, sex, and seasonal effects frequently play a role in resistance to infection. Nutritional deficiencies and hormonal imbalance play a role in susceptibility to infection. Vitamin deficiencies and certain amino acid deficiencies tend to reduce the natural resistance of the host, thereby inviting infection. It is not uncommon to see an individual with uncontrolled diabetes suffering from numerous boils and abscesses. This suggests hormonal effects on tissue or local resistance to infection. Cortisone reduces inflammatory processes and in many instances increases host susceptibility to a disease.

HYPERSENSITIVE CONDITIONS

When a foreign protein (antigen, preferably fluid or soluble type) is injected only once into the body (preferably intravenously) the antibody-producing mechanism begins to function. The cells responsible for antibody production begin to form moderate quantities of *raw* (attached to cells) antibodies and a few scattered *mature* antibodies (those cast off into the blood stream). If no further antigen injections are given for twelve or more days, the antibody-producing cells become irritated due to these attached antibodies (like burs on a dog's fur) and become hypersensitive to a reintroduction of antigen. The same antigen on reinjection (intravenously) after the above elapsed time combines with these cell-attached raw antibodies. The cells then undergo a violent reaction, many breaking up or disintegrating and liberate a substance called histamine or histamine-like chemical. Such liberated substances then cause smooth muscle constriction of the lungs, the animal or man being unable to inhale or exhale; also there is a marked fall of blood pressure; this anaphylactic shock may result in death. This shock, under the above conditions, produced by the antigen-antibody combination is called *anaphylaxis* or *anaphylactic* shock. Little danger of such shock phenomena is encountered with ordinary bacterial vaccines.

If repeated doses of serum (foreign protein antigen) are given in the same area intramuscularly daily for several days a local anaphylaxis is produced. This is called the *Arthus phenomena*. Redness, swelling, and local tissue death (necrosis) may result.

Serum sickness, which is due to the intravenous or even intramuscular injections of large quantites of a foreign serum, may be evident in five to ten days after injection, due to a generalized antigen-antibody combination. The patient develops a fever, body rash (hives), joint pains, and may be quite ill for several days. Antihistaminics may reduce these states of hypersensitivity in certain cases.

Allergy, whether natural or acquired, is a form of hypersensitivity. Hereditary allergies, called *atopy*, are such conditions as pollen and dust allergy, asthma, and certain food and drug allergies. Predisposition to these allergies (atopic) are transmitted by genes according to the Mendelian law of heredity trait transmission.

Acquired allergies such as skin rash due to soaps, chemicals, drugs, or other contactants are also examples of hypersensitive states.

Selected Reading References

Burnet, F. M., and Fenner, F.: The Production of Antibodies, ed. 2, Melbourne, 1949, The Macmillan Company.

Mayer, M.: Immunochemistry, Ann. Rev. Biochem. **20**:415, 1951.

Nungester, W. J.: Mechanisms of Man's Resistance to Infectious Diseases, Bact. Rev. **15**:105, 1951.

Smith, Theobald: Parasitism and Disease, Princeton, N. J., 1934, Princeton University Press.

CHAPTER 12

BLOOD GROUPS

Bloodletting was a commonly practiced procedure many centuries ago, accomplished either by cutting a vein or allowing leeches to engorge blood, as a method of treating many ailments.

In the eighteenth century the idea developed that blood could be replaced in those individuals who had lost blood by hemorrhage. It was at this time that various experiments were tried, transfusing animal blood (sheep, cow, etc.) into humans. First transfusions were frequently successful, but second transfusions of *foreign species* blood were frequently fatal. This practice was abandoned. Experiments were then performed transfusing humans with human whole blood. A few trials were successful; many trials, however, produced severe reactions in the recipient, and many deaths resulted.

From 1880 to 1900 extensive work was carried out on the above problem and it was found that foreign blood (serum of a cow, sheep, etc.) when injected into a human *sensitized* the person, and that after twelve or more days this sensitization was at its height. Injections of the same foreign protein would, after the above elapsed time, when injected into the sensitized person, produce a severe or fatal shock— the anaphylactic shock. However, human blood injected into humans could not be relegated to this same phenomena. In 1901 Landsteiner[1] produced experimental results explaining the cause of severe reactions in certain humans when transfused with blood from another human.

It was discovered that human blood could be divided into distinct groups called antigenic types—each group possessing its own antigenic peculiarity.

[1]Landsteiner, K.:   Wein. klin. Wchnschr. **14:**1132, 1901.

GROUPS O, A, B, AB

Human blood is now divided into four groups, called O, A, B, AB (international classification suggested by Von Dungern and Hirszfeld).[2]

This classification is based on the antigens found on the surface or in the stroma of the red blood cell walls and the antibodies found in the blood serum.

The group O cells have no antigen of the blood grouping type in the red cells. The serum, however, contains antibodies against both A and B cells, or alpha and beta antibodies. Group A cells contain A antigen; the serum contains beta antibody. Group B cells contain B antigen and alpha antibody in the serum. Group AB cells contain A and B antigen, but no antibody for either antigen in the serum.

These blood group antigens (A,B) and antibodies (alpha, beta) are inherited, according to Mendelian law. These *isoantigens* and *isoagglutinins* therefore play an important role in heredity in siblings from matings. If both parents are pure O, offsprings can therefore be only O blood group. If both parents are A, siblings may be either O or A group, etc.

TABLE IX

INTERNATIONAL CLASSIFICATION OF BLOOD GROUPS

GROUP	ANTIGEN IN RED CELL	ANTIBODY IN SERUM	BLOOD CELLS AGGLUTINATED	PER CENT DISTRIBUTION (U.S.A.)
O	O	Anti A and Anti B (alpha and beta)	A, B, and AB	45
A	A	Anti B (beta)	B and AB	41
B	B	Anti A (alpha)	A and AB	10
AB	AB	None	None	4

Since the antigen and antibodies of blood groups behave as any specific antigen-antibody reaction, the specific antibody will cause agglutination when mixed with the specific antigen. Thus a group A

---

[2] Dungern, F. Von, and Hirszfeld, F. Z.: Immun. Forsch. **4**:531, 1910; ibid **6**:284, 1911.

individual who may be accidentally transfused with group B blood will have a severe transfusion reaction. Group B blood contains B antigen in the blood cells and alpha antibody (anti A) in the serum. Group A blood contains A antigen in the red cells and beta antibody (anti B) in the serum. If we transfuse a person of A group with B group blood the beta antibody in the group serum A blood will cause the B group cells to agglutinate. At the same time the red cells will be lysed (c.f., complement fixation) since complement is present in the blood.

TABLE X

TRANSFUSION REACTIONS

BLOOD GROUP OF PATIENT TO RECEIVE BLOOD		BLOOD GROUP OF PATIENT TO GIVE BLOOD		REACTING ANTIGEN-ANTIBODIES	TRANS-FUSION REACTION
Antigen	Antibody	Antigen	Antibody		
O	alpha and beta	O	alpha and beta	None	None
O	alpha and beta	A	beta	"A" donor cells and antibody A (anti A) of recipient	Severe
O	alpha and beta	B	alpha	"B" donor cells and antibody B (anti B) of recipient	Severe
O	alpha and beta	AB	None	"AB" donor cells and antibody A and B (anti A and anti B) of recipient	Severe

It is therefore vital that blood must be grouped (typed) and cross matched before any transfusion. Cross matching is done merely as a safety measure in event that there might have been a mistake in the original typing procedure. To cross match, merely mix a dilute physiologic sodium chloride mixture of recipients' blood cells and donor blood serum and determine if the red blood cells agglutinate.

To initially type blood, make a dilute suspension of red cells of the patient's blood with known A and B antibody and place them in hollowed spaces on a spot plate. You then rotate the plate gently and allow the cells to settle. If the cells are agglutinated, the small groups of clumped red cells appear as many grains of red paint pigment dropped on the cement, with clear spaces between the grains. If no agglutination takes place, the red cells settle to the bottom in a smooth, unbroken cap or disc.

Group A cells have been further subdivided into two subgroups, A_1 and A_2 being the subgroups.   These subgroups are not too important insofar as blood transfusions are concerned.

OTHER BLOOD CELL ANTIGENS

Two additional human blood cell antigens M and N were discovered[3] in 1927; everyone has either M, N, or MN.   These antigens are unimportant in relation to transfusions.   They serve to identify further a given blood and this knowledge is helpful in medicolegal cases and as identification aids in genetic studies.

THE Rh ANTIGENS

By immunizing rabbits with red blood cells from monkeys (*M. rhesus*) it was found that the resulting red cell antiserum would agglutinate certain human red blood cells.   Landsteiner and Wiener[4] called their newly discovered factor the Rh (rhesus) factor.   This Rh antigen may be present on the red cell wall of any blood group, even in the group O.   This Rh antigen is inherited similarly to the blood group antigens, and according to Race three closely linked paired genes are capable of genetic transmission.

WIENER TERMINOLOGY	FISHER-RACE TERMINOLOGY	
Rh_o	Dce	
Rh_2	DcE	
Rh_1Rh_2	DCE	—Rh positive
Rh_1	DCe	
rh'	dCe	
rh''	dcE	
rh'rh''	dCE	—Rh negative
rh	dce	

[3]Landsteiner, K., and Levine, P.: Proc. Soc. Exper. Biol. & Med. **24:**600, 941, 1927.
[4]Landsteiner, K., and Wiener, A. S.: Proc. Soc. Exper. Biol. & Med. **43:**223, 1940.

The Rh antigen is extremely important in relation to transfusions and in the production of *erythroblastosis fetalis* in a fetus in utero. About 85 per cent of the white race is Rh positive. The D (Rh$_o$) antigen is responsible for about 95 per cent of immunizations to the Rh factor, and about 50-75 per cent of D negative (Rh-) persons may be sensitized by a single transfustion of D positive blood cells.

If an Rh-negative mother produces an Rh-positive baby, the Rh-positive antigen from the red cells of the fetus may pass through the placenta and sensitize the mother, who will in turn develop antibodies against this Rh$_o$ (D) antigen. The antibodies of the mother's blood may then pass the placental barrier to the fetal circulation causing a typical antigen-antibody reaction in the fetus. The antired cell antibodies will clump and hemolyze the fetal red cells, and, if severe, will produce a disease in the fetus called *erythroblastosis fetalis*. In this disease, the fetus is water logged or edematous and frequently jaundiced and death of the fetus may ensue.

Only a few Rh-negative (D−) women who bear an Rh-positive child have any difficulty in relation to an erythroblastotic fetus. Sensitization of the mother must take place, and comparatively few are sensitized at the first pregnancy. Some never become sensitized, probably because the placental barrier is not permeable to the Rh antigen of the fetus. Once a mother is sensitized and shows antibody for the Rh antigen there is a possibility that in subsequent pregancies, if an Rh-positive fetus develops, erythroblastosis in the fetus may be evident. Sensitization to the Rh antigen may also be brought about by direct transfusion if the female is Rh negative and receives Rh-positive blood. An Rh-positive fetus may then encounter difficulty as a result of the development of *erythroblastosis fetalis*.

An Rh-positive female will have no difficulty in relation to pregnancy. In fact, an Rh-positive person may be transfused with Rh-negative or Rh-positive blood with no fear of sensitization, provided the correct blood group is used. Both male and female Rh-negative individuals will become sensitized to the Rh-antigen if transfused with Rh-positive blood, and a subsequent transfusion of Rh-positive blood may produce severe reactions or even death.

THE BLOOD BANK

In most large communities there is at least one blood bank, run either by the Red Cross or by a hospital or private group. Here blood is taken from donors, stored in the refrigerator, typed, Rh determinations made, and serological tests are run on the blood. After these tests are completed it is ready for use.   Both whole blood and plasma may be used.   In large blood banks the globulin left over after purification for albumen may be used other than for transfusions.

One globulin fraction contains most of the antibodies and this is called the *gamma globulin* fraction.   After concentration and purification it is used for the prevention of measles and certain other diseases.

Selected Reading References

Boyd, W. C.: Genetics and the Races of Man, Boston, 1950, Little, Brown and Company.

Levine, P., Katzin, E. M., and Burnham, L.: Isoimmunization in Pregnancy; Its Possible Bearing on the Etiology of Erythroblastosis Fetalis, J. A. M. A. **116**:825, 1941.

McCurdy, R. M. C.: The Rhesus Danger; Its Medical, Moral and Legal Aspects, London, 1950, Wm. Heinemann, Ltd.

Race, R. R., and Sanger, R.: Blood Groups in Man, Springfield, Ill., 1950, Charles C Thomas, Publisher.

CHAPTER 13

SAFE AND UNSAFE WATER SUPPLIES

MICROBIOLOGY OF WATER AND SEWAGE

WATER SUPPLIES

It is trite to say that it would be impossible for us to live without water. However, in the United States and many other parts of the world we take for granted the safe, adequate water supply which we are privileged to enjoy. To appreciate fully the complete dependence of an urban center on the water supply around which it is built, try to picture the chaos and suffering which would result if New York City, Chicago, or Los Angeles were suddenly deprived of its water supply or if that water supply were to suddenly become unsafe to use.

The high standard of living in America has been developed, in part at least, because we have provided safe water supplies for a substantial part of our population.

The average city dweller will use more than 100 gallons of water per day and nearly 70 gallons are likely to enter the sewerage systems as sewage. Thus water creates a double problem; that of providing an adequate, safe, dependable supply and disposing of the sewage (secondary to such a water supply) in such a manner that it will not contaminate other water supplies. Because of its very nature, water can become a potential transporter of disease from one man to another.

With the vast increase in population in the world, prevention of pollution and contamination of waters which are potential or actual sources of urban supplies presents a very real problem which is rapidly assuming serious proportions.

Water is a major natural resource. Each of us needs to realize this and recognize the problems and dangers involved. Rain or snow represents the ultimate source of all water. This is distilled water and as such is theoretically pure water. What it contains as it reaches the earth in the way of bacteria, chemicals, dust, etc., will naturally depend on what is present in the atmosphere through which it passes. This in turn is dependent on the proximity of the falling rain to cities, plant-deficient desert areas, etc. A remarkable array of substances may be washed out of the atmosphere by falling rains. Dusts may add substantial numbers of the microorganisms which are normal soil inhabitants; pollen and fine bits of decomposed organic matter may also represent biological contributions. Chemicals such as sulfuric and metallic fumes may be present in the air over cities. Salt flats may contribute quantities of fine salt and halophilic bacteria. Even radioactive substances may find their way into some falling rain.

As soon as rain strikes the earth it becomes water of the next class, i.e., *surface water*. Such water is represented by streams, ponds, and lakes and probably even very shallow wells. Surface water is potentially dangerous as a carrier of disease wherever it comes in contact with man or his habitations. It represents a natural mobile highway which can easily carry disease producing microorganisms from an active case or carrier of a disease to any user of untreated water. The statement that surface water is never "safe water" while not entirely true is a good rule to follow. Appearance may be deceptive. The sparkling brook may be dangerously contaminated without its taste or appearance being impaired.

Ground water is represented by water from deep wells or deep springs. Being free from immediate surface contact its temperature is about the same both in winter and in summer. Generally ground water is safe water but contamination of ground water can occur through fissures in rocks, coarse gravel substrates, broken well casings, etc. The line of demarcation between surface and ground waters is not too clear cut.

Water as a Disease Carrier

Disease-producing microorganisms in water supplies do not arise spontaneously. They come from pre-existing cases of the same disease or from a carrier of such organisms. Typhoid does not develop as a result of filth in the water. It comes from a case of typhoid or a carrier of this organism. If we could eliminate all cases and carriers of typhoid the disease could become nonexistent and disappear just as cholera has disappeared in America.

Water may be described as (a) "potable," i.e., clean, safe, pleasing in appearance and taste; (b) polluted, i.e., having substances added which impair color, odor, or taste; and (c) contaminated, i.e., rendered unsafe through the addition of discharges from human or animal intestines or rendered dangerous by the addition of poisonous chemicals. Frequently the term polluted is used to imply a combination of both terms.

Diseases which are transmitted through water supplies are primarily those present in intestinal discharges. The reason for this is that through sewage, these intestinal pathogens find their way into water, where they generally have greater survival power than most of the other pathogens so introduced.

The bacterial diseases most commonly transmitted through sewage are typhoid, paratyphoid A and B, bacillary dysentery, amebic dysentery, and cholera. Some virus diseases are believed to be transmitted through contaminated water supplies. These include infectious hepatitis, (also known as infectious jaundice), and perhaps poliomyelitis, since the virus of poliomyelitis is discharged in great quantities from the intestines. However, the evidence in the case of poliomyelitis is not clear-cut. At least one epidemic of another type of infectious jaundice, caused by the spirochete *Leptospira icterohemorrhagiae*, whose chief reservoir is the common rat, has been reported.

A number of years ago, Russian workers reported an unusual epidemic of tularemia transmitted from a colony of "water rats" to workers on a communal farm who drank water from a brook.

It is often hard to prove definitely that a given disease is water borne because of the difficulty of isolating the causative organism from the water supply. However, epidemiological evidence often gives strong proof that a disease is water borne. We have known definitely for only about a century that water may serve as a carrier of disease. The story of this discovery is as follows:

In 1854 an epidemic of cholera flared in London. Peculiarly enough it seemed to be most intense in the Broad Street region. John Snow studied the outbreak, which raged in the area from mid-August until late in September and discovered that most of the victims had drunk water from the Broad Street well. Examination showed the well to be contaminated from a faulty sewer pipe. Nearly 700 deaths occurred in this area. After the pump was removed the intensity of the epidemic dwindled rapidly. Snow described his findings in a monograph on the "Mode of Communication of Cholera."[1] It represented the first clear-cut proof of the role of water as a carrier of disease and is a classical scientific paper.

Many other epidemics were also reported subsequently and the role of water as a carrier of disease amply authenticated.

A classical and tragic epidemic occurred in Hamburg in 1892 which first established the efficacy of filtration of water supplies. The city of Hamburg, which had a population of about 640,000 drew its water supply directly from the Elbe River where it flowed into the mains following a brief storage in settling basins. Altona (143,000) an adjacent city also drew its water from the Elbe at a point below Hamburg. This water was grossly contaminated with the sewage from Hamburg. Altona, however, subjected its water supply to slow sand filtration before using it. In the late summer of 1892 an epidemic of cholera occurred in Hamburg and before it subsided some 17,000 cases of cholera had occurred; of these, 8,605 persons died. Altona went through the epidemic virtually unscathed, a spectacular proof of the value of water filtration.

On the other hand, the well-publicized epidemic of amebic dysentery which occurred in Chicago during the World Fair of 1933, illustrates the need for constant vigilance and the dangers of faulty sewerage construction or mechanical failure of equipment.

No longer do the water-borne diseases make autumn a melancholy season which carries many people to their graves as a result of epidemics of typhoid fever and similar diseases. The tremendous strides we have made in water treatment is evidenced by the following statistics regarding the most important of the water-borne diseases:

The death rate from typhoid in 1900 was 31.3 per 100,000. Epidemics were frequent throughout the country. Of course, water was not the only carrier of the disease, since it shared this dubious honor with milk, food, fingers, and flies; however, it was the most deadly source of the dread disease.

In 1948 the typhoid death rate had dropped to a mere 0.2 deaths per 100,000, a decrease of more than 150 times.

This represents added years of life in which we all share and a monument to the proper application of bacteriological principles.

---

[1] Reprinted by Commonwealth Fund, 1936.

One of the greatest benefits that has come from adequate treatment of water supplies has been the amazing reduction in infant mortality which has occurred in recent years. However, recently the discovery has been made that certain surface well waters which are high in nitrates may be dangerous for infants. The nitrates apparently unite with the hemoglobin of the blood, changing it to methemoglobin, in which state its oxygen-carrying power is greatly impaired. Such infants become cyanotic or "blue babies" and a number of deaths from this cause have been reported. The nitrates are produced as a result of the bacteriological transformation of sewage or other polluting materials in the water.

Bacteriologically Safe Waters

Water as it occurs in nature is rarely entirely free from bacteria although water from deep wells may often show less than one bacterium per milliliter. Numbers of bacteria in surface water may vary from a few to thousands per milliliter depending on the nature and amount of external matter which enters the water supply.

If water is not contaminated with human or animal feces, the bacteria present will be harmless saprophytes whose normal habitat is water or bacteria which have been introduced from the soil.

Most of the normally occurring water forms will be members of the genera *Pseudomonas*, *Flavobacterium* and similar forms together with a variety of bacterial species which cannot be grown on nutrient agar or other common media. These latter organisms can be demonstrated by placing clean glass slides in water and allowing them to remain until the bacteria have become attached to the surface of the glass. The numbers and varieties of bacteria thus found are frequently great; however, they probably have little bearing on the sanitary quality of the water.

In addition to bacteria, large numbers of algae may also be present, especially during the spring and fall months. There is no evidence that these forms are pathogenic to human beings although sickness and death in cattle has been attributed to drinking water heavily laden with algae. Humans would be unlikely to drink such water since it would have a heavy scum of microscopic green plants. However, certain of the algae are notorious as producers of undesirable tastes and odors in water. A blue-green algae, *Nostoc*, which resembles an over-sized streptococcus, has the peculiar honor of producing a "pigpen" odor in waters. Others produce fishy, cucumber, geranium, and similar exotic odors or tastes.

In waters rich in iron salts, the sheathed iron bacteria, belonging to the *Chlamydobacteriales* may grow so rapidly that they sometimes rather effectively clog pipe lines.

Detection of Contamination in Water

Because many of the major cities of the United States must secure their drinking water from rivers or lakes which in turn have served as carriers or repositories of sewage from other cities, methods for determining the bacteriological safety of water supplies are of vital importance to the life and well-being of all of us. Also, the constant threat of mechanical failures in waterworks equipment makes imperative the frequent testing of water supplies.

The problem of developing a method for determining the safety of water supplies has been a difficult one to solve. The reason has been that it is almost impossible to isolate intestinal pathogens from any except heavily contaminated water supplies and the presence of any fecal contamination, even in the absence of specific disease producers, is itself highly objectionable.

Early workers such as Koch and others attempted to count the numbers of bacteria in water as a means of determining its safety. However, it was discovered that the kind of bacteria was far more important than mere numbers of bacteria.

Examination of human intestinal discharges revealed the fact that several kinds of bacteria were commonly present in the intestines. Three of these, now known as *Escherichia coli, Streptococcus faecalis,* and *Clostridium perfringens,* were found to be constantly present in these discharges.

For example, 200 to 400 billion *Escherichia coli* cells are excreted per day by an average person.

Therefore, the presence of these bacteria in a water supply is evidence of contamination with fecal material since their normal habitat is the intestinal tract.

Escherichia coli and closely related forms, known as the coliforms, have, therefore, been selected as suitable indicators of sewage pollution. While they will persist for a period of time in water, they apparently are not able to grow in water and they will ultimately die off. They are more resistant than the intestinal pathogens, so that when the coliforms are no longer present it may be assumed that the intestinal pathogens have also disappeared.

Streptococcus faecalis has also been used as a sewage contamination indicator but it does not lend itself quite as readily to the role of "indicator organism."

The objection to *Clostridium perfringens* lies in the fact that it is a sporeformer and its spores could readily survive such treatments as chlorination. Its presence would, therefore, not be a satisfactory indicator of recent sewage pollution.

For many years attempts have been made to develop laboratory tests which would afford information as to safety of water supplies by providing a reliable means of detecting the presence of bacteria characteristically found in the intestines. Out of the studies have grown the *Standard Methods of Water Analysis* prepared by representatives of the American Public Health Association and American Waterworks Association. This publication is now in its ninth edition and is undergoing constant revision.

Three tests or groups of tests are recognized (a) the presumptive test, (b) the confirmed test, and (c) the completed test.

Presumptive Test.—*Escherichia coli* and the related bacteria associated with the intestinal flora make up a group of organisms known as the coliforms. These bacteria are gram-negative rods which have the ability to ferment lactose (milk sugar) with the production of acid and gas. The medium used is essentially nutrient broth plus 0.5 per cent lactose. It is placed in test tubes (or bottles where large amounts of water are being tested) and a smaller test tube is inverted in the medium. When gas is formed it collects in the inverted tube and its presence can therefore be easily recognized.

Varying quantities of water may be used, ranging from as little as 0.001 ml. up to 100 ml. These are added to broth, the culture incubated at 37°C for 24 hours and the tubes examined for the presence of gas. The development of gas in the inverted tube is taken as "presumptive" evidence of the presence of coliforms. This medium serves as an enrichment medium since it is especially favorable to the coliforms. As a result, theoretically, at least, it is possible to detect a single coliform organism in 100 ml. of water. This represents an unbelievably sensitive test.

Levine[2] gives a graphic picture of the delicacy of the test in the following statement: "....The detection of one member of the

[2]Army M. Bull. **68:**178-194, 1943.

colon group per 100 ml. of water would be analogous to finding an object which weighs a millionth of an ounce in a 10,000 ton pile of coal or a stack of 40,000 bales of cotton. One should, therefore, not be surprised if the organism is occassionlly missed but rather that detection is so frequently successfully accomplished."

Since several bacteria, in addition to the coliforms, have the ability to produce gas from lactose, the test gives merely "presumptive" evidence of the presence of coliforms.

The bacteria responsible for false presumptives are primarily gram-positive sporeformers of which *Clostridium perfringens* and *Bacillus aerosporus* are major representatives. On the other hand, it has recently been demonstrated that the presence of fair amounts of nitrates in water may prevent a positive presumptive test even when coliforms are present. Presumably the lactose fermenting enzymes are blocked by the nitrates.

In spite of these limitations the presumptive test represents a relatively satisfactory method of detecting sewage pollution at least in a preliminary way.

Negative presumptive tests are considered as evidence of the absence of sewage pollution. A modified lactose broth medium which is said to give fewer false positives is now used in some laboratories. It is lauryl tryptose broth. It contains as its inhibitory agent sodium lauryl sulfate.

Confirmed Test.—Several differential media have been proposed which are designed to separate the presumptive tests caused by coliforms from these produced by other bacteria. These are: (a) Streak plates using Endo agar or eosin-methylene blue agar. Both of these media contain dyes and other components which permit the formation of characteristic, easily recognized *Escherichia coli* colonies.

The procedure consists of streaking on the surface of the agar, a small amount of material from a tube of lactose broth showing gas.

This culture is then incubated for 24 hours at 37°C and then examined for typical *E. coli* colonies.

Tubes of brilliant green broth with gas tubes may be used as an alternative culture medium. In this test, a loopful of culture from a tube of lactose broth showing gas is transferred to a tube of brilliant green bile broth. The production of gas in 24-48 hours at 37°C. is considered confirmatory for *Escherichia coli*.

The combination of brilliant green and bile effectively inhibits the growth of the gram-positive bacteria which produce gas from lactose.

Completed Test.—Occasionally samples are found in which the confirmed test may be in doubt. The completed test is then used. It consists of isolation of the organism, making a Gram stain and culturing in lactose broth.

The Membrane Filter.—Shortly after World War I, certain German investigators developed a commercially feasible membrane filter which could be used to filter bacteria from water. These could then be grown by placing the filter on a suitable medium in a Petri plate. However, in spite of the fact that a U. S. patent was issued for it in 1922 it was not given any special consideration until 1945, when Dr. A. Goetz of the California Institute of Technology published a report on investigation of its use in Europe. The United States Public Health Service and other interested parties have since made an extensive study of this device and indications are that it may serve to bring about a marked change in water testing procedures.[3] By this method large volumes of water may be filtered through the membrane, which is then placed on a selective medium such as eosin-methylene blue and characteristic colonies will develop. It is also possible to isolate pathogens, such as the typhoid-paratyphoid-bacillary dysentery organisms directly from water by this method since as few as one or two bacteria per liter could probably be detected. The membrane filter promises to have many applications in other fields.

Purification of Water Supplies

Less than one-half of the cities of the United States and less than one-fifth of the entire population secure their drinking water from underground sources. For them, water purification is not a serious problem because the water is generally potable at the time it is pumped. However, in cities of any size chlorination is generally practiced as a protective measure against chance pollution. Private wells or similar water sources are not generally chlorinated, although this can be done rather easily with a minimum of equipment.

On the other hand, practically all of the major cities use water in such great quantites that only large rivers, lakes, or reservoirs will

---

[3]See Clark, H. F., et al.: Pub. Health Rep. **66:**951, 1951.

afford ample supplies. These waters, being surface waters, are generally contaminated or polluted, often grossly so. As a result about four-fifths of the population of the United States is dependent on water supplies which are or should be carefully purified before they can be safely used for drinking and other household purposes.

At this point the reader should be cautioned against the idea that "running water purifies itself." While standing water tends to undergo a relatively slow self-purification, due to sedimentation and the operation of certain biological processes, in running water these activities are greatly reduced. Transmission of diseases through water taken from flowing streams is ample proof of the half-truths involved in such a concept.

The purification of water supplies has two objectives (1) to destroy or remove any pathogenic bacteria which may be present and thus render the water safe and (2) to remove all substances which impart turbidity, color, bad tastes, or odors to a water supply. When properly done, the water will be clear, sparkling, and desirable to drink. This combination of safety and good aesthetic qualities constitutes a "potable" water supply.

Purification Methods.—The process of purifying water supplies is usually accomplished by some scheme involving all or most of the following steps:

Sedimentation.—Water which is stored in reservoirs or natural lakes undergoes a substantial amount of natural sedimentation which greatly improves the quality of the water. Copper sulfate in small amounts must sometimes be added to prevent the growth of algae which impart undesirable tastes and odors. In many instances, water is merely drawn from rivers directly into sedimentation basins where it is allowed to stand for some hours before treatment begins. With very turbid waters, a great deal of the clay and silt carried by the water may be dropped during this preliminary holding period. The raw water is often given a preliminary chlorination at this point.

2. *Coagulation and Filtration.*—The common practice in many large cities is to introduce a coagulating agent before sedimentation. The coagulants most commonly used are aluminum sulfate or ferrous (iron) sulfate. In alkaline waters, these compounds are rapidly converted to a colloidal gel or "floc" which forms into small jellylike aggregates scattered throughout the water. If the water is allowed to stand in sedimentation basins, the floc soon settles to the bottom

taking with it the suspended matter including a fairly large proportion of the microorganisms.

In some cases, without preliminary sedimentation following coagulation, the water is pumped directly over sand (or gravel) filters where the floc accumulates at the surface and aids in the filtration process.

Fig. 14.—Automatic chlorinator.

Filter beds are of two types: (a) Slow sand filters which can be used with relatively clear waters but which are not adapted to handle turbid waters, and because of their relatively slow action large filter bed areas are required; a number of cities including Philadelphia, Pittsburgh, and Washington D.C. use slow sand filters. (b) Rapid sand filters are constructed with coarser grained sand and are adapted to filtration of water immediately after chemical coagulation.

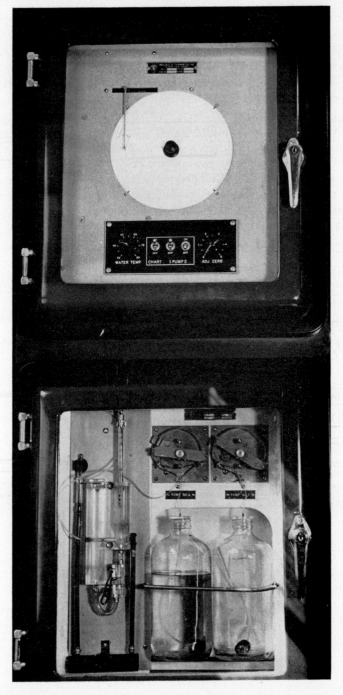

Fig. 15.—Control mechanism for automatic chlorinator.

In addition to filtration as described, compounds such as activated carbon may be added to absorb undesirable tastes, odors, or colors.

3. *Chlorination.*—It has been found that prechlorination, before water treatment actually begins, cuts down the load of bacteria and reduces growth in the filters. However, after the water is filtered it is again chlorinated. This has two purposes: (a) to destroy any pathogens which may have escaped treatment given up to this point (not a great likelihood in properly treated water) and (b) to remove objectionable tastes and odors. Most of these latter will disappear as a result of oxidation; and chlorine is an active and economical oxidizing agent. However, if inadequate amounts of chlorine are used, taste or odor-producing organo-chlorine compounds may be formed. To secure chlorination adequate to remove such tastes or odors one of the following processes may be employed:

(a) Breakpoint chlorination

This consists of the addition of enough chlorine so that the organic matter will all be oxidized and a small amount of residual chlorine will remain.

(b) Superchlorination.

This method involves the use of an excess of chlorine followed by removal with some dechlorinating agent.

(c) Chlorine dioxide.

This is a very unstable gas which has about two and one-half times the oxidizing power of chlorine. It is now being investigated extensively and is being employed in a practical way for treating water supplies in some cities at the present time.

These three methods have replaced in many cities the chlorine gas plus ammonia gas mixture formerly used. This latter combination forms chloramines which remain germicidally active for a longer period than chlorine alone.

Aeration is sometimes used to remove volatile compounds which may cause undesirable tastes and odors.

By the application of the methods outlined, high quality potable waters are made available to much of the American public. However, continued pollution of our surface water supplies constitutes a major problem which is constantly assuming greater proportions as our population increases.

SEWAGE AND SEWAGE DISPOSAL

Since man first began to build cities of any size he has been plagued with the problem of what to do with sewage, which represents the "used" portion of the water supply. Some 2,000 years ago Rome had the outstanding drain sewer of the times, the great Cloaca Maxima. Archeologists have found many of the ancient cities equipped with house drains and sewers. London, Paris, and other old European cities have had sewerage systems of some sort for centuries.

The water closet was invented in 1596 by Sir John Harington. As a gay young blade in Queen Elizabeth's court, he was "banished" to the city of Bath for translating a portion of a salacious poem. He did penance by translating the remainder of the poem! The foul odors emanating from a privy near his lodgings insulted his aristocratic sense of smell and prompted the invention of this great sanitary device. However, it did not assume much importance until well along in the nineteenth century.

We now recognize that sewage represents a dangerous and undesirable waste material which can be at once a prolific carrier of disease and a spoiler of our great natural resource, *pure water*. Dumped promiscuously into a stream, sewage fouls the area which borders the stream and contaminates the drinking water of the cities below. Fish may be killed, swimming and boating made hazardous, and property values along the stream destroyed. Improperly treated sewage is robbing us of millions of dollars every year through the impairment of our heritage, *pure water*.

Classes of Sewage

1. Domestic or Residential Sewage.—Heavily laden with human wastes, it represents a real peril because of the disease-producing bacteria and viruses it may carry. It is generally very low in solids, usually being more than 99 per cent water. The widespread use of the new home garbage disposal units will create a marked increase in the amount of solids carried by domestic sewage. This in turn may complicate the process of sewage treatment.

2. Industrial Wastes.—While sewage of this type is not likely to contribute any appreciable number of pathogens, it is often relatively rich in organic matter which may be attacked by microorganisms.   Such readily decomposed matter is spoken of as putrescible material.   Its decomposability is due to a high content of sugars, proteins, etc.   The amount produced by a single factory may be equivalent in terms of decomposable matter to that produced by a relatively large city.   For example, one brewery which was investigated in Chicago was pouring into the sewer as much putrescible material as a city slightly larger than Des Moines, Iowa, or Salt Lake City, Utah.

The following statistics indicate the great amount of waste water resulting from industrial operations: a ton of rayon requires the expenditure of a quarter of a million gallons of water; for every gallon of whiskey, eighty gallons of waste water is dumped into the sewer; and for every gallon of gasoline a thousand gallons of water are used.

The major problem with all types of sewage is a similar one. The two classes of sewage may differ in the number of disease-producing bacteria they may potentially carry, but both have substantial amounts of organic matter which is subject to ready attack by microorganisms.   It is not too farfetched to look upon sewage as a crude culture medium in which an amazingly complex mixture of microorganisms are growing vigorously.   In this mixture of bacteria, actinomycetes, yeasts, water fungi, protozoa, and phages a constant series of actions and interactions are in play.   As this mixed culture breaks down the organic components, it produces the foul odors of hydrogen sulfide and other malodorous compounds; objectionable colors and tastes develop in the water which carries the sewage; partly decomposed organic solids line the stream bed or shore line and these may be deposited in the form of filthy gray-black sludge bars which burp great quantities of sewer gas (methane).   Swimming beaches may be ruined with grease and defiled with half-decomposed body wastes. In such polluted waters fish and desirable vegetation are killed, concrete structures are damaged by the chemicals present or produced by microbial action, metal is corroded and paint chips and peels through contact with hydrogen sulfide.

Because of the pathogens present, the water becomes extremely dangerous for drinking, while swimming and boating become both health and esthetic hazards.

Aside from the transmission of disease-producers, the effect of sewage in water is primarily one of disturbing the oxygen balance of the water.

Pure fresh water normally contains some 8 to 12 parts per million of dissolved oxygen. (Recently boiled water tastes "flat" and lacks "sparkle" because heating has driven out the dissolved oxygen.)

When sewage is dumped into fresh water, readily decomposable organic matter is immediately attacked by the microorganisms present and the dissolved oxygen begins to be used up. When the oxygen is reduced to about 4 parts per million (ppm) game fish are killed. A further reduction gives rise to stale sewage which is milky in appearance. When all of the oxygen is used up the sewage becomes "septic," will be dirty gray in color, and have a foul odor.

Ultimately, if given time, decomposition of the putrescible material will be complete and water will again return to its original state so far as oxygen content is concerned. When this has taken place the putrescible carbohydrates will have been changed to carbon dioxide and water, while the proteins will have yielded these and also nitrates, sulfates, and phosphates. These are all relatively stable molecules which represent the ultimate in the oxidation of putrescible materials.

A picture of what takes place when sewage is dumped into a stream is shown in Table XI.

TABLE XI

STATE OF STREAM	DISSOLVED OXYGEN	FISH
Stream before pollution	Normal	Game fish present
Sewage introduced, stream becomes stale	Begins to be reduced Substantially reduced	Coarse fish present
Stream becomes septic	All D. O. used	No fish
Decomposition proceeding	D. O. increasing	Coarse fish present
Decomposition essentially complete	D. O. normal	Game fish present

A change of the type indicated above will take place in a relatively short distance in a slow-moving stream but may require miles to complete if the water is moving rapidly and little or no sedimentation is taking place.

Sewage Treatment Methods

Pollution of streams is becoming so marked that the cities of the United States are spending millions of dollars for the installation of sewage treatment plants which are designed to improve the quality of sewage before it is dumped into streams, lakes, or the ocean. The tremendous cost of making the streams, lakes, and ocean waters of the United States relatively free from pollution is indicated by the fact that it costs from $10 to $50 per person under a sewerage system to pay for an adequate sewage treatment plant.

Up until 1948 stream pollution control was largely left to the discretion of the individual states. Under the Taft-Barkley Act of 1948 the United States Public Health Service was authorized to assist the various states in setting up their programs for the control of water pollution.

While we may never be able to restore our streams to their native unpolluted state, continued public support will do much to alleviate the present deplorable condition.

Sewage treatment methods have two objectives: (1) to reduce to a minimum the settleable solids present in the raw sewage and, (2) to bring about as much decomposition as possible of the putrescible matter.

It is relatively difficult to determine the amount of decomposition which has taken place. The most satisfactory method appears to be the B.O.D. test. These letters stand for Biochemical Oxygen Demand and the procedure consists of diluting the sewage 1:1,000, or 1:100 times for strong industrial wastes down to 1:4 or even 1:1 for moderately polluted river waters. These dilutions are incubated in bottles with ground glass stoppers arranged to prevent any uptake of oxygen from the atmosphere. The dissolved oxygen present in the dilution water is determined at the beginning of the experiment. The sewage dilution is then incubated for 5 days at 20° C. (68 F.) and the dissolved oxygen is again determined. The loss in dissolved oxygen times the dilution represents the Biochemical Oxygen Demand. Thus if 3 ppm of oxygen were used in 5 days and the dilution was 1:100 the B.O.D. would equal 3 × 100 or 300 ppm.

During World War II an investigation of the B.O.D. of the sewage from various Army camps showed that it averaged about 250 ppm. On the other hand, sewage from a pea cannery in northern California has been reported as having a B.O.D. of 800 ppm.

Contamination of sewage with human wastes is estimated by the coliform count, while changes in the bacterial flora may be studied by means of bacterial counts.

Disposal of sewage may be accomplished with or without treatment.

Disposal Methods Not Involving Treatment

(a) **Dilution.**—This is the simplest and by all odds the most objectionable method. Sewage is dumped directly into a stream, lake, or the ocean. Sedimentation and oxidation by means of the dissolved oxygen present in the water represent the means of destroying the suspended solids and putrescible material present in the sewage. Large quantities of fresh water are required in proportion to the sewage added, and in dry seasons when the stream flow may be low, a profusion of foul odors may serve as an olfactory exclamation point to emphasize how bad this method really is.

(b) **Sewage Farming.**—In some cities of the world, of which Paris is a notable example, sewage disposal is accomplished by spreading the sewage over sandy soils as irrigation water. These soils are planted to crops.

A variety of procedures have been used in order to improve the quality of sewage before dumping it into a body of water. These methods may be divided into the following categories.

Disposal Methods Involving Treatment

(a) **Primary Treatment.**—Primary treatment is designed to reduce the total solids in sewage by sedimentation or to render it relatively free from pathogenic bacteria by chlorination or both.

Sedimentation.—Several modifications of this process are used. The simplest consists of running sewage into a basin then draining off the supernatant liquid after a retention period of 2 or 3 hours. Primary settling reduces the dissolved solids about 50 per cent and the B.O.D. about 30 per cent. The settled solids (sludge) are then pumped into a closed chamber or digester where they undergo anaerobic decomposition. This may require as long as 30-40 days and during this period most of the decomposable components of the sludge will have been broken down into simple compounds. The organic matter which remains is thus "stabilized" and not easily decomposed further. It is black and without any appreciable odor. Stabilized sludge may be dried, and sold as fertilizer or it may be burned or buried. During the digestion of sewage, an appreciable amount of gas may be produced which is trapped in the dome of the digester. The gas contains 65-85 per cent methane (CH_4). The rest is carbon dioxide and a small amount of hydrogen sulfide (H_2S).

The Imhoff tank, designed by Karl Imhoff in 1907, combines the settling basin and sludge digestion tank. It is arranged so that the sludge settles to the bottom and the sewage is allowed to flow slowly through. A baffle arrangement permits trapping of the gas.

The septic tank resembles the Imhoff tank in that the sewage is allowed to flow slowly through, depositing a portion of its solids in the process. The sludge which remains is subject to anaerobic decomposition.

With the settling tank, the Imhoff tank or the septic tank the outflowing sewage will have lost perhaps half, more or less, of its suspended solids and perhaps a third of its putrescible matter. It still contains much of its putrescible material and most of its load of pathogenic bacteria. (It should be remembered that the effluent from septic tanks is still potentially dangerous.)

Chlorination.—As an adjunct to primary treatment, chlorination is very useful. This may consist of prechlorination (before sedimentation) in which 15-20 ppm of chlorine is added or postchlorination of the effluent following settling in which 5-10 ppm of chlorine are added. While chlorination is designed to destroy pathogens it also serves to alleviate some of the unpleasant odors associated with raw sewage.

Following primary treatment the effluent is then generally dumped into some water channel for final disposal.

Primary treatment is sometimes considered sufficient if the sewage is not too high in solids and an ample amount of diluting water is available.

A modification of simple sedimentation is sometimes used. This consists of adding certain aluminum or iron salts to the water in which the alkalinity present (or added) causes the formation of a floc (colloidal masses of iron or aluminum hydroxide) which settles out and carries much of the settleable solids and a fair proportion of the putrescible material with it. This process, which is also applicable to water purification, has not been very widely used, probably because of the cost of labor and materials involved.

(b) **Secondary Treatment.**—While primary treatment improves the quality of sewage, the effluent or liquid portion may still be in need of further treatment for safety if stream pollution is to be avoided.

The combination of primary and secondary procedures is termed "complete treatment." It may produce an effluent which still has some small amount of solids and putrescible materials or it may even be continued to the point where the water would be of drinking quality. This later type of treatment, while assuring complete absence of any stream pollution, is usually not economically feasible.

The sewage is first passed through bar screens then through grinders or comminutors. It then flows into settling basins where the coarser materials settle out and the fats are skimmed off the top. During World War II a number of treatment plants were able to pay most of their labor costs by the sale of the skimmed fats.

The sludge which has settled out is now pumped into the digestion vats and the effluent then undergoes further treatment. This may involve the use of (a) intermittent sand filtration, (b) contact beds, (c) trickling filters, (d) high capacity trickling filters, (e) activated sludge, and (f) flotation processes.

Intermittent Sand Filters.—Construction consists of a bed of sand about 2-1/2 to 4 feet deep with drains underneath. The beds are filled with sewage to a depth of 3 or 4 inches and the sewage is allowed to filter through. In some instances, the sewage may not have had primary treatment but most generally it has been so treated. The sand soon develops a gelatinous coating which effectively filters out much of the solid putrescible material and bacteria. The bed is dosed only once or twice a day since a time interval must be allowed for decomposition of the organic matter. While this process yields a high grade effluent the capacity of any bed is small and many acres of filter beds would be required to handle the sewage of a sizeable city.

Contact Beds.—These beds consist simply of a tank filled with crushed gravel, slag, or coke. The pieces used vary in size from about a half to three inches and the bed is 4 to 6 feet deep. Drains for removing the filtered sewage are arranged at the bottom of the tank. In operation, the contact bed is filled with the effluent from a sedimentation or Imhoff tank. The sewage is allowed to stand for an hour or two. The solids and putrescible materials tend to be retained by the rocks which are coated, after the first few loads, with a film of stalked bacteria, other bacteria, water molds, and protozoa which are actively engaged in decomposing the organic matter. The effluent is then drawn off and air is drawn in producing active aerobic decomposition which is allowed to continue for several hours; then the process is repeated. Several beds may be set up to operate in tandem. If properly operated up to 90 per cent of the solids and 80 per cent of the B.O.D. will be removed. Contact beds are not widely used at present because their capacity is relatively low compared with some other processes.

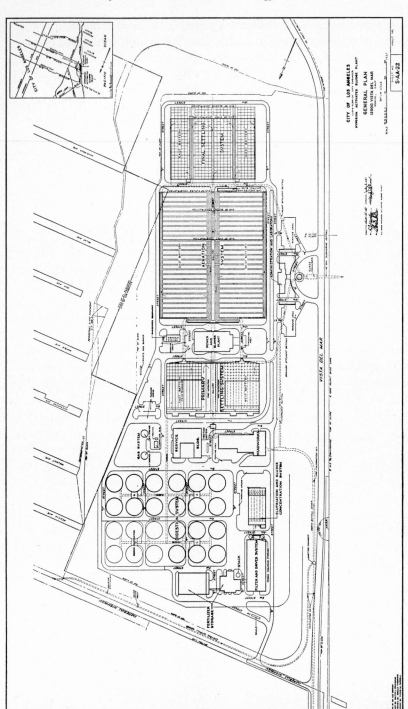

Trickling Filters.—The trickling filter is based on experiments performed at the Lawrence Experiment Station in 1894. It is essentially a process of exposing the sewage to active microbial decomposition under aerobic conditions. The trickling filter bed is filled with rather coarse pieces of stone, slag, etc., varying in size from 1 to 3 inches. This leaves a great many open spaces into which air can penetrate. After initial operation the rocks build up a coating of various microorganisms with the stalked bacteria forming an important part of the flora. Sewage is sprayed over the surface either from fixed pipes or from rotating metal pipes with holes at intervals along the pipe. These sweep over the surface like the spokes of a wheel and spray sewage onto the rocks. The practice is to operate the spray for a few minutes then turn off the spray and allow the sewage to trickle through. This insures the aerobic conditions so necessary for rapid decomposition.

Greater effectiveness and a higher capacity are obtained with the high-rate or bio-filter which is a modification of the trickling filter. Frequently, sewage is recirculated through at least two of these filters to produce a relatively high grade of effluent. Sedimentation tanks or "clarifiers" plus chlorination may be used as part of the secondary treatment.

Activated Sludge.—As has been previously mentioned, aerobic decomposition proceeds much more rapidly than anaerobic decomposition. The activated sludge process developed in 1912 at the Lawrence Experiment Station is a vigorous application of this principle. After the sewage has been sedimented the effluent is pumped into large open vats where great volumes of air under pressure are blown through the sewage from vents at the bottom of the tank. This effectively removes most of the odor, results in an active decomposition of the putrescible material and causes the finer particles to clump into floccules or masses which settle out rapidly when the sewage is run into a final sedimentation tank after activation.

The settled effluent may then be chlorinated before it is dumped into a stream, lake, or the ocean. This is true also of the filtrate from the other types of treatment.

A new development in the cleaning field, the detergent, has added complications to sewage disposal.

In the new, ultramodern Hyperion Activated Sludge Plant of Los Angeles City a great deal of difficulty has been experienced in the activated sludge beds. The detergents carried in the sewage have caused a profuse foam development in the aeration tanks. Encouraged by the heavy stream of air which is blowing through the sewage, foam often several feet thick forms at the top.   The aeration tanks at that time resemble Gargantuan beer mugs.   Several men with hoses must be kept constantly at work breaking down the foam masses.   On windy days, large chunks of foam are often swirled into the air and carried away.

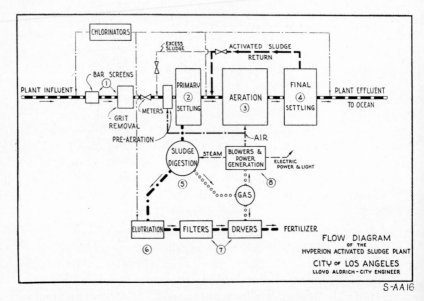

S-AA 16

Fig. 17.—Diagram showing the flow of sewage through the Hyperion Activated Sludge Plant and the various steps involved in the treatment process. (Courtesy City of Los Angeles.)

Flotation.—A number of years ago it was found that gold and other minerals could be separated from their ores by suspending the finely ground ore in water, adding a "flotation oil" which picked up the gold, and agitating the mixture in vats.   The gold could then be skimmed off in the froth which developed at the top.   This process is now being applied to sewage treatment.

In the various types of microbial decomposition which takes place in sewage, it appears that antibiotics and similar compounds may develop which tend to markedly reduce the numbers of pathogens present.   This, coupled with chlorination of the treated effluent can do much to not only improve the quality of our streams, but also greatly lessen the likelihood of future epidemics of water borne diseases.

The sewage disposal problem as has been indicated is a very real one and in the years ahead must be given immediate and ever increasing attention if we are not to be robbed of our heritage, *pure water*.

Selected Reading References

Hopkins, E. S., and Elder, F. B.: The Practice of Sanitation, Baltimore, 1951, Williams and Wilkins Company.

Maxcy, K. F., et al.: Rosenau Preventive Medicine and Hygiene, ed. 7, New York, 1951, Appleton-Century-Crofts, Inc.

Phelps, E. B.: Public Health Engineering, New York, 1948, John Wiley & Sons, Inc., Vol. I.

Standard Methods for the Examination of Water and Sewage, ed. 9, New York, 1946, American Public Health Association.

CHAPTER 14

SAFE SWIMMING

Swimming Pool Sanitation

The Greeks and Romans looked with great favor on swimming as an effective and highly desirable form of exercise which they included as an important part of their military training. Swimming was practiced in the open waters and no doubt in the great Roman baths since their pools were often rather spacious in construction. For example, the ruins of Aquae Sulis, the great Roman bath at Bath, England, discovered some two centuries ago, include a fairly large lead-lined swimming pool which still holds water and in which the original inlet and outlet conduits still function.

During the middle ages swimming was discouraged almost to the point of extinction as a sport because of the belief that the great epidemics of the period were caused in part by swimming in open waters and because of the active religious opposition to public bathing stirred into indignant action by the immorality of public bath houses.

In Europe and the United States swimming was resumed as a form of competitive sport only about a century ago. It is now being recognized as a most valuable form of exercise and bathing beaches contribute much to the physical well-being and mental relaxation of an apartment-confined urban populace.

Because contaminated open waters or improperly operated swimming pools have definite disease-spreading potential, bathing beaches and swimming pools have come under the same type of scrutiny by public health workers that is given drinking water and sewage disposal. In fact, swimming is often designated as a "tertiary" use of water.

Along with recognition by the public of the importance of a potable drinking water supply, we must also concern ourselves with the quality of water in which we are immersed while swimming. Only the thoughtless or uninitiated will drink water of questionable quality or fail to be critical of swimming pools or bathing beaches where the water is obviously subject to contamination.

It is a relatively easy matter to determine what disease or diseases may have been transmitted by a contaminated drinking water supply. On the other hand, it is difficult to prove that diseases are transmitted as a result of swimming in contaminated waters. Some of the factors which contribute to the problem are: (a) Cases may often be scattered or sporadic rather than epidemic as in the case of drinking water. This is occasioned by the great variability in the degree to which various swimmers may be exposed and the "dilution factor" which may spread the contamination in small "clouds" rather than uniformly throughout the water. (b) The incubation period may represent days or even weeks before the disease becomes evident. During this lapse of time, the causative organisms may have been carried down the sewer by draining the swimming pool, killed by heavy chlorination or currents may have swept an open beach clear of the offending contamination. Also, especially with open beaches, swimmers often come from such widely scattered areas that it is difficult to trace down infections that might be contracted by swimming.

In spite of the lack of clear-cut evidence as to the role of bathing waters in the transmission of disease "there is a strong suspicion on the part of medical and public health authorities that such transmission does occur."[1]

The kinds of diseases which are likely to be transmitted will be determined largely by the type of swimming facilities being used. In open waters, such as the ocean beaches or fresh water lake or river beaches, where a great deal of dilution occurs, the greatest source of danger is represented by the intestinal diseases because these bodies of water are often used as dumping places for sewage. Fortunately, the danger is not likely to be as great for the swimmer as it is for the person who may be using such water as a drinking water supply. It is recognized that only poor swimmers drink much of the bathing water while good swimmers who are thirsty apparently drink only widely advertised beverages. The danger is also lessened by the fortunate circumstance that typhoid rarely occurs in large cities today and sewage, therefore, seldom carries this dread organism. On the other hand, the milder salmonellas and shigellas and the amebic dysentery organism

---

[1]Maxcy, K. F., et al. Rosenau Preventive Medicine and Hygiene, ed. 7, New York, 1951, Appleton-Century-Crofts, Inc., p. 1231.

are still present in disturbing amounts and infection-producing organisms are often waiting to invade any open cut or abrasion.

Open waters are sometimes polluted with chemical substances which may cause injury to the skin or to mucous membranes of the eyes, nose, or throat.

Any open water swimming beach should, therefore, be situated as far away as possible from sewer outfalls delivering untreated sewage or sewage which is given only primary treatment and should be free from contamination with dangerous chemical wastes.

Ocean beaches have certain advantages over other forms of open waters in that the great quantities of water present afford a marked degree of dilution. A number of different investigators have also demonstrated that ocean water possesses a considerable amount of bactericidal action against *Escherichia coli* and probably against other related intestinal bacteria. However, this action requires a period of time to operate so that it does not serve as an immediate protection against any massive or recent sewage contamination. It does, however, apparently hasten the purifying action of ocean waters at least against the intestinal bacteria. As an evidence of this, although polluted harbor waters may be teeming with thousands of bacteria, many of them intestinal forms, a few miles from shore the number will be reduced to a few hundred and coliforms will be absent. Although this rapid reduction is attributed by various workers to a number of different factors, one of the most likely explanations is that it is a result of the action of heat sensitive antibiotic-like substances produced by the native microflora. This conclusion is based on the observation that autoclaved ocean water possesses practically no bactericidal action against *Escherichia coli*.

While fresh water lakes lack the salinity of the ocean, waters which are well removed from sewage contamination are likely to be entirely acceptable for bathing. On the other hand, with rivers, especially if they flow rather rapidly, a high degree of sewage contamination may be carried very considerable distances down the river, distances which might well be completely adequate for safety in the case of lake waters.

With the expanding of our cities and the great need for recreational areas such as our beaches, both public health workers and the general population are being awakened to indignation because of the great amount of valuable beach area adjacent to many of our cities that is rendered unsafe for swimming and undesirable to the senses by the accumulations of sewage wastes. As a result many large cities are now making concerted efforts to redeem these important recreational facilities by installing secondary treatment sewage disposal plants.

Swimming pools represent a completely different problem since the water with which the pool is initially filled is generally of drinking water quality. Organisms which enter the water are, therefore, those brought in by the swimmers on their bodies or on their swimming suits. Improperly sanitized towels may likewise be an incidental source of infection.

The pathogenic microorganisms which may be transmitted in improperly treated swimming pool waters are largely those which attack the upper respiratory tract, the eyes, and the skin. Indications are that the organisms responsible for intestinal infections may also be occasionally transmitted through swimming pool waters.

The bacteria responsible for upper respiratory infections such as sinusitis and which may be transmitted through swimming pool water, include the streptococci of which *Streptococcus pyogenes* would be the most dangerous, the pathogenic micrococci and similar forms. It appears likely that the viruses of the common cold and influenza may also be so transmitted.

Certain authorities in the past have incriminated the swimming pool as a transmitter of poliomyelitis. However, evidence of this is at best fragmentary, and considerable disagreement exists relative to the role of swimming pool water in transmitting the disease. The intestinal microorganisms which might be involved include the typhoid, paratyphoid, bacillary dysentery, and amebic dysentery organisms. However, the likelihood of a widespread transmission of these diseases through swimming pool waters appears to be fairly remote.

The possibility of transmitting venereal diseases through swimming pool water or perhaps improperly sanitized swimming suits has been mentioned. However, the evidence is far from conclusive. Skin infections such as impetigo, which involves pyogenic streptococci and micrococci, and of ringworm, caused by pathogenic fungi, have also been attributed to swimming pool contacts. The evidence here is apparently more conclusive.

Athlete's foot, a fungus disease of the foot, is probably not acquired any more often from swimming pool walkways or showers than in any other common public showers.

One of the great dangers likely to be encountered is overloading. Normally the facilities of properly operated swimming pools have sufficient capacity to afford protection against the infectious material likely to be introduced by the bodies of a normal load of swimmers due to the combined effect of chlorine and the dilution factor.

TYPES OF SWIMMING POOLS

(a) Fill and Draw Pools

These are pools which are filled and after a period of use the pool is completely emptied and refilled. The length of time between refills is generally determined by the bather load. A rule which is often followed is to allow 500 gallons of water per swimmer for each refill. Fill and draw pools are undesirable as public swimming pools because they are difficult to chlorinate, since they are often over-chlorinated immediately after being filled and under-chlorinated by the time the pool is drained. Pools of this type are frequently considered unacceptable by public health authorities for public use.

(b) Flow-Through Pools

Flow-through pools operate on the basis of maintaining a steady flow of water into one end of the pool while drawing out an equal amount at the other end. A safe concentration of chlorine can be maintained if the water is properly chlorinated as it enters the pool and a relatively large amount of water is used. The suggested amount is 500 gallons of water for each swimmer. Manifestly a large volume of water must be used if the water is to be maintained in a satisfactory condition. As a result, flow-through pools are wasteful of water, somewhat difficult to control and uneconomical to operate.

(c) Recirculating Pools

Recirculating pools represent the most desirable type of swimming pool design. If properly operated they afford satisfactory control of the sanitary quality of the water and are economical in their use of water. In operating a recirculating pool the water is pumped through a high efficiency filter and then chlorinated before it is returned to the pool. Proper operation calls for a complete overturn of the water in the pool every eight hours, making three complete filtrations in 24 hours. The filters used often combine sand with chemical coagulation although diatomaceous earth filters are becoming popular.

A daily water addition of at least 10 per cent of the pool's capacity is often required. Standard methods of construction and operation have been carefully worked out by sanitary engineers. One of the requirements is that 35 to 45 square feet of surface be allowed for each bather.

METHODS OF TREATING BATHING WATERS

No effective method has been developed for the disinfection of open waters used for swimming because of the tremendous water volume represented by the ocean or a large lake or river. However, where waters are artificially impounded in small lakes or reservoirs the water may be disinfected as it flows into the basin. This at least has the effect of controlling any contaminating organisms which may be present in the initial water source.

Swimming pool waters, on the other hand, can be rather easily disinfected. A variety of chemicals have been or are used for this purpose. In addition to chlorine gas, or chlorine plus ammonia, the hypochlorites or chloramines are also used for disinfecting swimming pool waters. Bromine has also come into fairly wide use in certain parts of the United States while cationic silver (silver ions propelled into the water from electrodes under the force of an electric current) has been used in some parts of Europe. In addition to controlling the pathogenic microorganisms which are present, it may be necessary to take measures to control the growth of algae in the water. Algae growth results in a layer of slime on the sides and bottom of the pool which creates a hazard for swimmers and the organic matter present reduces the effectiveness of the chlorine and unites with it to produce undesirable tastes or odors. Proper chlorination is generally sufficient to control the algae although under some circumstances it may be necessary to add small amounts of copper sulfate as an algacide.

The recommended chlorine concentration is 0.4-0.6 parts per million residual chlorine when free chlorine is used. When chloramines are used a combined available chlorine residual of 2 ppm has recently been recommended. Breakpoint chlorination is being employed in some pools.

The purpose in chlorinating bathing waters is to maintain a sufficiently high concentration of disinfectant to protect the swimmers continually against any pathogens which may find their way into the water.

The continual addition of chlorine to water may create an acid pH which is definitely irritating to the eyes and the membranes of the nose. In many instances it is necessary to add lime or some other chemical to maintain the water slightly on the alkaline side (pH-7.4-7.6).

With careful chlorination, water can be maintained in a safe condition without becoming especially irritating to the eyes or nasal mucosa, particularly if the pH is slightly alkaline.

Filtration is also a definite safety factor because it serves to maintain the water in a clear state which lessens the likelihood of drowning accidents and also improves the aesthetic appeal of the water.

In addition to maintaining the water in a safe condition, daily sanitization of the walkways, benches, etc., with strong hypochlorite solutions is also an important sanitary measure. Vacuum cleaning the bottom of the pool is also important. Suits and towels should likewise be treated adequately to render them free from all pathogens.

METHODS OF TESTING BATHING WATERS

Bacteriological tests for the safety of bathing waters involve the same procedures that are used for analyzing drinking waters. These were described in some detail in the chapter on Water and Sewage. The coliform index is taken as the primary yardstick for measuring the safety of bathing waters since it indicates the presence of or freedom from sewage contamination in the case of open waters or affords a clear indication of the effectiveness of disinfection in the case of swimming pool waters. In addition, bacterial counts are used to provide a quantitative measure of water quality or the effectiveness of water treatment. Standards for judging the safety of bathing waters vary considerably from city to city or state to state. However, a count of 200 or more bacteria per milliliter indicates unsatisfactory quality of open waters or inadequate treatment of swimming pool waters. If water from open beaches contains more than 50 coliforms per 100 ml its quality is subject to some question and if the coliform index is more than 1,000 coliforms per 100 ml the water is classed as unsafe.

Laboratory tests alone may not afford adequate information relative to the safety of a water supply. Therefore, a careful sanitary survey should always be made along with the laboratory tests. This may sometimes reveal that sewage contamination may be expected even though a bacteriological test on a particular water sample may fail to so indicate. Certainly, the presence of sewage grease or human fecal matter on any beach is conclusive evidence of sewage pollution and is sufficient to condemn any bathing beach.

With swimming pools the contaminating organisms are primarily the gram-positive cocci rather than the intestinal organisms. The streptococci and micrococci are considerably more resistant to chlorine than are the coliforms such as *Escherichia coli*. A number of sanitary bacteriologists now propose that the test organism be *Streptococcus faecalis* rather than *Escherichia coli*. However, complete agreement does not exist as to the merits of using *S. faecalis*, because some bacteriologists maintain that the enterococcus test is unnecessarily sensitive and imposes unjustified high standards for swimming pool water quality. The media used for isolating enterococci are those containing some inhibitory substance such as sodium azide or tellurite. These compounds rather effectively stop the growth of gram-negative bacteria while allowing the gram-positives such as the enterococci to grow rather readily.

The question as to the bacterial standards which may be used as a dependable measure of the safety of bathing waters has never been satisfactorily answered. A careful study of some bathing beaches on Lake Michigan near Chicago and some beaches along the Ohio River is being made by the Environmental Health Units of the United States Public Health Service. Results on at least one of the beaches indicate that a relationship apparently exists between the safety of a water supply and its bacteriological quality. In other words, at those sampling periods when the bacterial content increased significantly, an increase in illnesses also occurred among those bathing on that particular day. Until the problem is solved, relatively high bacteriological standards for the safety of the water supply coupled with adequate chlorination of swimming pool waters and sanitary surveys of open waters, will serve to protect the swimmer.

With swimming pool waters, a test for available chlorine represents a rapid method of estimating the condition of the water itself. If the available chlorine concentration is maintained at a safe level, the water is properly filtered and the pool is not overloaded, the water may also be expected to be bacteriologically safe. However, this must be checked periodically by bacteriological tests.

Chlorine content is generally determined by means of the ortho-tolidine-arsenite test. This involves the use of an organic compound, orthotolidine, which develops a characteristic yellow color if chlorine is present. The color varies in intensity with the amount of chlorine which is present, ranging from a faint yellow at 0.1 ppm to a bright canary yellow at 1.0 ppm.

SAFE SWIMMING

Swimming can be an exhilarating, safe, healthful, and highly desirable sport if a few simple common sense principles are applied.

When swimming in the ocean, lake, or river select those beaches which are approved by the local public health authorities or those which are manifestly free from any possibility of sewage contamination. "Swimmer's itch" which is caused by the free swimming cercaria of certain flukes which are not essentially human pathogens may sometimes be acquired as a result of swimming in contaminated waters. The cercaria are harbored by snails and these minute cercaria penetrate the skin of the swimmer and cause an intense itching which may last for about a week. "Swimmer's itch" has been reported from the lake regions of the north central states and recently from a lake in the state of Washington. Control measures are as follows: (a) waters stored for 36 to 48 hours, free from contact with snails, will be safe for swimming, because the cercaria will have perished, and (b) in snail infected waters, treatment with lime and copper sulfate will destroy these animals. A beach which has thus been freed from snails by such chemical treatment will remain so for two to four years.

In the swimming pool the precautions are of a different nature, because the water initially used for filling the pool is generally safe and often of drinking water quality. Since the primary sources of pathogens are persons who are in the swimming pool or the bather himself, the protection against infection revolves around the following rules:

1. Shower before entering the water.
2. Wear only properly cleaned and sanitized bathing suits.
3. Exclude persons with visible skin infections, open sores, etc.
4. Do not enter the pool if you have a cold or other type of upper respiratory infection since the water may carry organisms from the nasal membranes into the sinuses and even the middle ears. This precaution also applies to swimming in open waters. Infections can be induced which may be incapacitating or even dangerous to the life of the swimmer. The fact that a person may spend long periods of time in a swimming pool with a resultant chilling of the body and repeated washing of the nose and throat membranes adds to the problem.
5. Make it a practice to patronize only modern well-operated swimming pools which are equipped with recirculating and chlorinating systems.

Selected Reading References

Gainey, P. L., and Lord, T. H.: Microbiology of Water and Sewage, New York, 1952. Prentice-Hall, Inc.

Hopkins, E. S., and Elder, F. B.: The Practice of Sanitation, Baltimore, 1951, Williams and Wilkins Company.

Maxcy, K. F., et al.: Rosenau Preventive Medicine and Hygiene, ed. 7, New York, 1951, Appleton-Century-Crofts, Inc.

Phelps, E. B.: Public Health Engineering, New York, 1948, John Wiley and Sons, Inc., Vol. I.

CHAPTER 15

SAFE AND UNSAFE FOOD SUPPLIES

Microbiology of Food

We must compete constantly with microorganisms for the privilege of eating the food we raise. Our whole system of food preservation has been built up as a safeguard against the inroads of these ever-hungry microbes. Canning, pickling, preserving, dehydrating, refrigerating, and quick freezing all serve a dual purpose, to prevent the microbes from eating foods before we can consume them and to protect us against those which might poison us.

If there were no microorganisms, food preserving processes would be virtually unnecessary because foods would dry up without any spoilage taking place except perhaps overripening due to the natural enzymes in the food. On the other hand, if there were no microorganisms there would be no food because microorganisms are necessary in order to make available the nutrients required for the growth of food-producing plants. Also the so-called fermented foods, i.e., sauerkraut, dill pickles etc., would be nonexistent.

The food spoilage organisms are the same organisms which, in the soil, are decomposing organic matter and making the chemical constituents of organic matter available to higher plants. For them organic matter is food and they do not differentiate between that which is choice food for man and that which is merely waste organic material. For example, molds find the apple in the box in the grocery store just as palatable as an apple lying on the ground in an orchard.

Because of the technicalities now involved in food processing, the industry is coming to rely on the specialists for much of the required technical advice. As a result, a number of colleges and universities are now sponsoring training programs which lead to such specialties as food microbiology, food chemistry, and food technology.

With the food supply of the world definitely reaching the point where it is inadequate for the world population any process which will prevent the microbial spoilage of already produced food will auto-matically add to the world's precious food supply. Food preserving processes also have the very important function of making foods available throughout the year rather than only during the season in which they were produced.

Food spoilage also has a tremendous impact on our economy, be-cause losses of food due to spoilage are automatically reflected in an increase in the price of that food. While insects and rodents may cause a part of this food loss, microorganisms have a lion's share of the dubious honor of robbing us of our food supply.

The present chapter will deal with foods, other than dairy prod-ucts, since they will be discussed in the following chapter.

Foods vary greatly in their susceptibility to microbial breakdown. Wheat, beans, and similar food materials low in moisture may be kept successfully for many years without microbial spoilage pro-viding they are kept completely dry. The moisture content of wheat is about 15 per cent, while flour may have a somewhat lower moisture content.

Dehydrated foods likewise will keep for a fairly long period of time provided they remain dehydrated, unless oxidation due to atmos-pheric oxygen takes place.

On the other hand, foods which are high in moisture, particularly the high protein foods are extremely susceptible to microbial action.

The factors which determine the susceptibility of foods to micro-bial spoilage include moisture content, pH (an acid reaction inhibits bacterial growth while favoring the growth of yeasts or molds), temper-ature, oxygen (mold growth is inhibited by a lack of oxygen), and salt or sugar concentrations which produce high osmotic pressures.

Microbial action on foods may be conveniently divided into three categories:

(a) Microbial changes in foods which result in an impairment in the quality of the food by producing undesirable changes in taste, odor, flavor, or appearance without necessarily making the food unsafe to eat.

(b) Microbial changes in which the quality of the food may or may not be visibly impaired but which cause illness or death when the food is consumed. These products of microbial growth produce the so-called "food-poisoning."

(c) Microbial changes in foods which result in the production of a desirable product. These could hardly be termed "food spoilages," since the food quality is unimpaired or definitely improved; however, the microbial action is often rather similar to that involved in food spoilage since some constituent of the food is attacked enzymatically by the microbial enzymes.

MICROBIAL SPOILAGE OF FRESH FOODS

Fresh foods may be conveniently assigned to several classes on the basis of the kinds of microorganisms which commonly cause spoilage. Type examples are given below:

(a) Fruits, because of their moisture, acidity, and high sugar content are likely, if their protective skin with its waxy bloom is broken, to fall victim to the enzymatic onslaught of molds and yeasts. Indeed, there is a suggestion that certain molds appear to be partial to certain fruits. For example, some types of the common blue mold, *Penicillium*, seem to have a special affinity for apples and citrus fruits. Tomatoes on the other hand may be attacked by *Alternaria* in the field with the production of a characteristic "black rot" while *Fusarium* may grow on tomatoes to produce a white microbial growth, tinged with pink as the spores mature. The cottony molds, *Mucor* and *Rhizopus*, produce the "leaker" type of spoilage. The tissues of such tomatoes become softened and if the skin is broken the juice leaks out to infect the basket or soil the hand of the handler. In a basket of strawberries which has stood too long on the fruit stand, the fruit on the bottom of the basket may be held together by a mat of cottony threads which make up the plant body of the molds, *Rhizopus* or *Mucor*. Yeasts may likewise attack various kinds of fruit and bacteria may sometimes produce a foul smelling type of spoilage, especially in the case of tomatoes.

(b) Carrots, potatoes, parsnips, cabbage, lettuce, celery, and other nonacid vegetables are attacked to some extent by certain bacteria, at least when the skin has been broken or the tissue damaged. The result is characteristically slimy-soft and often malodorous breakdown of the plant tissues, especially the softer leafy ones. Molds and yeasts may also attack these plant materials rather vigorously.

(c) Raw meats because of their high moisture and protein content represent an excellent medium for the growth of microorganisms. If the animal has been properly slaughtered, the deep tissues will have few organisms in them. Because of this most of the microbial growth occurs on the surface of the carcass. Cutting of the meat lessens the keeping qualities by exposing more surfaces to ready attack by aerobic bacteria and molds. The kind of organisms which predominate depends to a large extent on the temperature and humidity under which the meat is held.

At refrigerator temperatures, molds may form characteristic mold masses on the surface or psychrophilic bacteria may appear as small pink, yellow, or white colonies. At somewhat higher temperatures, the mesophilic bacteria may produce a marked sliminess on the surface of the meat. Ground meats are subject to bacterial spoilage because the bacteria on the surface of the original meat are distributed throughout the entire mass of ground meat. If the temperature is raised to any extent, bacterial growth goes on at the rapid rate and microbial deterioration may quickly occur. Since anaerobic conditions prevail, the decomposing meat may develop a foul, unpleasant odor and if *Clostridium botulinum* is present it can grow rapidly and produce its lethal toxin, if the meat is held at temperatures much above 40° F. Ground meats should be carefully handled and well cooled especially when the temperatures are high, as in the summer time.

(d) Eggs, over a period of time are susceptible to microbial deterioration as all can attest who have had first hand experience with "rotten eggs." Whether or not microorganisms, capable of causing eggs to rot, may be introduced into the eggs before they are laid is a question which has not been satisfactorily answered. Certainly during the process of being laid or immediately thereafter, the surface of the egg becomes contaminated with various intestinal and soil bacteria, molds and actinomycetes spores. However, the egg is covered with a thin, membrane-like coating which keeps these organisms from penetrating the shell at least for a period of time. However, if this coating is washed or rubbed off or the egg is allowed to become moist, entrance of microorganisms is hastened. Highly humid atmospheres may provide enough moisture to speed the penetration of the shell by aiding microorganisms to decompose the protective mucinlike armor.

The kind of spoilage which may occur in eggs will depend on such factors as temperature, moisture, and the kinds of microorganisms involved. Usually spoilage is due not to one organism but several and the resultant deterioration may produce physical changes which have been described as "black-rots," "white-rots," "water-rots," etc.

If eggs are refrigerated in a dry condition they will keep for a considerable period of time without any marked deterioration in quality, although such "cold storage" eggs break down quickly after being removed from cold storage. Also, the natural enzymes within cold storage eggs produce a slow deteriorative change which reduces the quality of the egg.

(e) Fish are susceptible, particularly to surface spoilage, due primarily to the growth of organisms found in the water or which have been picked up from contaminated surfaces during the handling process. Under refrigeration, psychrophilic bacteria, which are frequently present in sea waters or the colder fresh waters, may grow and produce spoilage.

MICROBIAL SPOILAGE OF PROCESSED FOODS

The processing of foods tends to modify the flora by killing off the forms sensitive to the particular process, permitting the resistant forms to predominate. Whether or not these resistant forms grow will depend on the chemical composition, pH, temperature etc., of the processed food. Some typical examples are as follows:

Bread

The various constituents which go into the making of bread may each contribute some microorganisms; however, the flour itself may be the major source of microorganisms. These forms originate in the soil and flour may often carry thousands or even millions of such bacteria, yeasts, and mold spores per gram. Of these, the only microorganisms likely to survive the baking process are the bacterial spores and possibly the mold spores.

Certain aerobic sporeforming rods belonging to the genus *Bacillus*, mostly *Bacillus mesentericus* or *Bacillus subtilis*, are very heat resistant. As a result they are not killed by the baking process. Under certain conditions, particularly if the weather is warm, they grow rather vigorously in bread or cake and synthesize a considerable amount of bacterial gum or slime. When the bread or cake is broken apart, slimy threads are visible which may stretch to a considerable length before breaking. "Ropy" bread exhibits an unpleasant odor which may make it unpalatable. Since the "ropy bread" bacteria are sensitive to acid a small amount of lactic acid is sometimes added to the dough to inhibit the growth of these organisms or the presence of lactobacilli in the yeast may contribute some lactic acid. If bakeries become infected with large numbers of "ropy" organisms, difficulty may be experienced in ridding the bread-making equipment of this unwelcome guest.

Under some circumstances, the lactobacilli which are present in the commercial yeast preparation used may grow too vigorously and produce an excess of lactic acid. Such bread will be too acid to be palatable hence may be considered "spoiled." The lactic acid which forms in the dough is not volatilized during the baking process and as a result remains to give the baked bread a sour taste.

Molding may occur, especially if bread is kept under warm, moist conditions. The mold spores are primarily those which have survived the baking process although they may be derived from a bread can or other container which has been previously contaminated with mold spores.

While any one of several different kinds of mold may be responsible for moldiness, *Penicillium*, the common blue mold, is probably most frequently encountered and *Rhizopus*, often spoken of as the black "bread mold," may also be seen rather often.

A fermentation product, calcium propionate, has been used in the baking industry as a mold inhibiting agent.

Canned Goods

The canning process results in the development of a "practical sterility," "essential sterility," or "virtual sterility" in which the spoilage organisms are generally killed without canned foods necessarily being sterile. The organisms which escape the canning process are practically always bacteria in the spore form. In some instances they may cause spoilage, but frequently no change occurs, because they are not able to grow under the conditions extant in the can. The reason for this lack of complete sterility is that while it is necessary to use a temperature high enough to kill the spores of *Clostridium botulinum*, yet it must not be high enough to overcook the food. When overcooking occurs, the food becomes soft and the desirable flavors may be lost or greatly impaired. Under these conditions the bacterial spores which are more heat-resistant than *Clostridium botulinum* may escape destruction.

These heat-resistant forms are either members of the genus *Clostridium*, the anaerobic sporeforming rods, or facultative members of the genus *Bacillus*. These heat-resistant bacteria may also be thermophilic. As a result they may grow in the canned goods during the cooling period immediately following the process of sterilization. The true thermophiles will grow only at relatively high temperatures while some bacteria, the facultative thermophiles, which grow at either high or low temperatures, may also develop during the cooling period. A third group of common spoilage bacteria, the mesophiles, which are mostly anaerobes, will grow at ordinary temperatures. They are primarily putrefactive forms of which *Clostridium botulinum* is a representative form.

Thermophiles are of two types: (1) those causing "flat sours" are so-called because the growing bacteria produce acid and impairment softening of food without producing any gas. As a result the can of spoiled food will be normal in external appearance. (2) "Swells" result from the growth in foods of bacteria which produce gas which is about one-third carbon dioxide, two-thirds hydrogen. As a result, if a small hole is carefully punched in the top of the can, the escaping gas can often be ignited with a match.

With certain fruits a chemical reaction may take place between the fruit acid and the metal of the can resulting in the production of hydrogen gas. Such "swells" are usually found in highly acid foods while microbial "swells" are more likely to be encountered with foods that are low-acid to neutral or alkaline.

In either case, the amount of swelling which occurs will range from the barely discernible gas pressure to enough pressure to burst the can.

Another type of spoilage which may occur especially with canned corn or peas is the so-called "stinker" spoilage. This term is descriptive since the organism, *Clostridium nitrificans*, produces hydrogen sulfide or "rotten egg gas." Since this gas dissolves in water no gas pressure develops and the can remains "flat." However, the food has the unpleasant hydrogen sulfide smell and the food and metal of the can may be blackened.

With protein foods such as canned meats of various kinds, spore-forming anaerobic mesophiles sometimes develop. The proteins are strongly attacked with the production of a mixture of foul smelling compounds and often enough gas to produce a "swell."

If can closures are broken in some fashion so as to admit air, molding of foods, especially the acid fruits, may occur.

While commercially canned goods rarely show spoilage, except by the spore-forming bacteria previously mentioned, in home canned foods the heat used may have varied considerably and as a result forms that are sufficiently heat sensitive to be killed in commercial canning may escape and cause spoilage.

Proper precautions applied to home canning will greatly reduce spoilage and also reduce the danger of botulism. Some of the most important precautions are as follows:

(a) Make sure that you are not introducing large numbers of microorganisms by using only firm, unspoiled fruits or vegetables or carefully cutting out any small spoiled spots in any food. Use care in washing, blanching, and peeling to avoid contaminating the fruits and vegetables which are being prepared.

(b) Sanitize the cans or bottles by boiling before filling them.

(c) Use only the pressure cooker for heating nonacid foods and be sure the pressure gauge is accurately set. In many counties, home demonstration agents employed by the county, state, or federal government have the equipment necessary to test such pressure gauges for accuracy.

(d) The highly acid foods such as tomatoes, most fruits, fruit juices, pickles, sauerkrauts can generally be canned successfully by open kettle methods.

(e) In foods showing high concentration of solids and an acid reaction, such as jellies, jams, catsup, syrups, concentrates, etc., molds and yeasts are the most common causes of spoilage. Both mold spores and yeasts are quite sensitive to heating. A temperature of 170° F. for a few minutes will usually kill even the most resistant of these forms. When jars or glasses are filled with jelly or jam, care should be exercised to clear the "head space" of mold and yeast spores. This can be accomplished by filling the containers with hot fruit, capping, then inverting. This is also desirable when canning fruit in glass jars by the open kettle method.

With any canned foods it is important to discard any cans which are "swells" and to boil all home canned, nonacid foods for 10 minutes or longer. This will serve as insurance against the possibility of botulism. Foods which might be suspected of containing botulism organisms should be boiled before they are fed to chickens, hogs etc.

SPOILAGE OF OTHER PROCESSED FOODS

Chocolate candies with soft centers of the fondant type are sometimes "blown up" by gas produced by sugar tolerant yeasts and bacteria. Such forms are able to withstand the high osmotic pressures found in the soft candy, and for this reason are termed "osmophilic" forms.

Syrup and molasses sometimes exhibit surface molding after a long period of time in an unsealed container which affords prolonged contact with the air. Likewise, yeasts occasionally grow in honey.

Bottled beverages are sometimes subject to yeast and mold spoilage. The combination of a high citric or phosphoric acid content plus carbon dioxide probably inhibits bacteria or may even kill them. Spoilage can be largely controlled by preparing the ingredients under sanitary conditions.

FOOD POISONING

Food poisonings may be due to poisons inherent within the foods themselves, to inorganic compounds introduced into the foods or to the action of microorganisms. This last-named variety of food poisoning is by all odds the most common and most important of the three.

Following are some examples of inherent food poisoning: (a) Ergotism is due to eating rye bread made from grain infected with ergot, a type of fungus. Recently, a rather spectacular and widely publicized outbreak of this condition occurred in France. (b) Lathyrism, which comes from eating quantities of the chick pea, is due to a toxic agent in the seed. This condition is found among certain poorly fed groups in Italy, India, Africa, etc. (c) Mushroom poisoning is a notorious example of inherent poison. (d) Even the lowly potato, when the tuber becomes green, may produce an alkaloid, solanin, which causes an acute gastrointestinal upset. (e) Abraham Lincoln's mother is said to have died of "milk-trembles," a type of poisoning brought on by drinking milk from a cow which had been eating the rayless goldenrod.

A number of inorganic chemical compounds may cause acute intestinal upsets. These include salts of antimony, zinc, and cadmium which might develop as a result of mixing acid foods or beverages in galvanized containers or containers plated with cadmium, etc. Barium carbonate, sodium fluoride, or arsenicals used as rodent or insect poisons may cause serious or fatal poisonings when they find their way into human foods.

The so-called bacterial food poisonings apparently differ from food-borne infections in three or more ways: (a) The organisms produce toxic metabolites or growth products which are the probable cause of the acute symptoms which develop. (In certain cases the organisms themselves may have little or no invasive power.) (b) The organisms must grow for a period of time before enough toxin is produced to cause the symptoms of food poisoning. (c) The incubation period, after ingesting the food, is very short and the symptoms develop in an explosive fashion. On the other hand, in food-borne infections no proliferation of the bacteria in the food need occur and the incubation period is relatively long. With some organisms the combination of toxic effect and a true infection may occur. Some salmonella and shigella infections are of this nature. In these there is no clear-cut demarcation between a food poisoning and a food-borne infection.

Recently Feig[1] discussed the outbreaks of food poisonings and intestinal infections involving foods which had been reported to the United States Public Health Service in 1945-1947. More than 500 such outbreaks were reported. However, it should be recognized that those reported were only the more serious ones and probably represented only a relatively small part of the outbreaks which did occur because such infections are often not severe enough to require the services of a physician and thus are not reported. In most cases of food poisoning or infection the causative organism is not isolated. Estimates made in 1944 of staphylococcal enterotoxin gastritis cases alone suggested that there were between 570,000 and 1,000,000 such cases.

The interesting point is that the staphylococci (primarily *Micrococcus pyogenes* var. *aureus*) are by all odds the most important of the food-poisoning organisms. They caused a total of 14,988 cases, about 80 per cent of all cases reported, while the *Salmonellae* were a poor second with 3,430 cases, *Shigellae* were third with 2,883 cases and the streptococci were fourth with 1,157. *Escherichia coli* caused 302 cases, *Proteus* 78 cases, and *Aerobacter aerogenes* accounted for 52 cases. Since these cases dealt with gastrointestinal diseases only, no report was made of botulism cases.

A significant finding in this study was the observation that the length of incubation might be of value in indicating the causative organism involved. The median incubation period for 9,084 cases of staphylococcus food poisonings was 3.8 hours while the median for 2,284 cases of salmonella infections was 19.9 hours and an equal number of shigella infections showed a median of 53.4 hours. In the small number of streptococcus poisonings (629) the median was 10.3 hours.

The most common foods involved in staphylococcus outbreaks were meat products, 35.3; bakery products, 35.3; poultry, 22; potatoes, 8.6; and milk products, 6.6 per cent, respectively, of the outbreaks.[2]

---

[1]Feig, M.: J. Am. Pub. Health Ass'n **40**(11):1372-1394, 1950.

[2]More than one food was involved in some outbreaks, thus a percentage greater than 100.

Ham was the most frequent offender. Since micrococci are so frequently involved in food poisonings the knowledge of a few important facts may well serve as a safeguard against such poisonings:

1. The chemical nature of the causative substance is not known, but it is a toxic metabolite or product of growth which seemingly has an affinity for the intestinal tract.

2. The organism involved is usually derived from some active human infection such as a boil, carbuncle, wound infection on the hand, or a nasal infection such as sinusitis. The organism is usually *Microoccus pyogenes* var. *aureus* although the *albus* variety has been reported.

3. Although a staphylococcal food poisoning was first reported in 1914 from milk, its significance was not appreciated until this knowledge was revived by Dack, et al. in 1930, as the result of a study of a food poisoning outbreak.

4. The enterotoxin withstands boiling for 15-30 minutes.

5. Symptomologically, the disease is characterized by a short incubation period, generally about four hours, little or no temperature or a subnormal temperature, violent and acute symptoms of nausea, vomiting, and diarrhea, while complete recovery generally occurs within 24-48 hours.

6. As a preventive measure susceptible food should not be held at room temperature for any period of time, especially during the summer months. An incubation period of 8-10 hours at room temperature or warmer is conducive to the growth of sufficient numbers of organisms to cause poisoning symptoms. Refrigeration is important since the organism apparently will not grow or develop toxins at 40° F. The practice of reheating soon after preparing of foods such as cream puffs which cannot be refrigerated, serves to pasteurize them and thus destroy the staphylococci. Persons who have boils, hand infections, or active discharges from the respiratory tract should not prepare food.

Salmonella infections may come from human intestinal discharges since apparently 2-5 per cent of population are carriers, also dogs, cats, and other domestic animals may harbor the organisms, while flies may serve as efficient transporting agents. Powdered eggs have been incriminated in a number of cases. Essentially the same precautions should be observed as with the staphylococci although the organism can grow slowly at 45° F.

Shigella infections are much rarer and not so explosive in their symptomatology. The organisms primarily involved in this country are mostly *Shigella paradysenteriae* or *Shigella sonnei*. The precautions observed should be the same as with the *Salmonellae*. Asymptomatic human carriers of *Shigella* seem to be fairly common.

Compared with other types of food poisonings, botulism is rare indeed. However, it must be constantly guarded against, because the organism is present in soils in most parts of the United States and improper food handling or processing, particularly in home canning, may permit the growth of this organism. The highly lethal toxin which it produces is probably one of the most highly poisonous substances now known.

The major precautions against botulism are:

1. All nonacid foods should be canned only with a properly adjusted and operated pressure cooker following the manufacturer's directions carefully.

2. Such home canned foods should be boiled for at least 10 minutes before they are consumed.

3. Proper refrigeration of salads, etc., made with raw materials which must be prepared in advance, is a wise precaution since the organism will not grow or produce toxin at 40° F.

You will note that the words "ptomaine poisoning" have not been mentioned. These words, as suggested a good many years ago, are "good ones to forget" since food poisonings are apparently not due to ptomaines. The ptomaines are apparently products resulting from the partial breakdown of proteins and their malodorous nature would preclude any likelihood that such foods would be eaten. In addition, their toxicity has not been fully established.

METHODS OF FOOD PRESERVATION

Food preservation methods may be classed for convenience into those involving the use of heat to kill microorganisms, the use of cold to inhibit their growth, inhibition of growth by means of chemical compounds and preservation by removal of moisture. Of these methods canning is commercially the most important.

Canning

The History of Canning.—The basic procedure involved in canning consists of heating a food-filled, hermetically sealed container until all of the spoilage organisms have been destroyed. This principle was discovered by Spallanzani in 1765.   However, he failed to realize the great practical value of his discovery.

Napoleon's ambitions to conquer the world served as the mainspring for the development of commercial canning.   As the supply lines of the French armies became longer and longer it became apparent that some method other than salting, pickling, or drying was necessary to provide a healthful and palatable food supply.   Accordingly, in 1795 the Directory of France set up a prize of 12,000 francs to be awarded to the developer of a new and satisfactory method of preserving food.

The prize was not claimed until 1809 when a Parisian confectioner, Appert, demonstrated the method we now designate as canning.   He used wide mouth bottles which were filled with food, corked, and heated in a bath of boiling water.

The following year, canning became a reality when Peter Durand, an Englishman patented the idea of substituting tinned steel cans for bottles.

The idea caught on and in 1820 the first American canneries were put into operation using the Appert process by William Underwood in Boston and Thomas Kennett in New York.   Within the next 20 years the use of tin containers became widespread in the United States.

During this early period spoilage was probably commonplace and the first step toward improving keeping qualities came in 1861 when the practice of raising the boiling temperature by adding calcium chloride to the water bath was initiated.

The pressure cooker or "steam pressure retort" was patented in 1874 and was soon rather widely used.

In spite of the microbiological nature of canning, little was learned regarding the bacteriology of the process until the turn of the century.

The first open top, sanitary can was put into use in 1900.   In 1907 the National Canners Association was formed and soon after World War I they initiated a vigorous research program to develop procedures which would permit them to control *Clostridium botulinum* and the other organisms responsible for spoilage of canned foods.   By 1928 these procedures had been largely standardized and since that time, botulism from canned goods has become practically nonexistent in spite of the untold billions of cans of various foods which have been consumed by the American public.   Today studies are being made of canning procedures which involve higher temperatures but shorter periods of time.[3]

---

[3]For a complete history see Canned Food Reference Manual, American Can Co., 1939.

Operations Involved in Modern Canning.—In modern canning methods, the following basic steps are involved.

1. *Cleansing.*

Fruits and vegetables of various sorts generally carry on their surfaces considerable numbers of spoilage producing microorganisms. By greatly reducing the numbers of such organisms, successful canning can be more easily accomplished. The number of microbes may be reduced by washing the fruits or vegetables under strong sprays or flowing streams of water, by floating off chaff and other extraneous matter or by screening or using strong blasts of air to "dry clean" the foods which are to be canned.

2. *Blanching.*

In this process the raw fruits or vegetables are immersed for a short time in hot water or subjected to live stream. This makes the tissue softer so it will fill the containers properly; removes the respiratory gases from the plant cells (which otherwise might cause the can to bulge), and destroys many of the enzymes which might impair flavor.

3. *Peeling and Coring.*

After being blanched the fruits or vegetables are peeled and cored or given any other required treatment before being placed in the cans.

4. *Filling and Exhausting.*

The prepared fruits or vegetables are now carefully packed into cans and the cans are "exhausted." This consists of placing the cans in an "exhaust box" in which they are heated to about 130° F. This serves to drive out the air which may be trapped between the pieces of the fruit or vegetables.

5. *Sealing.*

After filling and exhausting, the cans are sealed to prevent contamination and in preparation for the heating process.

6. *Sterilization.*

The length of heating and the temperature used depends largely on the pH of the food being canned. If the food is more acid than pH 4.5 it is usually sufficient to raise the temperature to 200° F. in the center of the can. The pH is important for two reasons: (a) killing of microorganisms proceeds much more rapidly at a fairly acid pH and (b) *Clostridium botulinum* does not grow at a pH more acid than pH 4.5.

Nonacid foods such as beans, peas, carrots, meat, etc., must be heated to at least 240° F. for safe processing. *Clostridium botulinum* spores must be killed because they will be able to grow successfully in nonacid foods.   These spores are killed by holding at a temperature of at least 240° F for 10 minutes.   The relation of temperature to pressure is an autoclave is shown in Table XII.

TABLE XII

RELATION OF STEAM PRESSURE IN AUTOCLAVE TO TEMPERATURE

PRESSURE (LBS. PER SQ. INCH)	TEMPERATURE	
	°F	°C
0	212	100 Boiling Point
5	226	108
10	240	116
15	250	121
20	260	127

Note: Above temperatures can be secured at sea-level and only if air is completely replaced with steam.

It is important to recognize that as the altitude increases, the temperature at a given pressure is lessened.   This occurs because the temperature increase due to pressure is superimposed on the boiling point of water.   Thus, while at sea level the boiling point is about 100° C, at 5,000 feet altitude it is in the neighborhood of 95° C.   For successful use of the pressure cooker or autoclave at higher altitudes a higher pressure must be employed or a longer sterilization time utilized.

7. *Cooling.*

After an adequate heating period, which depends on the pH of the food, size of container, etc., the cans are cooled as quickly as possible to stop the cooking process and thus prevent undue softening of the food. Even after the initial cooling, the center of the can may still be relatively hot and this heat may not be dissipated for some time.   If thermophilic bacteria are present in the food, they may grow and spoil the food during this cooling process.   Attempts are therefore made to prevent an unduly long cooling period.

Pasteurization

While the temperature used in pasteurizing milk and food products is not necessarily high enough to kill all bacteria it produces a marked reduction in the bacterial content since it may kill as many as 99 per cent or even more of the mesophilic bacteria present. Thermoduric bacteria and bacterial spores largely resist pasteurization while thermophiles may grow rather rapidly at pasteurization temperatures. The process of pasteurization is not only useful for milk but the keeping qualities of many other food products is also improved by pasteurization. The process is used with fruit juices, especially grape juice and apple juice and is commonly practiced in the beer and wine industries. While the temperature varies somewhat, 140-150° F for 30 minutes is rather commonly employed.

Use of Cold

Within the common temperature ranges of bacterial growth (between freezing to slightly above body temperatures), the rate at which common microorganisms grow is reduced two or three times for every 10° C reduction in temperature. Refrigeration at above freezing temperatures owes its effectiveness to this marked reduction in the rate of bacterial growth. In addition, with some organisms, commonly employed refrigeration temperatures may be below the minimum growth temperature of an organism. In such an event, the organism will not only fail to grow but may die out at a slow or rapid rate. On the other hand, the psychrophilic bacteria present will continue to grow and produce spoilage of the food product. In the final analysis, refrigeration above freezing temperatures is essentially a method for deferring rather than preventing spoilage. Another advantage of refrigeration is that foods remain for a time in a state similar to that of the fresh product since both microbial growth and the action of the food enzymes is slowed down.

In certain syrups etc. which do not solidify at freezing temperatures, slow growth has been observed even at −2.2 to −3.8° C. (25 to 28° F.).

As has been previously suggested, the household refrigerator is best kept at about 40°F. except in the frozen food compartment which should be well below the freezing point (0 to−30°F. is recommended). The length of time that fruits and vegetables can be held in cold storage varies from about 10 days for cucumbers and berries to 8 months for apples.

With the introduction of mechanical refrigeration the preservation of foods by freezing was attempted. This did not meet with any degree of success because the slow freezing resulted in the development of large ice crystals which ruptured the cells and the foods became soft and watery as soon as they were thawed. With the introduction of the "quick freezing" or "sharp freezing" method which was based on the observation that in rapid freezing only small ice crystals develop, the quality of the food is essentially unimpaired. However, a few precautions must be observed with this process. Since many bacteria are not killed by quick freezing and some, like the typhoid organism, may remain viable for some months in quick frozen foods, care should be used if quick freezing is to be practiced in the home. The microbial load on the food to be quick frozen should be as low as possible. This may be accomplished by carefully washing the food and blanching it before it is placed in the containers for freezing. Blanching not only reduces the bacterial load but also destroys some of the enzymes which might otherwise cause a slow deterioration in the flavor and quality of the foods.

After they have thawed, quick frozen foods are susceptible to rapid spoilage by microorganisms. They should, therefore, be used immediately after thawing. *Thawed food should never be refrozen.* The danger from botulism is not too great because the food will probably be soured by the growth of certain lactic acid streptococci or made undesirable for use by other spoilage microorganisms before any quantities of toxin are produced by *Clostridium botulinum*. However, complete safety is assured if frozen foods are cooked or eaten as soon as they have thawed. Dry ice is effective for maintaining frozen foods during shipment because it is approximately twice as effective as ice as a cooling agent since the temperature of a block of dry ice is between -75 to $-100°$ C.

Chemicals for Food Preservation

Common salt (sodium chloride) is one of the oldest preserving agents used by man. Its preserving action appears to be largely due to the high osmotic pressure present in strong salt solutions. When used for "dry salting" of meats and other products, water is withdrawn from the food and the high salt content effectively inhibits microbial growth. In pickling brines used for curing hams and similar types of meat, the salt concentration used is about 25 per cent which is virtually a saturated solution. Sugar (10-15 per cent) and sodium nitrate may also be added to the "pickle." While most saprophytic bacteria are inhibited by the high salt concentration employed, certain halophilic (salt-loving) bacteria may continue to grow. These may reduce the nitrates to nitrites. This serves to intensify the red color of the meat. Occasionally the halophiles may cause spoilage of meat products or other salt-preserved products. Halophiles usually come from the salt itself.

Salt concentrations of 10-15 per cent will eliminate most of the common saprophytic bacteria especially the gram-negative rods. *Clostridium botulinum* is generally inhibited by a salt concentration of 10 per cent. Certain molds are inhibited by 15 per cent salt although a few molds and wild yeasts have been isolated which can grow slowly in a saturated salt solution.

Sugar, likewise, exercises its preserving action because of the high osmotic pressure of strong sugar solutions. Jellies and jams are generally 50 per cent sugar. This concentration, together with the acidity, inhibits the growth of bacteria, yeasts, and most molds. Some molds, especially penicillia, have been found which can grow very slowly even in a sugar concentration of 70 per cent.

Vinegar, which is about 4 per cent acetic acid, owes its preserving action to the combined effect of a highly acid reaction and the possible additional germicidal effect of the acetic acid. Lactic acid of the same concentration does not appear to have as much germicidal activity as that exhibited by acetic acid. These two acids are active preservative agents in the various types of pickles or pickled meats.

Certain spices and related substances tend to exert some germicidal action which makes them function not only as flavoring agents but also as preservatives. Cinnamon with its cinnamic aldehyde content and cloves with the active agent, oil of cloves (eugenol), are the most strongly germicidal of the spices. Garlic and onions also exert an antibiotic effect and freshly crushed horseradish has an inhibitory effect on many microorganisms. On the other hand, thyme and lavender exert only feeble germicidal actions.

Smoking of meats of various sorts has long been used, often associated with salting, as a means of preservation. The process of smoking is essentially a destructive distillation of the wood in which the various cresols, formic and acetic acids, formaldehyde, etc., are distilled out on the surface of the meat, effectively preserving it against microbial spoilage as long as it is kept dry. At the same time a desirable aromatic flavor and odor is imparted to the meat.

Dehydration consists of removing most of the water from a product. This has a dual effect, that of greatly increasing the osmotic pressure of the dried food and of robbing the microorganisms of the moisture necessary for growth. During dehydration many of the gram-negative rods appear to be destroyed with the result that the gram-positive sporeformers become the predominating bacteria in the food. Mold spores may also be present in large numbers. However, this is not true in all cases, because with certain products, very notably occasional samples of powdered eggs, pathogenic Salmonella bacteria have been found to persist and to be capable of causing food poisoning in poorly cooked foods.

Dehydrated foods will keep only so long as they are kept dry. In some cases a slow deterioration may occur due to oxidation with atmospheric oxygen.

FERMENTATION INVOLVED IN FOOD PRODUCTION

Certain microorganisms in the course of their growth on foods produce chemical products which impart desirable properties to foods.

Leavening of bread is perhaps the most universally employed of these fermentations. It involves: (a) the change of starch in the flour to sugar by the action of enzymes in the flour and, (b) the fermentation of this sugar by yeasts into carbon dioxide and alcohol. The carbon dioxide causes the bread to rise and gives it a desirable consistency while the alcohol is driven off during the baking process. This alcohol is responsible in part at least for the pleasantly sweet odor of baking bread and the stinging sensation caused by the gases from the oven when the oven is opened and the hot gas strikes the eyes.

In sour dough bread, leavening is due primarily to the production of gases by such bacteria as *Escherichia coli* and *Clostridium perfringens*, the gas gangrene organism. Lactic acid bacteria present in this mixture may impart a sour taste to the bread.

Lactic acid fermentations form the primary basis for the microbial changes involved in the production of sauerkraut and dill pickles. In preparing these products, the foods are first washed, then packed in layers with considerable amounts of salt and spices. The mixture is then covered with water and the fermentation allowed to proceed anaerobically. The salt tends to draw some of the solutes out of the plant material. These extracted sugars and related compounds are attacked at first by yeasts but lactic acid producing organisms soon begin to act. Initially these organisms will be of the *Streptococcus lactis* type but as the acidity increases *Lactobacillus* will take the ascendancy. The lactic acid which is produced serves both as a source of flavor and as a preserving agent. Some small quantities of acetic acid may also be produced in the course of the fermentation. Commercially canned dill pickles and sauerkraut may have considerable amounts of vinegar artificially added.

Silage for animal feeding is also preserved by a lactic acid fermentation process resembling the dill pickle-sauerkraut fermentation.

Vinegar production represents a somewhat different type of microbial change since it involves the change of an already fermented substance (cider or wine) into a second product. Unlike the alcoholic fermentation which is anaerobic, the oxidation of ethyl alcohol (in wine or vinegar) takes place under aerobic conditions. The organism involved is classified in the genus, *Acetobacter*. Any one of several different species of this organism may be involved, depending on the product which is being changed to vinegar. These organisms constitute the "mother of vinegar" film often seen in bottles of vinegar. In commercial processes, the alcohol concentration of the liquor used is about 10 per cent while the resultant vinegar may contain up to 5-6 per cent acetic acid. A moderately high alcohol concentration is necessary because the organism may otherwise continue the oxidation of the acetic acid completely over to carbon dioxide and water. Some procedure which will give adequately aerobic conditions is used to hasten the rate of acetic acid formation.

Selected Reading References

Smith, David T., et al.: Zinsser's Textbook of Bacteriology, ed. 10, New York, 1952, Appleton-Century-Crofts, Inc.

Tanner, F. W.: The Microbiology of Foods, ed. 2, Champaign, Ill., 1944, Garrard Press.

Townsend, C. T., and Esty, J. R.: The Role of Microorganisms in Canning, Western Canner and Packer, June-July-August 1939, p. 1-8.

CHAPTER 16

SAFE MILK AND DAIRY PRODUCTS

The Microbiology of Milk and Dairy Products

Milk and milk products are probably the best and most important single group of foods in the American diet. They constitute 1/5 to 1/6 of the average diet. Milk is, likewise, an excellent culture medium for the growth of many kinds of bacteria. This affinity of bacteria for milk has provided the basis for a distinct branch of microbiology, i.e., dairy bacteriology, which concerns itself with milk and milk products. Biologists are interested in milk from three distinct angles: (1) It can be a potential carrier of disease-producing bacteria if not properly handled. (2) It is one of the most easily spoiled foods. (3) Many desirable fermented products can be produced from milk by the use of the proper microorganisms.

Milk is mostly water. It usually contains about 87 per cent water and about 12-13 per cent solids. These solids are divided about as follows: Butterfat 3-5 per cent, lactose (milk sugar) about 5 per cent, and proteins (mostly casein with some albumin) about 3.5 per cent.

The human digestive system is equipped with enzymes which readily break down the various constituents of milk for their effective assimilation by the body.

Many bacteria are likewise able to attack milk components. Frequently, however, a given bacterium may attack one constituent while leaving the others essentially unchanged. If bacteria do not posesses the proper enzyme systems for digesting a given constituent they naturally must ignore this compound. To them, the particular compound is a lock for which they have no key.

Lactose or milk sugar, when attacked, frequently yields lactic acid which you recognize as the buttermilk or sour milk acid. In other cases proprionic, acetic, other acids or mixtures of acids or other similar compounds may be produced by bacterial action on lactose.

Carbon dioxide and other gases may also be produced, resulting in a frothy type of spoilage.

The protein portion of milk is largely casein, a phosphoprotein. Bacteria may produce a number of changes in the casein: (a) The acid which they produce from lactose may curdle the casein in the same fashion that vinegar added to milk causes it to curdle. (b) This acid curdled casein may be virtually unchanged if the bacteria lack the enzymes necessary to break it down, or they may split the complex protein molecule into proteoses, peptones, or such simple constituents as amino acids or even carbon dioxide, water, and ammonia. When this occurs, the milk curd is completely dissolved leaving a curd-free whey. (c) Bacteria may produce rennin-like enzymes which will curdle the casein without preliminary souring. This is termed "sweet curdling." The curd may then be digested or remain unchanged. (d) Casein may also be digested without preliminary curdling.

Fat is not readily attacked by most microorganisms, but certain ones can break it down giving a rancid odor and flavor. The major cause of the rancid odor is butyric acid. Agents other than microbes may produce a similar effect.

MILK AND DAIRY PRODUCTS AS CARRIERS OF DISEASE

In the past, many epidemics, sometimes devastating ones, have been traced to contaminated milk supplies. Modern methods involving sanitary inspection, pasteurization, etc., have virtually eliminated this hazard in pasteurized products, but raw milk, dispensed by careless handlers, still represents a very real hazard to health. The consumer of raw milk might well remember that he is drinking one of the few animal products that is now consumed in the uncooked state. Explosive epidemics still occur as a result of drinking raw milk produced under uncontrolled conditions. On the other hand, the careful procedures used today in producing high quality pasteurized milk have made it one of the safest if not the safest of all foods.

The chief reason for the health hazard inherent in improperly handled milk lies in the fact that a few disease-producing bacteria introduced into a milk supply may grow into thousands or even millions of cells within a few hours. As a result, a glass of such contaminated milk may represent a massive inoculation of millions of pathogens. The size of the inoculum may bear a direct relationship to the severity of the resultant disease.

Epidemics clearly traced to contaminated milk and dairy products were reported rather frequently a few decades ago. The diseases most commonly involved were typhoid fever, the bacillary diarrheas and dysenteries, scarlet fever, septic sore throat, and diphtheria. Occasional sporadic epidemics of these diseases are still being reported. They are traced almost invariably to raw milk or in very rare instances to pasteurized products contaminated after pasteurization through improper handling.

Tuberculosis of bovine origin was once fairly common, particularly in children. However, tuberculin testing of cattle and elimination of infected animals plus pasteurization have practically freed us of the danger represented by this slayer of children. On the other hand, less than a decade ago, public health workers in Great Britain reported that some 600 children were killed each year from bovine tuberculosis transmitted through infected raw milk. Human tuberculosis may also be transmitted through milk infected by contact with human cases.

Brucellosis, for which undulant fever is another but unsatisfactory synonym, is still a problem. While the disease is transmitted most often by contact with infected animals, raw milk or milk products still represent a common means of transmission of this enervating disease. A study made in Iowa indicated that about three-fourths of the cases were due to contact with infected cattle or hogs and one-fourth to the consumption of raw milk. Pasteurization, plus testing of dairy cattle to eliminate infected animals, plus vaccination of calves could make this disease (from bovine sources) as uncommon as bovine tuberculosis. A new method for detecting udder infections in cows by the presence of antibodies in the milk is aiding in the fight against brucellosis.

In a few small outbreaks of poliomyelitis, a virus disease, the finger of suspicion has been pointed at the milk supply as the possible carrier. A small outbreak of another virus disease, infectious hepatitis or jaundice, was blamed on a milk supply.

Recently, an epidemic of Q fever, a rickettsial disease, involving some 300 cases and 3 deaths occurred in California. A milk supply was thought to be involved. Public health workers are making a vigorous study of this new hazard.

Pathogens find their way into milk directly or indirectly. Infected cows may shed the causative agents of tuberculosis, brucellosis, septic sore throat, and related streptococcus infections, staphylococcal food poisoning, and Q fever directly into the milk. While the streptococcal and staphylococcal infections may come from an udder infection due to organisms of the bovine type, on the other hand, a milker with sore throat or an infected hand may be the source of the organism causing the udder infection.   Cases resulting from organisms of human origin are usually more severe than those arising from organisms of the bovine type.

Human cases or carriers in the past have been the source of epidemics of typhoid, paratyphoid, and bacillary dysentery.   Again modern sanitation has virtually eliminated such milk-borne diseases.

It should again be repeated that properly handled and inspected pastuerized milk represents a safe and highly desirable food.   The quality of today's milk supplies in many of our major cities represents a real monument to the combined efforts of the dairy bacteriologists and the sanitary bacteriologists.   On the other hand, careless handling by the purchaser of a safe, high quality pasteurized milk may result in making it unpalatable or rendering it dangerous to the health of the members of the household who consume it.

SPOILAGE OF MILK—IMPAIRMENT OF QUALITY

Bacteria, by virtue of their varied enzyme systems, can sit down to a meal of dairy products and produce bizarre changes in appearance or flavor which often make the milk undesirable for consumption by persons with fastidious palates.   In a few instances, food poisoning may follow consumption of such "spoiled" products, but in most cases the only result will be a displeased sense of taste or a disgusted sense of smell which may spell financial ruin to an unfortunate dairyman whose milk supply is involved.

Such spoilage may occur with milk, evaporated milk, cottage and similar soft cheeses, the semi-hard or hard cheeses, and ice cream.

In every case, microorganisms grow on or in the product and produce by the enzymatic breakdown of some milk constituents, products which taste, smell, or look undesirable.   However, dairy products considered spoiled by some Americans are avidly consumed by many other national groups.

Many different kinds of bacteria, yeasts, or molds may produce spoilage of dairy products.   In every instance, relatively large numbers of organisms are necessary before the products appear "spoiled."   For example, milk does not ordinarily taste "sour" until the number of lactic acid bacteria present has reached several million per milliliter which would mean a good many billion bacteria per quart of milk.

Other spoilages generally involve equally astronomical numbers of microorganisms.

The common types of spoilage are as follows:

a. *"Normal" souring.*

This is the spontaneous souring of raw milk often observed on a warm day, in which formation of a firm curd or "clabbered" condition, with a pleasant, sour taste develops. The organism responsible is usually *Streptococcus lactis.* This organism is commonly found on hay, fodder, and similar vegetation. Carelessly cleaned utensils frequently harbor many of these bacteria. When *Streptococcus lactis* finds its way into milk, it begins happily to change the milk sugar or lactose into lactic acid. It grows more rapidly than other bacteria which may be present. When a certain degree of acidity has developed, the casein is changed from a colloid sol to a colloidal gel and the result is a firm "clabber." The organism soon stops its own growth due to the accumulated acid. *Strepococcus lactis* has little or no effect on the casein curd which may be broken down subsequently by proteolytic bacteria, yeasts, or molds.

b. *Gassy curdling.*

Raw milk, even when carefully produced will contain some *Escherichia coli, Aerobacter aerogenes* and related forms. These will be recognized as the coliform, or common intestinal bacteria. These organisms have the ability to attack lactose or milk sugar with the production of acid and gas. When the milk is held in a rather warm place the organisms may grow rapidly producing bubbles of gas. In some cases enough acid is produced to curdle the milk. The curd is then filled with gas bubbles and may be spoken of as a "gassy curd." In some instances the milk merely foams due to the accumulated gas. Generally *Streptococcus lactis* will grow rapidly enough to keep the coliforms well in check, but if the temperature is relatively high it may favor the coliforms over *Streptococcus lactis.* Proper pasteurization will kill these bacteria. Gassy fermentations due to coliforms are not expected in properly pasteurized milk.

c. *"Stormy" curdling.*

Certain of the sporeforming anaerobic rods may develop in milk freed from the acid-producing bacteria by pasteurization. These bacteria are unable to compete with *Streptococcus lactis* or other acid producers so the fermentation they produce would rarely be encountered in raw milk. These organisms may coagulate the casein enzymatically, then produce gas which results in a violent breaking up of the curd, i.e., the so-called "stormy" fermentation. It may occasionally be encountered in evaporated milk and on very rare occasions in pasteurized bottled milk.

d. *"Ropiness" in milk.*

Certain bacteria produce gumlike material which causes the milk to become slimy or viscid. Such milk may form sticky threads when poured out or may "stretch" like a rubber band when lifted with a needle. Such a condition is objectionable to the consumers and in market milk may result in a severe loss to the distributor due to returned or rejected containers of milk. Ropiness is sometimes so pronounced that the milk can be "stretched" several inches or even feet without breaking the strand. On the other hand, in some countries "ropy" milk is considered a delicacy.

Alcaligenes viscosus, *Escherichia* and *Aerobacter* strains and some strains of *Streptococcus lactis* are most commonly responsible for ropiness.

Occasionally two organisms must work together in order to produce this condition.

e. *Off-flavors and odors.*

Flavors and odors described as fishy, malty, bitter, medicinal, potato-like, etc., may occasionally be produced by microorganisms. However, microorganisms are not the most common cause of off flavors or odors in milk. Food consumed by the cow, physiological condition in the animal or odors absorbed from the environment may all be contributing factors.

f. *Sweet curdling.*

Some bacteria produce a rennin-like enzyme which coagulates milk without souring it. Bacilli, (aerobic, sporeforming rods) are the most common cause of this condition. Such sweet curdled milk may develop a bitter taste.

g. *Color fermentations.*

Under certain conditions bacteria may produce unusual color changes in milk. *Pseudomonas pyocyanea* when growing in the presence of a lactic acid former may produce a distinctly blue color in milk. Occasionally, certain bacteria produce yellow or red colorations in milk.

h. *Other types of spoilage.*

Yeasts may grow on certain dairy products and produce undesirable changes. Pink colonies of such yeasts are occasionally noted in cottage cheese. Molds, particularly *Oospora lactis*, may form wrinkled surface growth on cream, cottage cheese, and other products which have been stored for some time.

SOURCE OF SPOILAGE ORGANISMS

Bacteria responsible for spoilage generally find their way into the milk or dairy products from improperly cleaned utensils in which growth of the organisms may have occurred, from water used to wash utensils or even from contaminated milk canals in cattle. Such contamination has been ascribed to cattle wading in swamps or polluted water or to contaminated milking machines. Improper washing of the udder before milking, hair from the animal, or even dust in the air may also be sources.

Modern sanitary methods of milk production do much to safeguard dairy products against undesirable fermentations. The wise dairyman is constantly on guard against such undesirable bacteria.

Fortunately spoilage of carefully produced dairy products is now uncommon due to the use of such bacteria-controlling factors as careful washing of utensils with hypochlorites or other chemical disinfectants followed by pasteurization and refrigeration. In fact, many types of spoilage formerly common are now a rarity and seldom encountered in Grade A pasteurized milk, distributed by reputable dealers under adequate sanitary city or state inspection.

It is well to remember, however, that in a refrigerator, even the highest quality milk will not keep indefinitely. Certain bacteria, classed as psychrophilic (cold loving) forms will continue to grow until the numbers become relatively high. One of the authors recently placed two samples of Grade A pasteurized milk side by side in a refrigerator to determine their keeping qualities. While sample A had 14,000 bacteria per milliliter at the beginning of the experiment and sample B showed a bacteria count of 13,000, at the end of 7 days sample A showed 1,000,000 bacteria while in sample B the count was only 140,000. The difference was apparently due to differences in the kinds of bacteria present. Psychrophiles may produce objectionable changes in milk such as bitterness and off flavors.

Commercial production of market milk has become, in many areas, particularly in milksheds for large cities, a highly technical procedure. This calls for spotlessly clean and sanitary milking barns.

Most city and state laws require frequent inspection and grading of such facilities. Not only must the producer give strict attention to the physical plant, but also he must carefully guard against disease in his herd. Likewise, the milking equipment, preferably milking machines, must be sterilized and kept spotlessly clean. Cleansing of the udders and teats of the cows also materially decreases bacterial contamination of the milk.

After production of the milk on the farm, immediate cooling is required, so that modern milk producing farms have refrigeration not only to cool the milk after production but also to hold the milk prior to delivery. Demands of large cities require milk haulage for several hundred miles from the producer to the pasteurizing and bottling dairies. Precooled milk is hauled for long distances in insulated tank trucks.

PASTEURIZATION—THE WHY AND HOW

The process of heating a product to improve its keeping qualities was applied to wine by Louis Pasteur, 1860-1864 in an attempt to save the wineries of France from ruin due to the wholesale spoilage of the product. It proved eminently successful and its author has been honored by having the process bearing his name emblazoned on untold millions of milk bottle caps and cartons to say nothing of bottles of cider and other products.

In 1886, Soxlet suggested that milk which was to be fed to infants should be heated. Soon after this the dairy industry began the practice of pasteurizing market milk.

Pasteurization of milk has two objectives, (a) to destroy pathogenic bacteria and (b) to improve the keeping quality of the milk. Pasteurized milk is not sterile but the number of organisms is generally greatly reduced. Usually 90-99 per cent of the bacteria originally present in the raw milk are killed by pasteurization. Occasionally thermophilic (heat-loving) bacteria, which actually grow at the pasteurization temperature may be found. In such a case the bacteria count may actually increase during pasteurization. On the other hand, thermoduric (heat-enduring) bacteria, which resist the pasteurization temperature may also be encountered. Their presence is evidenced by the fact that the bacteria count is almost as high in the pasteurized as the unpasteurized milk. Neither have any public health significance but may produce off flavors.

Two pasteurization procedures are used: The first, the LTLT (long-time-low-temperature) or holding method consists of holding the milk in a vat at 142°-145°F. for 30 minutes.   The second, the HTST (high-temperature-short-time) or continuous method consists of heat-

Fig. 18.—Modern milking parlor. Note that the cows are on offset or staggered cement platforms. This decreases body discharge contamination from one cow to another. Such units are much more easily kept clean and sanitary. (Courtesy Artillo Guernsey Farm, Salt Lake City, Utah.)

ing the milk at 160°F. for 15 seconds.   In actual practice it is more common to use 161-165°F. and 16 to 17 seconds.   This method is ingenious in that the milk which enters the preheater is heated by the milk which is flowing from the heating chamber.   This in turn serves to cool the pasteurized milk quickly. The heating chamber is a battery of baffles through which the heat flows rapidly. The temperature is very accurately controlled and the design guarantees a full minimum temperature for the required time. This process functions successfully only when a high quality raw milk is used.

Fig. 19.—Modern milking machine. Note that the cow is on a sufficiently high platform so that the milker can stand, cleanse the cow's udder easily, and conveniently attach the milking machine. (Courtesy Artillo Guernsey Farm, Salt Lake City, Utah.)

Electric heating devices may be purchased on the market for home pasteurization of milk in the bottle.

The temperatures and times prescribed for pasteurization have been determined experimentally. *Mycobacterium tuberculosis*, the most resistant to heat of disease-producers likely to be encountered in milk, has been used as the test organism. Temperatures which kill this organism provide a sizeable margin of safety against other pathogens. *Streptococcus lactis* the common milk-souring organism is generally killed by pasteurization so that pasteurized milk will rarely sour spontaneously as does raw milk.

Fig. 20.—Truck and trailer milk tanks for Carnation Milk Company. Cork-insulated; 440-gallon total capacity in both units. Hauling milk from Bakersfield to Los Angeles, Calif., 365 days per year. (Courtesy C. E. Howard Company.)

The organisms which do survive are generally sporeforming rods, micrococci producing yellow colonies, *Streptococcus thermophilus* and probably *Streptococcus liquefaciens* and some especially resistant coliforms. This latter situation is rare since pasteurization practically always gives complete freedom from coliforms in 1 ml. portions of milk. The method used will be explained subsequently.

Higher pasteurization temperatures sometimes depress the cream line in glass bottles of unhomogenized milk. However, the cream line is not always an indicator of cream content since it can be increased by methods of processing while the actual fat content may be simultaneously reduced.

Spoilage, when it does occur in pasteurized milk, is most likely to be sweet curdling or the development of bitterness or off flavors.

Pasteurization has justified its existence by the remarkable decrease in milk-borne epidemics which has accompanied its widespread use. Indeed Rosenau has made the comment "milk-borne outbreaks of disease are always due to raw milk....There is no record of a milk-borne outbreak attributable to properly pasteurized milk."[1]

The improvement in keeping quality of milk in the past decade due to the combination of a high standard of sanitation in the producing and handling of milk, pasteurization, and refrigeration has given rise to the practice of every other day delivery or holding of milk in grocery stores for several days.

Any slight impairment in the nutritional value of pasteurized milk over raw milk, is apparently limited almost entirely to a slight reduction in the ascorbic acid (Vitamin C) content. This loss is completely outweighed by the remarkable degree of freedom from disease-producers of pasteurized as compared with raw milk. The loss in vitamin C is not a serious one because the amount present in milk is far too little to supply our needs adequately. No carefully controlled experiments have established any nutritional advantage in raw over pasteurized milk. However, these experiments do indicate a far greater incidence of milk-borne diseases among the consumers of raw milk.

Certain milk enzymes are destroyed by pasteurization, but this appears to be an advantage because certain of these enzymes tend spontaneously to produce some off flavors not encountered in pasteurized milk.

METHODS FOR ESTIMATING NUMBERS OF BACTERIA IN MILK

Since the presence of large numbers of bacteria in milk tend to indicate faulty handling or even udder infection in the cows producing the milk, methods for counting the numbers of bacteria are essential for the production of high quality milk. These bacteria counts not only indicate the quality of the raw milk used but also provide a measure for the effectiveness of pasteurization. Routine bacteria counts do not necessarily reveal the presence of disease-producing bacteria, except for the microscopic method which reveals certain

---

[1]Rosenau, Milton J., et al.: Preventive Medicine and Hygiene, ed. 6, New York, 1935, D. Appleton-Century Company, Inc., p. 792.

pathogens such as streptococci. However, a low count milk indicates care in producing the milk and a greatly reduced possibility of the presence of disease-producing bacteria.

Fig. 21.—A modern flash type (high temperature-short-time) milk pasteurizer. (Courtesy HiLand Dairy, Salt Lake City, Utah.)

The Standard Plate Count

This technique depends on the ability of the bacteria in milk to grow on a given medium at a certain temperature. It does not reveal all of the bacteria present because some bacteria will not grow on the medium used or at the incubation temperature employed. However, the procedure as standardized by the United States Public Health

Fig. 22.—Pasteurizer controls. (Courtesy HiLand Dairy, Salt Lake City, Utah.)

Service provides a very useful and relatively reproducible method of estimating the numbers of bacteria in milk. The method involves the following steps:

1. Milk is diluted 1:100 in sterile water (1 ml. of milk added to 99 ml. of sterile water).

2. This is shaken in a specified manner then 0.1 ml. is placed in a sterile Petri dish and 1.0 ml. is added to a second Petri dish. The first plate, therefore, represents 1/1000 ml. of milk, the second 1/100 ml.

3. Over this is poured sterile tryptone glucose extract agar which has been melted, then cooled at 45° C. The agar solidifies at about 40-42° C. (When the agar has solidi-

fied, each bacterium or cluster of bacteria will be held in one spot and begin to grow. The mass of growth forms a visible colony which can then be counted.)

4. The plates are incubated at 32 or 35° C. in an inverted position for 48 hours.

5. From the two plates, the one which has from 30 to 300 colonies is chosen for counting. (If both fall in this range, both are counted.) The number of colonies is then multiplied by the dilution. Thus 33 colonies on the 1:100 dilution would be reported as a standard plate count of 3,300 bacteria per milliliter. On the other hand 45 colonies on the 1:1,000 plate would be reported as a count of 45,000 bacteria per milliliter. The corresponding 1:100 plate would not be used since it had more than 300 colonies.

The plate count procedure is useful for estimating the number of bacteria in any liquid or any substance which can be dissolved or suspended. Modifications of this method are used for estimating numbers of bacteria in ice cream, water, sewage, ground meat, soil, etc.

For determining the number of thermophilic bacteria present in milk the same procedure is followed, however, an incubation temperature of 55°C. is used in place of 32 or 35°C.

True cold-loving bacteria can be counted by the same procedure except that the incubation temperature is 5-10°C. (41-50°F.) (which you will recognize as a characteristic refrigerator temperature) and the time is 10-14 days. This is an indication of the slow rate at which these organisms grow. Thermoduric bacteria (which resist pasteurization) can be counted by making a plate count on the raw milk, then after pasteurizing it for 30 minutes at 145°F. making a second count. If the count is nearly as high after pasteurization as before, the presence of thermoduric bacteria is indicated.

Raw milk normally contains a fairly large number of coliform bacteria such as *Escherichia coli.* These, however, are ordinarily killed by pasteurization. A method which counts the coliforms is, therefore, a measure of the effectiveness of pasteurization. This requires a special medium, violet red bile agar, and is carried on in the same manner as the standard plate count except that one milliliter of milk, undiluted, is added to the plate and mixed with the melted agar. After the agar solidifies, a second layer of melted agar is poured over the plate to insure subsurface colonies only. Coliform colonies are 1 mm. or more in size, red in color and display a zone of precipitated bile.

Properly pasteurized milk generally shows no coliforms in a milliliter sample of milk.

The presumptive test or similar tests used on water may also be utilized.

Direct Count

For raw milk the direct microscopic method of Breed and Brew affords a rapid and relatively simple method of estimating the numbers of bacteria in milk. The procedure is as follows:

1. One-hundredth milliliter of milk is placed on a glass slide and spread over one square centimeter.

2. After drying, the smear is fixed over boiling water, flooded with xylol to remove the fat, alcohol to remove the xylol, then stained with methylene blue. The spot occupied by the fat globules forms a lightly stained "ghost," while the bacteria and body cells are a deep blue against the lighter blue background of stained milk protein.

3. Using a microscope with a special 6.4X eyepiece the amount of area seen at any time, the so-called "field," represents about 1/300,000 ml. of milk. Therefore, to give a fair estimate, the number of individual bacterial cells are counted in 30 fields and the average number per field multiplied by 300,000 to give the "number of bacteria per milliliter by the direct method." If a 10 X eyepiece is used, only about 1/500,000 ml. of milk is seen at any one time; therefore, the average number of bacteria per field is multiplied by 500,000.

The direct count has the advantage of being rapid, requiring relatively little equipment and permitting the recognition of many of the kinds of bacteria present. For example, a streptococcus infection of the udder can be easily seen, while dirty utensils or improper cooling each give a characteristic picture under the microscope. On the other hand, the method is not applicable to low count milk samples or pasteurized milk. The direct count will be about four times as high as that obtained by the plate method.

Methylene Blue and Resazurin Reduction Tests

Certain dyes or stains change from a colored to a colorless form when reduced. In certain solutions such as milk, bacteria reduce methylene blue or resazurin to a colorless form by adding hydrogen to the dye molecule in oxidizing milk constituents in order to secure energy.

Tests have, therefore, been devised in which the time required for the bacteria in milk to render these dyes colorless is used as a measure of the bacterial quality of the milk. The methylene blue test is probably more commonly used than the resazurin test. Milk which changes from blue to colorless, pink to colorless, (depending on the dye) in less than two hours is considered poor, while excellent milk remains colored for more than eight hours.

Phosphatase Test

This test is an ingenious method for determining whether or not milk has been properly pasteurized. It is based on the fact that the enzyme, phosphatase, which is normal to raw milk is destroyed by proper pasteurization. This enzyme breaks down phosphoric esters into inorganic phosphate and alcohol.

For the test, disodium phenyl phosphate is added to milk. An indicator which gives a blue color if the phenyl group has been released is used. Properly pasteurized milk remains uncolored after incubation while improperly pasteurized milk shows a blue color.

BACTERIAL STANDARDS FOR MILK

In virtually all large cities and most of the small ones the market milk is sold as Grade A pasteurized or in rare cases raw milk. To be assigned this grade, the milk must be produced on farms which meet the prescribed minimum standards of sanitation and equipment, be processed in acceptable dairy plants and show bacterial counts below the maximum allowed by the given city.

Cities often operate under the standards suggested by the United States Public Health Service. In many cities raw milk to be used for Grade A pasteurized milk may contain not more than 200,000 bacteria per milliliter by the standard plate count, while the pasteurized milk, itself, may not have more than 25,000 bacteria per milliliter by the standard plate count at the time of delivery.

Cities often vary in their standards. Many maintain more rigid requirements than those specified by the United States Public Health Service. For example, Los Angeles requires that milk to be used as Grade A pasteurized milk may contain not more than 75,000 bacteria per milliliter by the standard plate count and the maximum allowed for Grade A pasteurized milk is 15,000 per milliliter.

As a result of our knowledge of milk and its sanitary aspects, milk has now become "our safest food" in the words of Doctor Feemster in his report on "Milk Borne Epidemics in Massachusetts."[2]

FERMENTED MILK PRODUCTS

Desirable fermentations represent the third and most directly monetary of all the applications of dairy bacteriology. The production, by use of selected cultures of bacteria, of butter with a fine aroma, cheeses with a fine bouquet, represents the foundation of the dairy processing industry.

Butter

The delectable flavor and aroma of high quality butter is derived primarily from the growth of lactic acid bacteria contained in the so-called "butter culture." This consists of a mixed culture of *Streptococcus lactis* and some other closely related lactic acid streptococci. Sweet cream butter is made by adding the "butter culture" to pasteurized sweet cream and holding it for a period of time at the proper temperature for growth of the bacteria. On churning, the butterfat is gathered into masses and the "churned buttermilk" remains. In addition to producing lactic acid, the bacteria also produce certain volatile compounds which give the desired aroma and flavor.

If raw cream is used, spontaneous souring will take place due primarily to the growth of *Streptococcus lactis*. However, no control on flavor is possible because undesirable bacteria, other than *S. lactis* may be present. In addition, the flavor-producing lactic acid streptococci may be absent.

Microorganisms may occasionally produce off flavors in butter. Some of these are (a) malty flavor produced by *Streptococcus lactis*, var. *maltigenes;* (b) yeasty flavor due to *Torula cremoris*, one of the yeasts which often grows vigorously in sour cream; (c) cheese flavor, resulting from excessive growth of the lactobacilli. Not all off flavors, however, are bacterial in origin. Certain volatile compounds, such as gasoline, are readily absorbed from the atmosphere by butterfat.

Molds may occasionally develop in butter, particularly tubbed butters.

------

[2]Am. J. Pub. Health **36** (7):777-781, 1946.

Buttermilk

Buttermilk may be of two types, churned buttermilk, previously described, and culture buttermilk. This latter product is made by adding a "butter culture" with its mixed flora of *Streptococcus lactis* and related streptococci to skim milk or whole milk. The fermentation which occurs results primarily in the development of lactic acid and certain flavor-producing compounds. Particular "cultures" vary in the flavor which they produce and many dairy plants use extreme care in handling their "cultures" so as to insure uniformity in flavor and quality.

Bulgarian Milk or Yogurt

The major organism in this "health food" which has recently become popular, is *Lactobacillus bulgaricus*. This organism is somewhat more difficult to grow than *Streptococcus lactis*. It requires a relatively higher incubation temperature. When grown alone it produces a flavor which may not be too pleasing to many consumers. As a result, cultures of *S. lactis* may also be included to improve the flavor.

Cheese Fermentations

The production of cheeses is a microbiological fermentation in which a considerable variety of microorganisms may be involved. The particular combination of salt, incubation temperature, pH and culture used will determine the kind of cheese produced.

The basic operations in cheese production are as follows: (a) Holding of milk, with or without added starter to promote initial souring by *Streptococcus lactis*. If pasteurized milk is used, starter must be added. If raw milk is used the *Streptococcus lactis* present may be relied on to produce a spontaneous souring. The lactic acid streptococci rather rapidly convert the lactose to lactic acid. This acidity is highly important in preventing the growth of putrefactive bacteria which would soon change the milk proteins into a foul-smelling, unedible mass. In addition, the acidity favors the action of the enzyme rennin which is added to the souring milk to promote coagulation of the casein.   (b) After formation of a firm curd by the added rennin the curd is "cut," i.e., divided into small pieces then pressed to remove the whey.

At this time salt, coloring matter, and mold or other cultures may be added. (c) The "hoops" of cheese are then set aside at a selected temperature to ripen. The temperature used has a direct bearing on the type of fermentative reaction which will take place.

In the early ripening of cheese, *Streptococcus lactis* is the primary organism present. It has been found to make up as much as 99 per cent of the bacterial flora during the first 2-3 months of ripening. The lactose by this time has been practically all consumed and the acidity has become inhibitory to the streptococci. Lactobacilli now begin to grow and within another couple of months may represent as much as 99 per cent of the flora.

In the ripening process the tough, rubbery protein curds are broken down. This is a hydrolytic process which yield caseoses, peptones, amino acids, and even some ammonia.

While the rennin and *Streptococcus lactis* may play some part in the ripening process, the lactobacilli (or molds) probably are the major agents in the ripening process.

The variety and flavor of cheese produced depends on the particular strain of organism involved.

The most important commercial varieties of cheeses are indicated in Table XIII.

TABLE XIII

IMPORTANT COMMERCIAL VARIETIES OF CHEESES (Adapted from Hammer)[3]

GENERAL GROUP OF CHEESES	VARIETY OF CHEESE	PRIMARY RIPENING ORGANISMS
Hard	Cheddar	Bacteria, not including eye formers
	Swiss	Bacteria, including eye formers
Semi-hard	Brick	Bacteria
	Blue	Mold
	Roquefort	Mold
	Gorgonzola	Mold
Soft	Limburger	Bacteria
	Camembert	Mold
	Cottage	Unripened
	Cream	Unripened
	Neufchatel	Unripened

[3]Hammer, B. W.: Dairy Bacteriology, ed. 3, New York, 1948, John Wiley and Sons, p. 491.

Cheddar Cheese is the most commonly sold variety of American cheeses. It is a hard curd cheese without "eyes". The organisms primarily involved are probably *Streptococcus lactis* and related streptococci and various lactobacilli.

Undesirable types of spoilage may occassionally be produced due to foreign organisms. These include gassy curd, bitterness, unclean taste (*Escherichia-Aerobacter*) and pink color due to yeasts.

Swiss Cheese.—Lactic acid streptococci and lactobacilli have a part in the ripening process. The "eyes" are produced by the growth of a species of *Propionibacterium*. This organism breaks down lactic acid into propionic, acetic acids, carbon dioxide, and water. The accumulated carbon dioxide forms the "eyes."

Brick cheese is softer than cheddar; *Streptococcus lactis*, some lactobacilli and salting are combined to produce this particular cheese.

Blue Cheeses are mold-ripened cheeses and include as the semi-hard cheeses, Roquefort (France), Gorgonzola (Italy), Stilton (England) and certain American Blue cheeses. Camembert, also mold-ripened, is classed as a soft cheese.

In making the cheese, mold spores of the desired species are added to the curd just before pressing or dusted on the cheese in the hoop. *Streptococcus lactis* produces lactic acid as in the case of other cheeses. Lactobacilli are often present but are not necessary for successful ripening. The major agent in the final ripening is the mold. In addition to breaking down the protein, fat hydrolysis also appears to contribute flavors and odors as a part of the ripening process.

The molds primarily involved are *Penicillium roquefortii* Thom (a mold quite often found on other dairy products) and *Penicillium camemberti*. Some other species may also be used.

Limburger Cheese is produced by the action of the lactic acid streptococci and a number of other bacteria and yeasts. High salt concentration, low pH and the temperature are combined to produce the limburger type of ripening.

Cottage Cheese is an unripened cheese. It is produced commercially by adding a culture of lactic acid bacteria to pasteurized milk followed by a small amount of rennet. After a firm curd forms it is cut, cooked and drained. Cream and salt are frequently added. Flavor is due to lactic acid and certain volatile compounds found also in high quality butter. Since it is not ripened, cottage cheese is very susceptible to spoilage, particularly by yeasts and molds.

The variety of cheeses produced is very large but most of them are related to certain of the cheeses described above. In addition to the products mentioned, a number of other fermented products may be produced from milk and milk products. These include koumiss and kefir, both of which contain ethyl alcohol produced by certain yeasts from lactose.

Selected Reading References

Hammer, B. W.: Dairy Bacteriology, ed. 3, New York, 1948, John Wiley & Sons, Inc.

Hopkins, E. S., and Elder, F. B.: The Practice of Sanitation, Baltimore, 1951, Williams and Wilkins Co.

Maxcy, K. F., et al.: Rosenau Preventive Medicine and Hygiene, ed. 7, New York, 1951, Appleton-Century-Crofts, Inc.

CHAPTER 17

SAFE EATING FOR THE PUBLIC

Restaurant Sanitation and Inspection

Today most of us in the United States eat occasional meals or all of our meals in public eating places. As a result the sanitation of such eating places has become important to everyone. Because public eating involves so many of us, public health agencies on every level are attacking the problem of restaurant sanitation with the primary objectives of preventing illness and possible death from food poisoning or food-borne infections or from diseases which may be transmitted from one person to another by improperly sanitized eating utensils. In addition, there is a growing awareness of the problem on the part of forward-looking food handling establishments and many of them are voluntarily initiating protective measures to provide safer and more pleasant eating.

Where this new awareness has gained a foothold, public eating is more secure and more aesthetically satisfactory than it has ever been before. However, there are still many carelessly operated eating places in the United States. These constitute a perennial and avoidable health hazard. On the other hand, properly managed eating places which apply the sanitary procedures described in this chapter provide safe eating, as safe or safer than that which you enjoy in your own home.

The public health aspects of the problem of restaurant sanitation are much more complex than those presented by milk and water control. It is usually difficult to prove that certain procedures such as improper utensil sanitization are giving rise to infections or to trace food poisoning outbreaks back to a certain meal in a certain restaurant. In addition, with a steady turnover among employees of eating places, it is often difficult to create among food handlers an awareness of their importance and responsibility to the public and a knowledge of the proper methods of handling foods and eating utensils.

The matter of restaurant sanitation has three separate aspects: (a) the food itself, (b) the food handler and his techniques in handling and serving food, and (c) the eating utensils themselves.

THE FOOD

Our concern with the food itself is occasioned by the possibility that it may produce food poisonings or food-borne infections. In addition, we should be protected against unwarranted microbial impairment of food quality.

The sources of contamination of foods may be any one of the following: (a) rats or mice may be carriers of salmonella infections from bacteria which may find their way into foods through the droppings of the animals. (b) Cockroaches and flies may likewise carry the organisms of food poisoning and food-borne infection organisms. (c) Polluted water is a perennial source of danger particularly from the *Salmonella* and *Shigella* organisms. (d) Food poisoning micrococci may come from infected wounds or boils on the hands of food handlers. (e) The salmonellas and shigellas may be transmitted by the carelessly washed hands of food handlers who happen to be carriers of the organisms at the time. (f) Dirty cooking or serving utensils also represent a threat. (g) Another important source of contamination is exposure of foods in the open where they will be subject to flies, airborne dust, or of respiratory droplets from customers.

Food-borne infections are most likely to be of the typhoid-paratyphoid group although it is possible that foods may be simple carriers of respiratory pathogens or similar organisms when the food is exposed unnecessarily to contamination. The common cold, influenza, etc., might be conceivably carried in this fashion, although improperly sanitized eating utensils might present a more direct and important route.

Proper refrigeration is of prime importance in keeping down the number of potential food poisoning organisms. If foods are prepared beforehand to be eaten cold they should be consumed as soon as possible or stored in refrigerators at 40°F. or colder. Overcrowding of the refrigerator should be avoided and the prepared foods should be stored in shallow pans, well protected from contaminating contact with raw foods. Steam tables, for keeping the food warm, should be maintained at more than 150° F. to prevent bacterial growth. Salads should be stirred with a spoon, not mixed by hand.

In well-operated eating places these precautions will be taken voluntarily and in carefully policed eating places they may be enforced.

THE FOOD HANDLER

The food handler has a twofold role in the transmission of undesirable microorganisms to the customer. If he has an infection he may be the source of food poisoning micrococci, salmonellas, shigellas, etc., or he may transmit some other infection from himself to the customer. This is probably not too serious a problem. On the other hand, he may serve as an active intermediary in the transmission of microorganisms from one customer to another because of the unimaginative manner in which he handles eating utensils.

In an attempt to rule out the food handler as a personal carrier of infection, periodic examinations have been tried. However, the procedure has never been especially satisfactory since it was often cursory and the result was a false sense of security.

On the other hand, a careful initial physical examination is being recognized as having great merit. It serves to eliminate active tuberculosis cases and those with previous records of diphtheria and typhoid who may be potential carriers of these organisms.

Coupled with this, the exclusion from work of those with respiratory or other infectious diseases or those with infected cuts or with boils will do much to eliminate the food handlers as personal carriers of disease.

On the other hand the procedure followed by the food handler is highly important. Some of the errors made by food handlers which might cause the contamination of food or the transmission of pathogens from one customer to another are as follows:

(a) The most flagrant disregard for common sense handling of utensils is the practice of picking up two or three used glasses at a time by the rims often dipping the fingers into the remaining water in the process, then picking up clean properly sanitized glasses in the same manner. Such a person would never think of putting his finger into the mouths of successive customers yet this is essentially what he is doing.

Picking up forks by the tines or spoons by the bowls also represents a potential source of hazard to the customer. Likewise, picking up a pat of butter with the soiled fingers instead of a fork not only is undesirable but leaves an incriminating finger print as well.

In an exclusive and widely publicized European restaurant an oversolicitious waiter seeking to impress the source of his gratuity was observed graciously dusting clean dishes with a filthy cloth before placing them before the customers.

It is apparent that proper food handling is largely a matter of proper understanding of the highly responsible position occupied by the food handler. Educational programs for the handlers are being found to be much more effective from a public health viewpoint than the periodic physical examination. Many cities have a compulsory food handlers' training course. This course must be completed before a person may be employed as a food handler. Many of the more progressive eating establishments in the United States have now voluntarily initiated such training programs. In addition to knowledge of proper technique, good personal hygiene with proper hand washing facilities and habits are also essential to safe food handling.

UTENSILS

Proper sanitization of eating utensils falls in the same category as pasteurization of milk and chlorination of water. The improperly sanitized eating utensil can transmit diseases as readily if not more so than polluted water. The eating utensil serves as a link in the chain of infection because it acts as a *fomite* or inert object which can carry infection. On the other hand, properly sanitized utensils make eating as safe as it would be in your own home, or more so since glassware is often washed in a much more satisfactory manner than in the home. The public and progressive restaurant owners alike are becoming increasingly aware of the importance of properly sanitized eating utensils. The public is sending back forks with food between the tines and restaurant owners have found that few men care to share lipstick with a woman whom they have never seen.

The diseases which may be transmitted by improperly sanitized eating utensils are essentially those of the mouth and respiratory tract. These include the common cold, influenza, measles, mumps, whooping cough, tuberculosis, trench mouth, septic sore throat and similar diseases. In a few rare instances syphilis has apparently been transmitted through grossly unsanitary eating utensils.

The protection against the hazard of improperly sanitized eating utensils lies in proper dishwashing.

Dishwashing Methods

Standard dishwashing procedures have been worked out by public health agencies which, if followed, will guarantee completely safe, properly sanitized eating utensils. The exact details of requirements set up by different health departments vary, but the procedures followed are essentially the same.

If the mechanical dishwasher is used it should be equipped so that the detergent will be automatically fed into the washing chamber. The washing should be done with water held at 140°-160°F. and should be continued for at least 40 seconds. Rinsing should be carried on for a minimum of 10 seconds at 180°F. Overturn of waters should be 9 gallons per minute at 15 pound pressure per square inch and 1.8 gallons of water should be allowed per person per day.

For washing dishes by hand a 3 compartment sink is required and the wash water should have a temperature of 120°F., the rinse water 140°F. and the final rinse is to be in water at 180°F. for 30 seconds. In place of the final rinse a hypochlorite solution with 50 ppm. or more of available chlorine may be used and the utensils immersed for at least one minute. The quaternary ammonium chloride compounds are also being used to some extent although their method of germicidal action and their effectiveness is still not completely understood.

If possible, drying with dish cloths should be avoided. Proper waste disposal and hand washing facilities should be provided adjacent to the dishwashing equipment. After washing, sanitized dishes should be protected against accidental contamination.

Bacterial Examination of Eating Utensils

In order to determine the sanitary condition of washed eating utensils, a standard method has been devised by the Sub-Committee on Food Utensil Sanitation of the American Public Health Association for the "Bacteriological Examination of Food Utensils." This is analogous to the standard methods devised for the examination of milk and water. Briefly, the procedure is as follows:

1. Cotton swabs are prepared on standard wood applicator sticks and placed in screw cap test tubes containing as many milliliters of buffered distilled water as utensils are to be swabbed (4 ml. for 4 utensils, 5 ml. for 5 utensils, etc.) Sterilize in the autoclave.

2. The utensils to be examined should "include at least glasses, cups and spoons, if used, and at least 4 of each shall be selected at

random from the shelves or other places where clean utensils are stored." "Use 1 swab for each group of 4 or more similar utensils." Excess water should be squeezed out of the swab, then it should be rubbed "slowly and firmly 3 times over the significant surfaces of 4 or more similar utensils, reversing the direction each time. After swabbing each utensil, return the swab to the container of dilution water, rotate (whip rinse) the swab in the dilution water, and press out the excess water against the inside of the container before swabbing the next of the 4 or more utensils of the group."

Note: Significant surfaces of the various utensils are as follows:

(a) Upper half inch of inner and outer rims of glasses and cups.
(b) Inner and outer surfaces of tines of forks.
(c) Inner and outer surfaces of bowls of spoons.
(d) Inner surfaces of bowls, halfway between bottom and rim.
(e) On plates the entire upper surface is a "significant surface." The sampling method used is that of swabbing 3 times reversing the direction of each stroke "completely across each of two diameters at right angles to each other."

After swabbing, break off the contaminated part of handle and drop the swab into the dilution water. Pack tubes in ice and return to laboratory.

3. For counting, shake the swab container rapidly 50 times, over a distance of 4-6 inches "striking the palm of the other hand at the end of each cycle and completing the whole in about 10 seconds."

4. Transfer 1 ml. with a sterile pipette to a sterile Petri dish and add about 10 ml. melted tryptone glucose extract agar (without milk).* Mix and incubate 48 hours at 32 or 35°C.

5. Count the number of colonies and report it as the number of bacteria per utensil.†

The United States Public Health Service recommends a standard of not more than 100 bacteria per utensil. Carefully operated establishments will generally show counts well under the maximum.

Many cities make a practice of designating restaurants on the basis of their willingness and ability to meet minimum sanitation standards. In such cities it is, therefore, a wise practice to patronize those establishments which display a Grade A placard. In all cases

---

*Note: This is the same medium as that used for making milk counts.

†Note: For more complete details see report of the Sub-Committee in American Journal of Public Health Yearbook 1947-1948, Part 2, **38:**(5) 68-70, 1948.

it is important to encourage the efforts of the public health authorities and the alert restuarant operators who are trying to give you safe eating.

SCHOOLS, CHURCH FUNCTIONS

In school, church, fraternal and other social affairs light refreshments or even full dinners are often served. In many instances a lack of care in the proper sanitation of eating utensils or in the preparation and handling of salads, etc., is often very much in evidence. For the safety of the group, common sense and the application of the procedures previously suggested may do much to prevent the outbreaks of food poisoning (and sore throats) which may follow these public functions.

In the home if these same principles are applied to food handling and eating utensil sanitation they will pay dividends not only in lessening the likelihood of food poisoning and reducing the respiratory diseases which would otherwise be shared by all of the family but also in preventing loss of foods due to spoilage, an important economic item today.

Selected Reading References

Hopkins, E. S., and Elder, F. B.: The Practice of Sanitation, Baltimore, 1951, Williams and Wilkins Company.

Maxcy, K. F., et al.: Rosenau Preventive Medicine and Hygiene, ed. 7, New York, 1951, Appleton-Century-Crofts, Inc.

Phelps, E. B., et al.: Public Health Engineering, New York, 1948, John Wiley and Sons, Inc. Vol. II.

CHAPTER 18

MICROORGANISMS WHICH WORK COMMERCIALLY

Industrial Microbiology

Microorganisms, because of their enzymatic versatility, are, literally, minature chemical factories capable of turning out in large or small amounts, an infinite variety of chemical compounds, many of which we have not yet been able to duplicate in the laboratory. In most instances, the growth products of a given microorganism may contain such a mixture of compounds that they cannot be put to any particular use. However, if the microorganism produces a relatively pure substance or a predominance of one substance or if the particular compound can be easily separated from the mixture, these microbial growth products may often be used to great advantage. In other instances, the enzyme systems of naturally occurring microorganisms may be utilized to produce a desired change in some raw material.

Examples of the uses of microbial enzymes have already been given. These are represented by the lactic acid fermentation involved in the production of certain dairy products and of foods such as sauerkraut and also in the ripening of cheese.

However, this chapter will concern itself primarily with those chemical substances or reactions produced by microorganisms which are utilized in the industrial field rather than in food production. It will not be possible to adhere rigidly to this differentiation because a good many microbial products are used both industrially and directly or indirectly as food.

Industrial uses of microorganisms tend to fall into two categories: (a) those which utilize a pure culture or pure cultures and involve large scale manufacturing techniques, and (b) those involving the use of naturally occurring mixtures of microorganisms under conditions that may be rather primitive, as a means of producing a desired change in some product of industrial value. The discussion will follow in that order.

MICROBIAL PRODUCTION OF COMMERCIAL SOLVENTS

Modern industry with its infinite variety of processes and products, uses great quantities of chemical compounds which will dissolve resins, oils, waxes, etc., which are immiscible in water. These dissolving agents are termed "commercial solvents" and they are mostly alcohols or related compounds. These compounds are also finding wide use in the production of many of the new plastics, in the synthetic rubber industry, etc. The solvents are essential for the production of many of the organic substances used in industry and medicine. Microbiological and biochemical laboratories could not operate without them.

Many of them may be produced either by microbial fermentation or by chemical synthesis. The cost of production naturally determines which method will be the more widely used. Since the raw materials most commonly employed in industrial fermentations are agricultural products, a great deal of time and effort is being expended to develop methods for the utilization of agricultural wastes in the commercial solvent industry.

For example, the Forest Products Laboratory of the United States Department of Agriculture at Madison, Wisconsin, is attempting to develop methods of using wood wastes of various sorts for the production of ethyl alcohol and similar fermentation products. The various Regional Laboratories are also working on the utilization of agricultural wastes for the same purposes.

Ethyl Alcohol

The following statement by Porter aptly describes the commercial position of ethyl alcohol:

"So extensive is the use of ethyl alcohol in industry and the arts and sciences that, with the exception of water, it may be regarded as the most important accessory chemical employed today."[1]

Ethyl alcohol or ethanol enters into the production of a legion of industrial products ranging from synthetic rubber to perfumes. It has even been used with some success as a motor fuel. While ethanol

---

[1] Porter, J. R.: Bacterial Chemistry and Physiology, New York, 1946, John Wiley & Sons, p. 899.

may be produced in part as a by-product of the distilling and brewing industries or may be synthesized from ethylene or acetylene gases, much of it comes from industrial alcohol plants.

In the production of ethyl alcohol in these plants, special strains of yeasts are used which have a high tolerance for alcohol. Some of

Fig. 23.—Barley malt production. The germinating drums have a capacity of 250 bushels. Moist barley kernels (44 per cent moisture) are placed in the aerated, revolving (1 revolution each 35 minutes) drum and maintained at 60-65° F. for five to six days. The acrospire denotes *diastase* formation. (Courtesy Becker Brewing Company, Ogden, Utah.)

these strains of *Saccharomyces* may tolerate alcoholic concentrations as high as 10 to 12 per cent and ferment nearly all of the glucose in the medium.

Various raw materials are used, depending to some extent on their availability and relative cost. Low grade molasses, corn, potatoes, wheat, sulfite liquors from the paper-making industry and even wood wastes have all been used. The first two are probably the most important. If starchy materials are used, they must first be converted to sugars before the yeasts can ferment them. This conversion may be accomplished by the use of acid and heat or by means of enzymes such as the starch-splitting enzymes (diastases) from barley malt, or similar malts or with enzyme preparations from diastase-producing molds.

Fig. 24.—Brewing kettle. The capacity of this large copper kettle is 130 barrels. The brewing kettle serves three purposes: (1) evaporates excess water to concentrate brewer's extract and to sterilize the wort; (2) coagulates proteins and develops a break in the wort; (3) extracts flavor of hops. (Courtesy Becker Brewing Company, Ogden, Utah.)

The cellulose in wood wastes must likewise be converted into glucose before it can be fermented. The combination of acid plus steam under pressure is generally employed to accomplish this change. The great potential source of energy represented in wood wastes when they can be used widely for the production of industrial alcohol is a fascinating subject to consider. However, the economics of the method have yet to be worked out. It has been reported that a ton of dry, bark-free wood will produce nearly 65 gallons of ethyl alcohol.

In the actual alcoholic fermentation, acidified mashes are used in order to inhibit the growth of undesirable microorganisms and heavy inocula of yeasts are also employed. The exact method followed depends on the initial raw material being utilized.

In addition to producing alcohol, the fermentation yields large quantities of carbon dioxide which may be compressed into cylinders and used for carbonating beverages, etc. The yeast cells themselves become valuable as sources of human food, vitamins, or animal feeds.

While yeasts are the most efficient producers of ethyl alcohol, many bacteria may also produce it but in such small quantities or so mixed with other compounds that they cannot be used for the industrial production of ethyl alcohol. *Pseudomonas lindneri* represents an exception. This bacterium was isolated by a German bacteriologist, Lindner, from pulque, a Mexican drink made by fermenting the sap of *Agave americana*, the century plant. Lindner's organism has been employed successfully in Germany for the production of commercial alcohol. It is said to be able to produce up to 10 per cent alcohol.

Of the fermentation products it produces by the breakdown of sugar, slightly more than 90 per cent will be alcohol and carbon dioxide, about equal amounts of each, with lactic acid representing about 7-8 per cent of these products. Small quantities of other compounds make up the balance. By way of comparison, in the yeast fermentation, alcohol and carbon dioxide represent about 95-96 per cent of the fermentation products with 2-3 per cent being glycerol and the remainder usually made up of lactic acid, succinic acid and fusel oil.

Glycerol Production

The microbial production of glycerol presents an interesting aspect of the alcoholic fermentation. Investigators made the discovery during World War I that by adding sodium sulfite during the fermentation process yeasts could be forced to produce 5 to 10 times as much glycerol as the 2-3 per cent they normally produce during the alcoholic fermentation. Utilizing this method, factories were put into operation for glycerol production by means of the yeast fermentation. The glycerol thus formed was used in the manufacture of nitroglycerine and other explosives. This process is no longer of any importance since the glycerol supply is now ample because glycerol-yielding fats are relatively abundant and glycerol can be produced synthetically.

YEASTS AS FOOD FOR MAN OR ANIMALS

While this phase of yeast utilization is not a fermentation process it represents a by-product of the fermentation industry or is closely allied to that industry.

The brewing industry produces as one of its products the familiar "brewer's yeast." Thousands of tons are marketed each year, primarily as a source of the Vitamin B complex or to a lesser extent as a secondary protein source. Brewer's yeast is used either as a human or animal food supplement.

Yeast cells themselves are an excellent source of protein since they contain approximately 50 per cent protein. In addition, they can produce this protein from such simple nitrogen sources as ammonium sulfate, ammonium phosphate, and related compounds. Although the American diet is able to secure a fairly ample supply of protein from milk and meat, many areas of the world suffer from a serious protein deficiency because of a lack of meat-producing animals. This is especially true during periods of war.

Yeast cells have, therefore, been utilized with considerable success, chiefly on an emergency or experimental basis, as protein sources. However, the primary difficulty is the cost of producing such protein. Cheap carbohydrate sources such as wood wastes plus large scale production may make the widespread use of yeast proteins economically feasible in the future. The use of such wood wastes for the production of "fodder yeasts" is being investigated extensively at the present time.

In the protein deficient West Indies the British have set up a plant for the production of yeast proteins. The process utilizes one of the false yeasts, *Torulopsis utilis*. This organism grows vigorously, and the resultant product is free from certain objectionable tastes and odors often produced by strains of *Saccharomyces*. The carbohydrate source used is molasses, however, experiments have been carried on in which an attempt has been made to utilize algae, which can manufacture its carbohydrates directly by photosynthesis.

In growing food yeasts, a low concentration of sugar is used, and aerobic rather than anaerobic conditions are maintained. This leads to a profusion of cell growth with reduced alcohol production.

Butanol-Acetone Fermentation

Acetone and butanol (butyl alcohol) resemble ethyl alcohol in their chemical make-up and likewise in their value as solvents and as generally useful chemicals. Their chemical structures are shown below for purposes of comparison:

$$CH_3CH_2OH \qquad CH_3CH_2CH_2CH_2OH \qquad CH_3CO\ CH_3$$

Ethanol   Butanol   Acetone
(ethyl alcohol) (Butyl alcohol)

Present industrial developments have resulted in a great demand for these compounds. The lacquer on your automobile may have been dissolved in one of these fermentations products. While acetone and butanol can be synthesized, a very substantial part of the supply is produced by means of fermentations. Unlike ethyl alcohol, these two compounds are the products of a bacterial fermentation.

The ability of bacteria to produce butanol was first demonstrated by Pasteur in 1861, while acetone was shown to be produced by the bacteria in 1905. During World War I, acetone became a critical item because of its use in airplane "dope" and in the production of explosives. Factories were, therefore, built and considerable amounts were produced by bacterial fermentation. After the war was over the industry was largely abandoned because of a lack of demand for acetone. Today, we are once more using large quantites of acetone but butanol which is produced in the same process is even more important. A large part of the total amount of butanol used probably comes from the fermentation industry.

The organism employed is *Clostridium acetobutylicum*, an anaerobic sporeformer. A number of different strains of this organism have been patented. It ferments carbohydrates rapidly to produce a mixture which is high in butanol, with a fair amount of acetone and relatively smaller amounts of ethyl alcohol. Quantities of carbon dioxide and hydrogen are also produced together with very small quantities of other organic compounds. The three major compounds may be rather readily separated by fractional distillation. The proportions of these three constituents produced will vary somewhat with the strain of bacteria employed and the raw material used. The gases are generally trapped and the H_2 and CO_2 may be combined catalytically to produce methyl alcohol (methanol) or the CO_2 may be compressed in tanks or used to make "dry ice."

The carbohydrate sources most commonly utilized are low grade molasses or to a lesser extent corn. With starch, no diastase needs to be employed since the organism apparently has its own starch splitting enzyme.

Other solvents such as isopropyl alcohol may also be produced by fermentation. However, the petroleum industry produces it more economically, thus discouraging attempts to utilize the fermentative process.

ORGANIC ACID FERMENTATIONS

Microorganisms of various sorts yield a large number of different organic acids. Of these, three are commonly produced by an industrial fermentative process. These are lactic, citric, and gluconic acids.

Lactic acid ($CH_3 CHOHCOOH$) will be recognized as the sour milk acid. It has a variety of industrial applications including its use in tanning hides, textile dyeing, and the manufacture of plastics. It is used in connection with a number of foods such as candies, extracts, pickle brines, etc., and to acidify beer worts to prevent the growth of undesirable organisms which would cause the beer wort to develop off-flavors or odors and thus seriously impair the quality of the beer.

A considerable number of carbohydrates may be used as raw materials for the lactic acid fermentation. Whey and molasses are commonly employed, although potatoes, corn, and other starchy substances may also be used after they have been hydrolyzed into sugars.

Several different microorganisms may be utilized, including certain species of *Lactobacillus, Streptococcus lactis, Leuconostoc mesenteroides* and even some molds. Since lactic acid may occur either in the dextro- or levo-form, i.e., it rotates the plane of polarized light to the right or left, the organism must be selected not only on its ability to produce lactic acid but also of d- or l- lactic acid, whichever may be desired.

It is customary to think of citric acid as being primarily a product of the citrus industry; however, this is not the case. As pointed out by Foster, about 80 per cent of the 13,000 tons of citric acid produced annually in the United States comes from the fermentation industry. The remainder is the citric acid produced from citrus fruits.

$$
\begin{array}{l}
H_2—C—COOH \\
\quad\ \ | \\
HO—C—COOH \qquad \text{citric acid} \\
\quad\ \ | \\
H_2—C—COOH
\end{array}
$$

Much of the citric acid is used for medicinal purposes, while smaller proportions go into the production of soft drinks, candies, etc.

Unlike the fermentations previously discussed, citric acid is a mold fermentation. Many different molds are able to produce citric acid and the high yielding varieties are selected for commercial fermentations.

High concentrations of sugar (14-20 per cent) are employed and the medium is distributed in shallow pans. The incubation proceeds at about 28°C for 7-10 days. The citric acid is then harvested by crystallization or removed as the calcium salt. Sulfuric acid is used to divorce the calcium, leaving free citric acid.

Gluconic acid is another organic acid which is being produced commercially by fermentation methods. Many molds and some bacteria are able to produce this molecular modification of glucose, which is essentially an oxidation of one end of the glucose molecule from the aldehyde to the acid form; however, gluconic acid is most commonly produced commercially by the use of molds.

Calcium gluconate, the calcium salt of gluconic acid is being used medicinally as a calcium source for expectant mothers. In the body, the glucose portion is oxidized as a readily available source of energy while the calcium is immediately available for bone building or other uses.

ANTIBIOTIC PRODUCTION

Although antibiotics per se are discussed elsewhere, the process of manufacturing antibiotics is a definite aspect of industrial microbiology. The techniques employed are similar to those used in the production of ethyl alcohol and other fermentation products. When penicillin was first produced on a commercial scale in the United States during the World War II years, the job was assigned to several of the major processors of alcoholic beverages because they possessed the large-scale equipment so urgently needed. Since that time tremendous new developments in equipment and methods have occurred in the antibiotic industry which have vastly increased the efficiency of the operations. In fact, we may well expect the new techniques which have been created for the antibiotic industry to make possible the use of microorganisms for the production of many products for which the use of industrial fermentation processes has not previously been feasible.

In addition to the commercial fermentation processes which have been discussed, other processes of this type are in industrial use; however, little is known concerning them since they represent closely guarded trade secrets.

INDUSTRIAL USES OF CRUDE CULTURE MICROBIAL ACTIONS

Many processes involving the use of crude, naturally-occurring cultures of microorganisms have been known and utilized for long periods of time. One of the most ancient of these processes is the retting of flax or similar fiber-producing plants. In certain of these plants the bast fibers are extremely hard to separate from the stems because they are glued in place with the pectins of the middle lamella. The practice of water or dew retting is, therefore, utilized to facilitate this separation. In water retting, bundles of the pulled flax plant are placed in water and weighted down with stones. In Ireland and Italy small ponds or "flax dams" are used. In Belgium and Holland the bundles are immersed in sluggish streams. As soon as the plant is placed in the water a vigorous anaerobic decomposition begins. This is marked by the development of a foul odor and the grayblack color of septic sewage. The pectins are attacked first, which serves to release the cellulosic bast fibers. Retting is continued just long enough to complete the liberation, but, before the cellulose is attacked to any extent, it is stopped, by throwing the plants on the bank to dry. The major organism involved appears to be *Clostridium pectinovorum*, an active pectin dissolver.

"Dew retting" is accomplished by allowing the pulled flax plants to lie on the ground subject to the effects of dew and rain until the fibers are released. A mixture of microorganisms probably carries on this reaction. After retting, the plants are subjected to breaking and scutching to remove the fibers. A number of research agencies are attempting to develop new retting processes which will be more rapid and more easily controlled.

The tanning of leather has also involved the enzymatic actions of mixed cultures of bacteria. In the past, removal of hair was aided by the action of bacteria in dissolving the material which served to "cement" the hair in the follicle. Preparation of hides for tanning was further continued by the action of bacterial enzymes from such crude cultures as pigeon manure infusions or bran allowed to ferment in contact with the hide. Today known enzyme preparations are often substituted. In all of these processes care must be exercised to produce the desired results without the hides being damaged by microbial action.

The curing of tobacco is accomplished by piling the tobacco leaves which have been moistened with water or a sugar mixture and allowing them to ripen. During the curing process starch and reducing sugars are broken down and a reduction occurs in the amount of nicotine, malic acid, and other compounds present. While the plant oxidase enzymes play an important part in the curing of tobacco, the microorganisms present probably exert some considerable effect.

Selected Reading References

Foster, J. W.: Chemical Activities of Fungi, New York, 1949, Academic Press.
Porter, J. R.: Bacterial Chemistry and Physiology, New York, 1946, John Wiley and Sons.
Prescott, S. C., and Dunn, C. G.: Industrial Microbiology, ed. 2, New York, 1949, McGraw-Hill Book Company, Inc.

CHAPTER 19

MICROBES, FERTILE SOIL, AND FOOD FOR ALL

Microbiology of Soil

The soil mantle which covers the earth is more than a mass of finely divided rock particles which functions as an anchor for plant life. It is literally the reservoir from which the life of the earth is drawn. On it all living things rely directly or indirectly.

Higher plants, the first visible product of this life reservoir, draw their essential mineral elements from the soil. On the death and return to the soil of these higher plants, the soil minerals imprisoned in the plant tissue are made once more available for subsequent generations of plants through the intermediation of soil microorganisms. Any productive soil is an inorganic-organic complex seething with life, a teeming mass of microbial and chemical activity. In it an innumerable host of living forms carry on the multiplicity of actions and interactions, mostly chemical in nature, which literally guarantee our continued existence on the earth. Without the microorganisms of the soil, life, as we know it today, could not long exist.

To understand and appreciate the vital role of soil microorganisms, one has only to contemplate what would happen if the soils of the earth were suddenly deprived of their microbes. If such a tragedy occurred, the limited store of precious life-giving nitrogen compounds would soon be tied up, unavailable to succeeding plants in the form of the complex proteins of undecomposed plant and animal remains. In the same manner much of the phosphorus, sulfur, and other essential elements would become less available for the growth of plants. Thus, deprived of the microbe-released plant food elements such as nitrates and sulfates, the earth would soon be covered with dead and dying plants, struggling to exist on the meager supply of nitrates produced by lightning and of sulfur from volcanoes. Animal life itself would eventually be doomed to hunger-ridden extinction, ending, no doubt, in utter destruction.

These vital life-givers, the soil microbes in functioning as the "digestive juices of the soil," remain largely unrecognized and unsung. Like a good citizenry they carry on their activities in such a quiet, unpretentious fashion that their beneficence is unrecognized except by the initiated.

A picture of the astronomical numbers and varieties of these soil inhabitants can be gained when one realizes that a handful of rich garden soil has in it more bacterial cells than the entire human population of the earth. If, to this handful of soil is added readily decomposable organic matter and adequate moisture, bacterial spores quickly germinate and rapid growth of previously quiescent nonsporeformers takes place. In a matter of hours this handful of soil may contain more bacterial cells than the total number of all the humans who have ever lived upon the earth.

Add to the masses of bacteria the thin twining threads of the actinomycetes; the budding yeast cells; the coarser filaments of the saprophytic molds seeking nutriment in this welter of life; protozoa busily engulfing bacteria; the metazoans engulfing both bacteria and protozoa and the result is a profoundly animate universe within a universe.

This little world is not without its catastrophies. Certain investigations indicate that viruses may attack the bacteria or actinomycetes, producing epidemics, on a grand scale; antibiotics produced by some microorganisms may poison to death their microbe neighbors; predatory protozoa may devour bacteria and may on occasion receive "food-poisoning" from the bacteria they engulf.

This picture suggests the fallacy of speaking of soil as inert dirt. Even dry dust is merely sleeping, to be awakened to full life by a moistening rain and a new influx of organic matter, such as dead leaves, a dead earthworm, or a dead animal.

While the numbers of microorganisms in the soil may be counted in millions or even billions of cells per gram they vary tremendously between soils. This depends upon the amount of organic matter such soils contain, their acidity or alkalinity, and above all their degree of dryness or wetness. Even in a given soil, the numbers may be increased a hundred-fold in a few hours as a result of a rain or the plowing under of easily decomposed plant or animal remains.

MICROBES OF POVERTY OR PLENTY

The presence in the soil of two general types of microbial populations was suggested by Winogradsky near the turn of the century. The two groups he termed the "zymogenous" and "autochthonous" floras of the soil. As the term indicates, the "zymogenous" microbes are the fermentation producers. They are creatures of prosperity showing intense activity when fresh organic matter or a fresh supply of moisture is added to the soil. They subside into an inert spore or quiescent state when the readily decomposable organic materials are broken down or the soil becomes dry. This group is probably represented by the sporeforming rods and some nonsporing bacteria such as *Pseudomonas*. Certain of the actinomycetes, yeasts, and molds may also fall in this group.

The "autochthonous" forms represent the "native flora" which carry on a quiet, steady type of existence during the microbial depressions as well as the lush periods which follow the plowing under of organic matter. Even the burying of a leaf can create a temporary microcosm of zymogenous microbial activity. On the other hand, the autochthonous forms go on quietly securing their energy through the slow breakdown of hard-to-digest organic matter.

The decompositions or molecular changes produced by microorganisms are brought about primarily by the action of enzymes, similar to the enzymes which function in our own digestive system. The changes produced by any microorganism will depend on the enzyme system it possesses. These enzyme systems appear to be inherited or genetically determined characters.

Because of the complexity of the soil microorganisms, we are almost as much in the dark as we were half a century ago in our understanding of the total functions and activities of the soil microorganisms. Our lack of knowledge results from the difficulties encountered in attempting to isolate the many and varied microorganisms from the soil, then attempting to interpret the multiplicity of interactions which probably go on in within this varied population. The use of radioactive elements may do much to alleviate this confusion by providing labelled molecules which can be followed through an entire series of microbial soil-plant changes.

It is now becoming apparent that the reactions occurring in the zone adjacent to the root, i.e., the "rhizosphere," are quite different from those taking place in the soil zone where no roots are present. In addition to contributing organic matter in the form of sloughed off cells, the plant roots may exude products which in turn influence the microorganisms.

In spite of the apparent complexity of these interrelations, one generalization can be made. In performing the function of the "digestive juices" of the soil the soil microorganisms become the world's greatest simplifiers. By an intricate and sometimes varied display of teamwork they: (1) Reduce proteins, the life stuff with its huge, intricate molecules of carbon, hydrogen, oxygen, nitrogen, sulfur, and perhaps phosphorous to the simple compounds of ammonia or nitrates, sulfates, phosphates, carbon dioxide, and water. (2) Break down the great complex cellulose molecules, which give woody tissue its properties, into carbon dioxide and water. (3) Split the big, resistant fat molecules into carbon dioxide and water. (4) Simplify a great many other compounds in the same fashion, including such improbable ones as tar, phenol, and paraffin.

The simple compounds thus produced, serve as plant foods or as solvents which in turn release the mineral nutrients from their rock prisons.

In addition to the inorganic or mineral elements made available from organic compounds, substances analogous to the vitamins may be produced by soil microorganisms. These may serve to stimulate plant growth as vitamins do the growth of animals. This particular aspect of soil microbiology has not been adequately studied, but investigators have long recognized the growth stimulating effect of organic matter added to the soil.

Organic matter which is difficult to decompose may accumulate in the soil as humus. This material gives soil its dark color, makes it "mellow" and greatly increases its ability to absorb and hold water. The darker color of the soil also increases its power to collect heat thus producing a "warmer" soil which is more conducive to plant growth, especially in the spring months.

PLANT FOODS IN CIRCULATION

The breakdown of complex organic materials into simple inorganic compounds is cyclic in character and part of the continuing building up and breaking down process which constitutes the interminable flow of life in the earth. These changes are generally designated as the nitrogen, carbon, phosphorus, iron, and sulfur cycles. This is an over-simplification since these cycles probably never function independently of each other but are meshed together to form a smoothly operating, beautifully synchronized effect. However, for convenience in describing the various phases of this welter of activities, the conventional scheme of cycles of the elements will be followed.

NITROGEN CYCLE

Life on the earth is a paradox of starvation in the midst of plenty. With about four-fifths of the entire atmospheric blanket of the earth made up of nitrogen gas, nitrogen is still the most critical of the plant food elements, the one most likely to be in short supply. The reason, of course, is apparent when one realizes that nitrogen gas is seemingly useless to all living things except for a very small group of microorganisms. Higher plants, living alone, are denied all access to this unlimited supply of life substance.

To be useful, either to plants or animals, nitrogen must be united with some other element or elements to form a compound. Nitrogen gas is itself extremely inert and difficult to change from the elemental to compound form. As a result, the nitrogen which is in the form of a compound becomes an extremely valuable part of a limited supply.

For practically all plants, the nitrogen must be in the form of ammonia (NH_3) or preferably the nitrate ion (NO_3). Animals use as their primary source of nitrogen the proteins of plants or of other animals. Since plants are the ultimate source of all nitrogen compounds it is apparent that for the continuation of life on the earth, the limited supply of nitrogen compounds must be kept constantly in circulation.

Aside from the small amount of compound nitrogen formed by lightning flashes, or spued from volcanoes or furnaces, the entire responsibility of distributing and maintaining the precious nitrogen compound supply of the living world rests on soil microorganisms or the synthetic processes developed by man.

When a complex protein molecule is added to the soil in the form of plant or animal remains, a series of microbial splittings of molecules results in the eventual production of the nitrates so essential to plant growth.

This is the familiar nitrogen cycle and in Table XIV its stepwise operation is indicated.

Starting with plant or animal proteins the steps in the cycle proceed as follows:

The protein molecule is a mosaic built up primarily of amino acid molecules. These amino acids are made up of simple or more complex

Table XIV

The Nitrogen Cycle and Balance Sheet

Cycle of Nitrogen

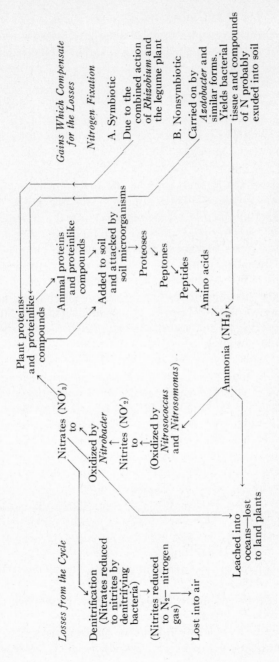

The breakdown of proteins to ammonia is often called *ammonification*.
The oxidation of ammonia to nitrates is frequently termed *nitrification*.

organic acids having the NH_2 or amino group.   The simplest of the amino acids is glycocol.   It is an acetic acid (vinegar acid) molecule on which the NH_2 group has been substituted.   The structure of the two molecules is compared below:

acetic acid

H
|
H—C—C=O
| |
H OH

amino acetic acid
or glycocol

NH_2
|
H—C—C=O
| |
H OH

Most of the 25 or so amino acids now known are much more complex than glycocol.   These amino acids are arranged in varying proportions and degrees of complexity to make up the protein molecules.   These molecules are huge, commonly ranging from molecular weights of about 10,000 to a million or more.   The size of this mosaic may be visualized when one remembers that water has a molecular weight of 18, while the molecular weight of glycocol is about 75.

In the simplification of the protein molecule the first step is the production of several proteose molecules.   These in turn are broken down into smaller masses, the polypeptides, then to the peptides and finally the amino acids.

One of the processes by which the amino acid is broken down is know as deamination.   This consists of splitting the amino acid into a modified organic acid and ammonia.   The organic acid is now separated from the nitrogen cycle and becomes part of the carbon cycle where microbes may simplify it to carbon dioxide and water.   This latter process probably proceeds simultaneously with or independent of the process which separated off ammonia.

The process of breaking down a protein occasionally may be accomplished by a single organism, a fact which can be demonstrated in the laboratory.   However, it is more likely that in the soil a good many microorganisms work together to accomplish this result.

Some of the ammonia which is produced may be utilized directly by certain microorganisms and by some of the higher plants.   However, the important change in ammonia is its oxidation by autotrophic bacteria to nitrates.

In a well-aerated fertile soil, ammonia does not seem to accumulate to any extent.   It may be absorbed or "fixed" for a time by the clay complex of the soil unless directly oxidized to the nitrate form.   In some instances a small part may be exuded into the air or washed away by percolating water to find its way eventually into the ocean. These two possibilities represent losses which may reduce the total amount of nitrogen compounds available in any given soil.

The process of oxidizing ammonia to nitrates is one of the most remarkable of the biological processes. It is carried on by a group of bacteria known as the chemosynthetic bacteria or autotrophes. These bacteria are "self sufficient" as their name indicates. They secure their energy by the oxidation of inorganic compounds (ammonia or nitrites) and derive the carbon for building their tissues from carbon dioxide. For them, our exhaled breaths would represent foodstuff.

These bacteria were discovered by Winogradsky in the early 1890's. He found that the process involves two major steps: (1) oxidation of ammonia to nitrites which is carried on by either *Nitrosomonas* or *Nitrosococcus*; (2) oxidation of nitrites to nitrates carried on by a rod-shaped form, *Nitrobacter*.

When higher plants take up these nitrates from the soil and use them to build up their own protein, the nitrogen cycle is completed. This represents the inner circle of the nitrogen cycle (Table XIV).

The process is not 100 per cent efficient since "friction" develops which results in a loss from a given soil of the nitrogen compounds originally added to it as protein or proteinlike compounds.

These losses are: (1) losses of ammonia due to effusion into the air, or leaching out in underground water chiefly as nitrates. The nitrate losses may be considerable because nitrates are highly soluble. Leached out nitrates may ultimately find their way into the ocean, completely unavailable to land plants. (2) In waterlogged soils, a more immediate loss occurs. Because of the lack of soil air, bacteria may attack the nitrates in order to secure the oxygen they contain. Nitrates are first reduced to nitrites and eventually to nitrogen gas. This process is know as dentrification. (3) Assimilation of ammonia or nitrates by microorganisms in competition with higher plants, will temporarily make these nitrogenous compounds unavailable until the microorganisms die and their proteins are again broken down. The native microbial flora of the soil is likely to be an active competitor with higher plants for these compounds.

Importance of Nitrogen Fixation

While the losses of nitrogen compounds just described are very real, the economy of nature can be relatively self-perpetuating in so far as nitrogen compounds are concerned unless dissipated by profligate agricultural practices. This is due to the replacement of the lost nitrogen compounds as a result "nitrogen fixation."

Briefly, nitrogen fixation is a physiological reaction peculiar to certain bacteria and closely related forms which enables them to transform the useless, inert nitrogen gas of the atmosphere into compounds which may ultimately become available to higher plants as nitrates. The nitrogen gas is apparently transformed into proteins or simpler nitrogen compounds in the bacterial tissues or in the tissues of leguminous plants. The bacteria or plant tissues must then undergo decomposition, to bring about the release of ammonia and its oxidation to nitrates. In addition, some nitrogen compounds may be excreted directly into the soil by the bacteria or from the roots of inoculated leguminous plants.

The nitrogen fixing bacteria are of two types: (1) the symbiotic nitrogen fixing bacteria which live in the roots of legumes; (2) the non-symbiotic nitrogen fixing bacteria which live free; i.e., without a host plant, in the soil.

Symbiotic Nitrogen Fixation

The term "symbiotic" may be translated, "living together." It aptly describes the relationship which exists between the leguminous plants and the root nodule bacteria. These bacteria belong to the genus, *Rhizobium*. The name means "root dweller."

For many centuries observant agriculturists have recognized that legume crops such as alfalfa or clover may have a beneficial effect on subsequent crops. This bacteria-plant relationship was first demonstrated by Frank in the 1870's and its fundamental importance established by the work of Hellriegel and Wilfarth in 1888.

Recent work indicates that symbiotic nitrogen fixation is limited rather definitely to members of the legume family, the *Leguminosae*. As pointed out by Allen and Allen,[1] the claims that nodule production and nitrogen fixation have occurred in nonleguminous plants have not, at least to date, withstood critical examination. Even in the legume family itself, the picture is incomplete. While the family represents some 10,000 species, less than 10 per cent of them have been studied carefully for nodule production. Of this group about 8 per cent have not shown nodule formation "under any known conditions."

---

[1]Bact. Rev. **14** (4):273-330, 1950.

Botanically, the members of the *Leguminosae* have several characteristics which set them apart from other plants. The seeds are borne in pods attached to the midrib of the pod and the flowers are papillionaceous or as the name indicates butterfly-like. This peculiar flower appearance is due to the great enlargement of two of the petals. The sweet pea flower is a familiar example. The leaves are generally pinnate. While the presence of root nodules is a distinctive character as has been indicated it is not shared universally by all members of the family.

Some legume plants will grow fairly well with or without *Rhizobium* being present. In other cases, alfalfa is a notable example, growth is extremely meager in the absence of the root nodule bacteria. The most significant difference between inoculated legumes and uninoculated plants, either legumes or nonlegumes, is the much greater nitrogen content of corresponding tissues in the inoculated leguminous plants. This may be as much as twofold greater in the inoculated legumes.

This increase in protein or fixed nitrogen in legumes has a profound influence on the permanency of agriculture and the nutrition of the human population it supports. Without doubt, the ability of the oriental civilization to persist for centuries on a meager diet depends in part on *Rhizobium* which is present in the roots of soybeans. The widespread use of beans as a staple item of diet by the population of Central America is justified by the protein producing ability of *Rhizobium* in the roots of the bean plant. Crop rotations which contain a legume which is plowed under as green manure are a great aid in maintaining soil fertility. However, when the entire plant is harvested, as in cutting alfalfa for hay, no appreciable gain in soil nitrogen occurs.

Bacteria of the genus, *Rhizobium*, are gram-negative, motile rods which can be grown easily on artificial media in the absence of the host plant. In such cultures, the bacteria greatly resemble *Escherichia coli* in shape. On the other hand, in the root nodules they often elongate and swell into forms resembling irregularly stained baseball bats or even swollen X or Y forms. These are termed bacteroids, i.e., "rodlike." The relationship of these bizarre forms to the actual nitrogen fixing process has never been adequately explained.

At the present time, seven species of *Rhizobium* are recognized. These are separated largely on the basis of the plants they infect.

The species, together with their host plant, as listed in *Bergey's Manual of Determinative Bacteriology* are as follows:

1. *Rhizobium leguminosarum*—produces root nodules on sweet peas, peas, vetches, and lentils.
2. *Rh. phaseoli*—produces nodules on beans.
3. *Rh. trifolii*—produces nodules on members of the genus *Trifolium* (the clovers).
4. *Rh. lupini*—produces root nodules on the lupines and serradella.
5. *Rh. japonicum*—produces root nodules on soybeans.
6. *Rh. meliloti*—produces root nodules on alfalfa, sweet clover, burr clover, yellow trefoil, and fenugreek.

Because *Rhizobium* is often absent in an agricultural soil, particularly if it is acid, it is frequently necessary to inoculate the legume seeds with the bacteria before planting. In acid soils, the soil must also be limed to neutralize the acid before the legume seeds are sown. Inoculation is accomplished by (1) securing soil directly from a field on which inoculated plants have been growing and broadcasting it directly over the surface of the field (this procedure is now little used because of the labor involved and the danger of spreading plant diseases); (2) applying bacteria directly to the seeds before planting. The cultures may be either agar slants of *Rhizobium* or finely ground peat or soil used as a carrier for the bacteria. After the seeds are planted, entrance of bacteria into the roots with the resultant formation of nodules, takes place rather early in growth of the plant.

The bacteria-host plant relationship is not a simple one. Some strains of *Rhizobium* may be extremely "effective" in stimulating nitrogen fixation, other strains of the bacteria may be relatively "ineffective." Some may even prove to be a burden to the host plant, like pampered sons of a rich father. The yield obtained from a given legume crop may often be closely associated with the "efficiency" of the bacteria in the roots.

Within the past two decades, another complicating factor has been recognized. It has been found that *Rhizobium* suffers from a virus or phage infection which may destroy many of the bacteria.

The "fatigue" shown by clovers and alfalfa grown for some time on the same soil has been attributed to the destruction of the effective strains of *Rhizobium* by the phage or virus.

While the mechanism of symbiotic nitrogen fixation is still not well understood, its practical significance in replenishing the losses in the nitrogen cycle remains one of the major contributions of the soil microorganisms to the economy of the world.

Nonsymbiotic Nitrogen Fixation

The second and less spectacular, but more universally present, form of nitrogen fixation is that carried on in the soil by nonsymbiotic or free-living microbes. These microorganisms possess the remarkable power of utilizing nitrogen gas in building up their own protoplasm. Following the death of such bacteria, their proteins are transformed into nitrates through the stepwise action of the ammonifying and nitrifying bacteria. In addition, the free-living nitrogen-fixing bacteria may apparently excrete small amounts of nitrogen compounds into the soil.

The most widely studied of the free-living nitrogen fixers is *Azotobacter*, a large gram-negative, nonsporing, coccoidal rod or diplococcus resembling a yeast cell in some respects. It is widely distributed in all except acid soils and can be easily isolated by adding soil to a nitrogen-free medium.

The production of a brownish pigment is characteristic of the most widely distributed of the *Azotobacter* species, *A. chroococcum*. The two other *Azotobacter* species, which are relatively less common, do not form a brown pigment.

The critical soil pH for the growth of most *Azotobacter* spp. appears to be about 6.0. In soils more acid than this, this organism may not be present.

In the laboratory, *Azotobacter* generally shows rather vigorous nitrogen fixation. It is doubtful that conditions exist in the soil which are conducive to such a high degree of nitrogen fixation. In certain alkaline soils, however, *Azotobacter* is present in large numbers and apparently makes vigorous growth.

In addition to fixing nitrogen, *Azotobacter*, through its gelatinous exudates, may aid in protecting soils against erosion.

Probably second in importance to *Azotobacter* is the anaerobic, spore-forming rod, *Clostridium butyricum* which has been shown to be capable of fixing nitrogen in the soil. It appears to be an important source of fixed nitrogen in certain acid soils devoid of *Azotobacter*. Other species of soil clostridia may also be able to fix nitrogen.

Rhodospirillum rubrum, one of the pigmented bacteria, which apparently functions photosynthetically under certain conditions, also appears to be able to fix significant amounts of nitrogen.

Certain algae are likewise thought to be able to fix nitrogen. Such fixation may be of a practical significance in rice paddy soils.

If left alone and undisturbed the soils in many areas show a definite positive nitrogen balance which results in a slow but consistent increase in nitrogen content. When this balance is disturbed by man's cultivation and harvesting, the fertile soils suffer a nitrogen depletion which soon forces an artificial replenishment of the supply of nitrogen compounds for successful crop production.

THE SULFUR CYCLE

Sulfur, like nitrogen, is essential to all life since it is a constituent of certain indispensable amino acids which are necessary for the building up of most protein molecules. It is not present in as high a concentration as nitrogen, but without it life could not exist. Proteins normally contain from 0.2 to 2.0 per cent sulfur. In addition certain important vitamins such as thiamine and biotin also contain sulfur.

Sulfur in the soil is primarily in the organic form, as proteins and related compounds, and secondarily in the inorganic form as sulfates. In most humid soils about 15 to 20 per cent of the sulfur present may be in the sulfate form. As aptly pointed out by Starkey,[2] "organic sulfur in soils is important as a reservoir of the element for plant development. Sulfate is readily leached from the soil, whereas the sulfur in organic compounds resists leaching and is gradually broken down to sulfate. It is generally assumed that sulfate may be the principal source of sulfur for plant development; it may not be the only source."

A wide variety of microorganisms may play a part in the transformation of sulfur and may in turn compete with the higher plants for this essential element.

Quoting Starkey further, "There is no simple sulfur cycle in the soil. Only in a general way can one refer to the sulfur cycle. When all the many microbial transformations of sulfur are considered one finds an interwined network of reactions."

A few of the transformations in sulfur compounds produced by microorganisms are of particular interest. When elemental sulfur is applied to most soils it is rather quickly oxidized to sulfuric acid. The most active of the sulfur oxidizers is an autotrophic form, *Thiobacillus thiooxidans*. This microbe is remarkable in that it tolerates the high acidity which results from the sulfuric acid it produces. It is of special interest in that it is apparently responsible for the high

---

[2] Soil Science **70**:55-65, 1950.

concentration of sulfuric acid present in drainage water from coal mines. These waters are rich in sulfur which is quickly oxidized to sulfuric acid. Starkey points out that the amount of sulfuric acid which enters the Ohio River and its tributaries from such coal mine waters approximates 3,000,000 tons per year. This often causes serious corrosion of metal or concrete structures and other types of damage.

Elemental sulfur is often added to alkaline soils to reduce the alkalinity or in some cases to create acid conditions necessary for the growth of acid-loving plants. Even when *Th. thiooxidans* is absent in a soil many other microorganisms are present which can oxidize the sulfur to the sulfate or thiosulfate form.

A reaction characteristic of waterlogged soils, sewage, etc., is the anaerobic reduction of sulfates with the production of hydrogen sulfide. The hydrogen sulfide or "rotten egg gas" odor found under such circumstances is the olfactory evidence that such a microbial action is taking place.

THE PHOSPHORUS CYCLE

Phosphorus is not a common element in the soil, but nevertheless it is one of the essential elements, indispensable to all life on the earth. One of its important life functions is the part it plays in the release of energy through the action of energy-rich phosphate intermediate compounds. As an example, the energy which is derived from the sugar you eat, which keeps your body warm and your heart beating, is released by the "phosphorylation" of sugar molecules. Many other vital structures or functions of the living tissue contain phosphorus or require it for their operations.

Unlike nitrogen, phosphorus always occurs in nature in the form of compounds. When separated into the element form it is highly reactive, uniting violently with oxygen when placed in contact with air.

The phosphate of the soil is highly insoluble and as such is unavailable to higher plants. Certain of the organic phosphorus compounds likewise contain phosphate in an unavailable form.

Microorganisms play a vital role in keeping phosphorus circulating in the world of living things. While this relationship is cyclic in nature it is not as direct and uncomplicated as the nitrogen cycle.

The major actions of microorganisms in relation to phosphorus are as follows:

1.   By producing such dissolving agents as carbonic acid, organic acids of sundry kinds and inorganic acids such as sulfuric or nitric, the insoluble phosphates of the soil are rendered soluble.   On entering the soil solution, these phosphates may be taken up by the plants or microorganisms or may "revert" back to the insoluble state.

2.   Plant and animal residues, incorporated with the soil, are the source of fairly large amounts of phosphorus-containing organic compounds.   These include such compounds as the nucleoproteins, phosphoproteins, and phospholipids.   The phosphorus they contain is unavailable to plants and must be changed to the phosphate form, i.e., "mineralized" before the phosphorus can be utilized by growing plants.

3.   The micoorganisms compete with the higher plants for phosphates.   However, if phosphorus is snatched from a plant by a microorganism, on the subsequent death of the microbe other microorganisms will mineralize the tissue and release the phosphorus it contains as phosphoric acid.

THE IRON CYCLE

Another element vital to plant growth is iron.   Without it plants become pale or chlorotic and die; animals become anemic and likewise succumb.   Iron ranks next to oxygen in its abundance, representing more than 5 per cent of the earth's crust.

Much of the iron of the soil is in a relatively insoluble mineral form and a lesser part in the form of unavailable organic compounds. By means of the acid which they produce, bacteria render a portion of this iron soluble and thus available to plants.   They may also compete to some extent with plants for these compounds.

In addition to the action of the decay bacteria on the iron compounds of the soil, the so-called "iron bacteria" are widely distributed in rivers, lakes, ponds, swamps, and bogs.   These bacteria, members of the *Chlamydobacteriales*, bring about a precipitation of ferric hydroxide in the slimy sheath which covers the filament of cells making up the bacterial mass.

These organisms may grow rather vigorously in waters containing any appreciable amounts of iron salts where they may accumulate as brown "slime" on the surface of the rocks.   In water supplies carried in pipes, these accumulations of iron bacteria may be sufficient to clog the pipes or seriously impede the flow of water.   This increase in organic matter may bring an increase in the numbers of saprophytic bacteria, discolor the water or produce objectionable tastes.

It is believed that these microorganisms may have played a major role in the development of certain deposits of "bog-iron."

CARBON CYCLE

Carbon represents the basic building material out of which the framework of all organic compounds is constructed. Approximately half a million such compounds are known. In addition, inorganic carbon in the form of carbonates (limestones, dolomites, chalks, etc.) contribute substantially to the rocky skeleton of the earth itself. Our own skeleton, is a complex mixture of calcium carbonate (chalk) and calcium phosphate.

Carbon may occur in nature in the elemental form. Diamonds are relatively pure crystalline carbon while charcoal made from certain pure materials may be relatively pure uncrystallized carbon.

The most universally distributed of the carbon compounds is carbon dioxide since the air is about 0.1 per cent carbon dioxide and we contribute to the supply with our every breath.

This carbon dioxide in turn represents one of the raw materials out of which we ourselves are made.

The process involved is the familiar photosynthetic process. All plant and animal tissues and even coal and oil were initially produced by photosynthesis. Without doubt this is probably the most important of all the life processes. By it, the energy of the sun is locked in organic compounds to become the energy animating all living things.

The green plant thus becomes the source of all life on the earth. The details of photosynthesis are shown in the following formula:

CARBON DIOXIDE plus	WATER plus	MINERALS plus	SUNLIGHT
(from the air, entering through the stomata of the leaf)	(absorbed from the soil by the roots)	(absorbed with the water)	(source of energy for the process)

	in the presence of CHLOROPHYLL	yields	CARBOHYDRATES
	(the catalyst which "triggers" the reaction)		(and indirectly other compounds such as proteins, fats and vitamins)

plus OXYGEN

From this discussion it can be seen that plants, from the smallest bacterium to the largest tree and all animals are made directly or indirectly out of carbon dioxide and water, under the impact of energy from the sun and through the magic of chlorophyll.

Carbon dioxide, therefore, becomes a vitally important compound which might conceivably be in short supply. However, the contribution of microorganisms to the interminable carbon cycle is much less important than is the case with the elements previously considered.

Volcanoes and the burning of coal, wood, oil, and other organic compounds probably contribute the lion's share of atmospheric carbon dioxide.

However, the greatest contribution of microorganisms to the cycle of carbon lies in supplying carbon dioxide to the soil where it becomes a major solvent in releasing inorganic plant foods from the minerals of the soil. Of almost equal importance is the operation of breaking down the vast amount of woody tissue or cellulose of plant origin which would otherwise leave much of the earth an impenetrable bramble of dead trees and smaller plants. This process likewise leaves, as a residue carbon dioxide, water, and the soil humus so important as an aid to vigorous plant growth.

SUMMARY

The soil is not inert, not dead. It is the abiding place of an innumerable host and variety of microorganisms. These may serve as the "digestive juices" of the soil, releasing the plant food elements bound in plant and animal remains, to be used by subsequent generations of plants; or by other microorganisms which in turn compete with plants for plant food elements thus released.

The microbiological processes which take place in the soil represent a complex of interactions, of cycles upon cycles which proceed in an apparently orderly fashion as yet not well understood. An adequate understanding of these vital interrelations can only be secured by the application of the newest devices for scientific research. When such knowledge is gained it may ultimately contribute much toward banishing the specter of hunger from an overpopulated world.

In this chapter no mention is made of the plant pathogens which may lurk in the soil ready to destroy susceptible plant life. However, these are of great importance and may have a profoundly detrimental effect on the food supply of the world.

Selected Reading References

Allen, E. K., and Allen, O. N.:   Biochemical and Symbiotic Properties of Rhizobia, Bact. Rev. **14** (4):273-330, 1950.

Conn, H. J.:   The Most Abundant Groups of Bacteria in the Soil, Bact. Rev. **12** (3):257-273, 1948.

Waksman, S. A.:   Principles of Soil Microbiology, ed. 2, Baltimore, 1932, Williams & Wilkins Company.

Waksman, S. A.:   Soil Microbiology, New York, 1952, John Wiley and Sons, Inc.

CHAPTER 20

GOVERNMENT AND PRIVATE AGENCIES THAT PROTECT YOUR HEALTH

The truism "public health can be purchased" expresses the role that both government and private agencies can play in providing more years of life for us. The purchase of public health involves first of all the acquiring of knowledge on which to base the procedures necessary to protect us against the hazards of our environment. Behind all the accomplishments of public health agencies has been the unselfish work of the public health employees, often underpaid and unsung plus the altruistic actions of private individuals who are willing to give their time and talent toward the control of some disease or adverse condition.

The net result of the activities of these agencies is that many of us are alive today and many of our children will survive or escape diseases which in the past or even in our own time have led only to disability or death.

In the past, control measures instituted by public health agencies have produced miraculous results. For example, cholera and yellow fever which were once dreaded killers are now unknown in the United States. Typhoid is now a medical oddity and diphtheria no longer chokes entire families in their beds. The battle, however, is still far from won and we must rely constantly on those who guard us against the evils of our environment and our own acts of ignorance or carelessness.

Official public health agencies are the primary forces which protect us against our environmental hazards. These include not only communicable diseases but also many others which have been an outgrowth of our modern way of life. Control measures involve a great variety of functions which have been integrated into our federal agency, the United States Public Health Service, and to more or less complete degree into the state and municipal health departments.

The major activities which bear on the control of environmental factors include control of drinking water supplies and stream pollution; sewage disposal; milk and food control; restaurant sanitation; air pollution control; pest control; industrial health; communicable disease laboratories and control of radioactive wastes.

The role of federal, state, and municipal agencies will each be considered separately.

THE UNITED STATES PUBLIC HEALTH SERVICE

This federal public health agency has had a history almost as venerable as the federal government itself. Although the constitution limits its legal powers to interstate commerce and similar interstate activities, the United States Public Health Service guards our shore lines, our insular possessions, and our borders against the invasion of diseases. It provides the states with much needed public health knowledge; on invitation it assists in setting up and financing public health programs; it provides model laws and regulations which can be used in the various states and local organizations and in cases of emergency is ready to step in and assist in combatting the effects of epidemics and disasters.

What is now the United States Public Health Service came into being on July 16, 1798, when the Fifth Congress passed an Act for the relief of sick and disabled seamen which was financed by a tax of 20 cents a month per sailor to be collected by the Treasury Department. As time went on, hospitals were erected at various ports including some situated on inland rivers.

In 1870 the Marine Hospital Service was reorganized and became for the first time a national health service. Since that time it "has operated programs in three major categories; namely, medical and hospital care for certain groups of the population; research in the biological, social and related sciences; and services designed to control disease and promote health in the general population."[1]

In 1873 the newly reorganized Service swung into action in order to bring under control the cholera epidemic which raged in the Mississippi Valley and New Orleans.

It stood guard at our ports and even in foreign ports to protect us against the cholera epidemic which circled the world about 1892.

---

[1] A Philosophy of Modern Public Health Services by Leonard A. Scheele, 1950, Reprinted by National Sanitation Foundation.

Yellow fever claimed its attention in Louisiana in 1895 and in the last epidemic in the United States which flared up in New Orleans in 1905.

When bubonic plague reared its ugly head in San Francisco in 1900 representatives of the Service were there. They were at first greeted with public derision but the same public remained to applaud their acts. Since that time the disease has appeared briefly in several other ports and even occasionally has been found in the rodent population of inland areas.

Rocky Mountain spotted fever, tularemia and psittacosis are among the diseases against which the men of the Service have matched their skills.

In 1935 the Federal Security Act provided "Grants in Aid" for the state and local health units which made them partners with the United States Public Health Service.

Research has been a vital function of the U.S.P.H.S. since the Hygienic Laboratory was organized in 1887. In 1930, this laboratory, now grown to major size, became the National Institute of Health. In 1948 it was rechristened, the National Institutes of Health since it embraced seven institutes including such widely varied fields as a National Institute on Arthritis, a National Dental Research Institute, and a National Cancer Institute.

This represented the "largest medical research organization in the world—the presently (1950) constituted National Institutes of Health"[2] which occupies nine large buildings at Bethesda, Maryland.

Added to this are the activities of the Environmental Health Center in Cincinnati, Ohio, the Communicable Disease Center of Atlanta, Georgia (which has an important function, the training of state and municipal laboratory workers), and the huge research grants administered by the U.S.P.H.S.

As a result of all this the U.S.P.H.S. has assumed a major role as a producer and dispenser of public health information and as an encourager of health research. Not only has this information concerned itself with the control of a great variety of communicable diseases but water, milk, food sanitation, sewage disposal, and related problems have all been subjected to extensive and fruitful investigation.

In 1953, the U.S.P.H.S. became an integral part of the new Department of Health, Education and Welfare, the chief of this department holding cabinet rank.

---

[2]Williams, R. C.: The United States Public Health Service, 1798-1950, Comm. Off. Assn., United States Public Health Service, 1951.

STATE PUBLIC HEALTH ORGANIZATIONS

While the public health departments in the various states may differ to a considerable extent in their functions and activities, they are practically all vigorous, functioning organizations which stand between the citizens of the state and the hazards which give rise to illness and death. They are not the health departments of fifty years ago, whose widespread quarantines and harsh fumigation and disinfection procedures often worked great hardships on the unfortunate citizens who fell prey to some communicable disease. Nor do they spend any great portion of their time abating the nuisances which may aggravate the more fastidious members of the community. Today, state and county and municipal health organizations all have definite and widespread functions which have a bearing on the whole broad aspect of the health of the individual states and the communities within their state.

These state and local agencies have the definite responsibilities of breaking chains of infection, improving the environment as it influences the general health, combatting chronic and even mental ills and educating the entire population so that its members can better protect themselves against these various health hazards.

From the legal viewpoint, each state in the United States is the sovereign power. In it rests the majority of the legal responsibilities and prerogatives. The Federal government possesses only those powers which are specified in the Constitution, such as the control of interstate commerce. All other rights are reserved to the various states. The cities, towns, and villages within a state have only those rights delegated by the state constitution and they can carry on no health (or other) activities at variance with state regulations.

In practice, however, an excellent working relationship generally exists between federal, state, and municipal health organizations.

State health organizations generally function actively in the unincorporated areas and smaller cities of the state while the larger cities generally assume the responsibility of operating their own health services. County or county-city health organizations are now being set up in certain states. In most instances these units work closely with the state health department.

Historical

Although some Boards of Health were set up prior to the Civil War they were short-lived. The first permanent State health organi-

zation was that set up the by State of Massachusetts in 1869.   In
the 1870's epidemics of cholera and yellow fever stimulated sufficient
interest in public health that during the decade between 1870 and 1880
a total of 18 state health organizations were established.   By 1900
state boards of health were functioning in thirty eight states and by
1913 all of the states except one had such organizations.

In the early state organizations, the boards of health were made up
of physicians who served without pay and carried on studies of various
epidemic diseases.   Their findings were transmitted to the state
legislative bodies with recommendations for action.   It should be
remembered that prior to the turn of the century our knowledge of
diseases in terms of public health was meager indeed.   Use of bacteriological procedures did not become widespread until about the beginning
of World War I.

Responsibilities

The present responsibilities of state health organizations include,
among others, enactment of sanitary and communicable disease regulations and related measures; diagnostic procedures as they relate to
communicable diseases and sanitation; inspection of dairies and eating
places; control of occupational hazards; examining and licensing medical
laboratory technicians; and enforcement of health laws and sanitary
codes.   In the discharge of these functions the state (and municipal)
health departments stand guard over the health and well-being of all
of us.

The environmental sanitation control activities include water
supply control, sewage disposal and stream pollution regulation;
restaurant sanitation measures; the control of insect vectors, inspection
of hotels, camps and bathing places especially outside of the larger
cities.

Of these functions the milk supply regulation may sometimes be
in the hands of the state department of agriculture rather than being
vested in the state health department.

Public health laboratories carry on a variety of tests which are of
great importance, particularly in the control of communicable diseases.

In the "Report of State Health Programs 1949"[3] the following
laboratory procedures are listed as performed by the various state
public health laboratories in the order of the frequency with which

---

[3]By Josephine Campbell and Clifford H. Greve, published 1950 by United States
Public Health Service.

they are carried on in the various states.   The first four are carried on
by decreasing proportion of the state laboratories:

> Diagnostic tests for tuberculosis
> Diagnostic tests for undulant fever
> Diagnostic tests for syphilis
> Bacteriological examination of foods
> Bacteriological examination of milk
> Analysis of water samples
> Examination of animal specimens for pathogenic micro-
>   organisms or viruses
> Development of new laboratory techniques
> Diagnostic service for rickettsial diseases
> Analysis of sewage samples
> Provision for urinanalysis
> Chemical examination of foods
> Preparation of biological products for free distribution
> Provision for tissue examinations
> Approval of local laboratories for general diagnostic
>   service

In addition, many special procedures not listed above are commonly performed including tests for gonorrhea, intestinal pathogens,
serological diagnosis of infectious diseases, blood typing, etc.

The control of communicable diseases is another important function
of state health departments.   It includes: (a) Measures against
tuberculosis which involve mass x-rays, laboratory tests of suspected
cases, BCG vaccine for children exposed to the disease in households
where the infection is present, and construction and operation of sani-
tariums.   (b)   The control of venereal diseases through large-scale
blood tests, especially prenatal and premarital tests, dark-field examinations, microscopic and cultural tests for gonorrhea, free clinic and free
treatment for indigent patients.   Such free treatment is important
because the incidence of venereal disease is highest among those who
cannot afford or will not take treatments from private physicians.
Venereal disease treatment is often a municipal rather than a state
responsibility.   (c) The control of acute communicable diseases in
general, including virus diseases, is a perennial problem.   This includes
laboratory diagnosis, epidemiological studies, and the institution of
control measures.

In addition to the functions which deal primarily with the control
of communicable diseases other activities include vital statistics, maternal and child health, public health education, school health programs,
and industrial hygiene.

LOCAL PUBLIC HEALTH ORGANIZATIONS

Municipalities as secondary political divisions within the state often exercise a specific and often rather effective control of public health hazards. Since cities vary greatly in their populations, the nature of their populations, and environment, the freedom for action provided by their charter, and even in the extent to which the population is interested in public health, it is difficult to give a very satisfactory picture of the protective functions performed by municipal health organizations.

One of the earliest public health boards was that established in Boston in 1799. Paul Revere headed the organization as its first health officer. However, very little interest was shown in local health boards since, according to a U.S.P.H.S. report of 1873, only five boards were established between 1800-1830.

Interest increased after about 1870, although no real strides were made until after the turn of the century.

The functions which most local health organizations share are vital statistics; communicable disease control especially as related to venereal diseases and tuberculosis; the operation of a public health laboratory; sanitary controls particularly of milk supplies, foods, and restaurants, water supplies and sewage disposal; child and maternal hygiene; public health nursing, and often other and perhaps unrelated duties. Cities may have in addition to or in place of a board of health, a lay advisory committee. County health organizations may operate in conjunction with a city health unit or may operate as independent units.

In 1949 approximately 30 per cent of the population of the United States had the services of a full-time city health unit, about 25 per cent were served by a full-time county unit, and about 15 per cent by a local or state health district and more than 25 per cent did not report a full-time health unit.[4]

INTERNATIONAL HEALTH ORGANIZATION

Recognition of the fact that the health of any nation is dependent on the health of the world prompted the development of health organizations which are world-wide in scope.

Included in such organizations are L'Office International d' Hygiene Publique set up in 1907 and financed by contributions of

---

[4]Greve, C. H., and Campbell, J.: Public Health Facilities and Services in Local Area, United States Public Health Service, 1950.

member states.   Its primary function was notification regarding the incidence and spread of such diseases as cholera, yellow fever, plague, typhus, and smallpox.   It was made part of the World Health Organization in 1947.

In 1902 the International Sanitary Bureau grew out of an Inter-American Conference.   It became the Pan-American Sanitary Bureau in 1924 and functioned as a regional office of the League of Nations. In 1947 it became a regional office of the World Health Organization.

The Institute of Inter-American Affairs, established by the Department of State in 1942, has included a Cooperative Public Health Service among its many active functions.

The League of Nations included a Health Section in its organization.   This functioned well in the active years of the League, particularly in establishing international standards of antisera, vaccines, etc.

The United Nations World Health Organization is the most recent of the international programs.   It was chartered in San Francisco in 1945 and formally established in 1948.   It embodies six regional organizations and is primarily concerned with the control of malaria, tuberculosis, venereal diseases, maternal and child health, and improved environmental and nutritional conditions.   It is financed by contributions of member states.

VOLUNTARY HEALTH ORGANIZATIONS

Public-minded private citizens have over the years organized groups which tend to supplement or implement the activities of the tax-supported public health organizations.   Many of them have been aimed at the control of certain diseases, at public health education in some field or the arousing of official interest in the problems of some phase of public health.   Some of these organizations have been rather markedly successful while others have achieved less in the way of tangible public benefits.

The National Tuberculosis Association grew out of the Pennsylvania Society for the Prevention of Tuberculosis which was organized in 1892.   It had as its express purpose the preventing of tuberculosis or caring for victims of the disease.   A considerable portion of its funds are now used to finance research in the field.   Institution of the Christmas seal sales in 1914 gave it new impetus by providing a substantial increase in available funds.

The American Society for the Control of Cancer, organized in 1913 has been increasingly active in recent years and has been able to supply substantial amounts for research on this vital problem.

The American Social Hygiene Association, organized in 1914, has had its special interest in the control of syphilis and gonorrhea.

The American Heart Association was organized in 1933 to aid in the fight against the disease condition which kills more Americans than any other single cause. Funds for research have been rather widely distributed.

One of the most recent and perhaps most successful organizations is the National Foundation for Infantile Paralysis which started its fund raising campaign on President Roosevelt's birthday in 1938. It has provided large sums for the care and rehabilitation of paralytic victims of the disease and for financing a very extensive research program.

Philanthropic foundations have also done much to improve the health of the public. The Rockefeller Foundation, the Milbank Memorial Fund, the Commonwealth Fund, the Kellogg Foundation, and many others have contributed much toward the improvement of the health of the world. Professional organizations such as the American Public Health Association, the American Medical Association, the American Dental Society, and other similar groups have aided directly the attempts of professional public health workers to improve the health and well-being of the public.

Selected Reading References

Hanlon, John P.: Principles of Public Health Education, St. Louis, 1950, The C. V. Mosby Company.

Maxcy, K. F., et al.: Rosenau Preventive Medicine and Hygiene, ed. 7, New York, 1951, Appleton-Century-Crofts, Inc.

Smillie, W. G.: Public Health Administration in the United States, New York, 1949, The Macmillan Company.

Williams, R. C.: The United States Public Health Service, 1798-1950, Commissioned Officers Association, United States Public Health Service, 1951.

THE DISEASE-PRODUCING MICROORGANISMS
—PATHOGENIC MICROBIOLOGY

CHAPTER 21

THE PYOGENIC COCCI

BOILS, SORE THROAT, AND PNEUMONIA
MENINGITIS AND GONORRHEA

THE GRAM-POSITIVE COCCI

All of the round to coccoid microorganisms discussed in the first part of this chapter are gram-positive. Although many of these types of bacteria are nondisease-producing, only those having disease-producing potentialities or exhibiting parasitism will be discussed. The saprophytic group will be discussed later.

MICROCOCCI   *Staphlococcus*

Micrococci were first observed by Leeuwenhoek in 1676. Pasteur in 1880[1] described their presence in pus, but it was not until 1884 that Rosenbach correlated the constant presence of micrococci to suppurative wounds, abscesses, and other similar disease processes. This group have, for many years, been named *Staphylococcus*, meaning cluster or seedlike clusters of organisms. More recently (1948) their generic name has been changed to *Micrococci* (small seeds). The most pathogenic one of this group is *Micrococcus pyogenes* var. *aureus*, undoubtedly the one described by Rosenbach. Micrococci are ubiquitous and are found on the skin, on the mucous membranes of the mouth, nose, sinus, respiratory tract, gastrointestinal tract, in water, soil, milk, food, air; in fact, from almost any source these organisms may be isolated.

---

[1]Pasteur, L.: Bull, d l'Acad. de Med. **9**:447, 1880.

Structure

Micrococci are coccoid or round and about 0.8 μ* in diameter or about one tenth the size of a human red corpuscle. They do not possess flagella or spores and only rarely are capsulated. They are easily stained and retain the gram-positive stain tenaciously.

Cultivation

These microorganisms are easily cultivated in the laboratory on ordinary simple nutrient broth or agar. Culture media devoid of thiamine and nicotinic acid, however, fail to support growth. Micro amounts of these vitamins are generally present in ordinary media preparations. Blood-containing media supports more luxuriant growth and a potent red blood cell hemolysin is easily demonstrated by observing the large area of clearing around the colony. A golden yellowish pigment is produced by *M. pyogenes* var. *aureus*, while the *albus* varient is nonpigmented or whitish. *M. citreus* produces a lemon-colored pigment, while other varieties may produce a red pigment. Colonies, on solid media, appear like round daubs of paint, varying in size from 3 to 5 mm. or larger. More luxuriant growth is obtained under atmospheric oxygen pressures, but many strains grow, although feebly, in the absence of atmospheric oxygen. Only a few strict an-aerobes have been described, the most common types being *M. an-aerobius* and *M. aerogenes*. An excess of carbon dioxide will enhance the production, by *M. pyogenes* var. *aureus*, of enterotoxin, particularly if starch is present in the media.

These organisms will grow over a wide temperature range, 10° to 40°C, but the optimum is 30°-37°C (86-98.6°F).

Many enzymes are produced by the pathogenic form, namely, gelatinase, maltase, rennin, etc. Many carbohydrates are fermented with the production of acid, but no macroscopic amounts of gas. Dextrose, mannitol, sucrose, and lactose are fermented. Glycerol is usually fermented by most strains. Inulin, raffinose, and salicin are not attacked. Gelatin is liquefied and milk is acidified and coagulated.

The resistance of the micrococci to heat, germicides, drying, and other physical and chemical agents is greater than the other gram-positive cocci. Most species isolated five years ago were susceptible to penicillin, but resistance to penicillin of many *M. pyogenes* var. *aureus* strains is becoming alarmingly frequent.

---

*μ = micron (1/1000 of a millimeter).

Exotoxins

Several exotoxins are produced by *M. pyogenes* var. *aureus* most of which play some role in disease processes due to this organism. These exotoxins are: Beta hemolysin, coagulase, small amounts of fibrinolysin and hyaluronidase, leucocidin, necrotizing toxin, and enterotoxin.

Diseases Produced by Micrococci

Besides several different disease processes in man, the *aureus* variety of micrococci produces diseases of lower animals, such as black head in turkeys, boils and abscesses on rabbits, and local skin infections in several species of animals. In man, boils, abscesses, and suppurative wounds are the most frequent diseases. However, the most serious disease is osteomyelitis, an infection of the marrow and bone substance. This organism may also produce a septicemia or blood stream invasion, with multiplication of bacteria in the blood stream, which is a very serious disease. Multiple metastatic abscesses may develop in many organs of the body due to a spread from a focal infection. Bronchial pneumonia and lung abscesses may be caused by this organism. It also is a frequent cause of sinus infections, and in the common cold it is probably responsible for the copious pus produced. Kidney and bladder infections may also be caused by this organism. The micrococci are frequently the cause of infected tonsils or tonsillitis, wound infections, and infected surgical wounds. The enterotoxin produced by this microorganism causes a severe gastroenteritis, with severe vomiting and diarrhea about two hours after eating food containing this toxin.

A fairly common characteristic of this organisms is pus production. Pus, being dead leucocytes, may be found in large amounts, if the infection is extensive. The pus is usually thick and fairly free of blood. Many dead leucocytes, if stained, will show the presence of ingested bacteria, indicating that the defense mechanism of the body is locally active.

How an Infection With Micrococcus Pyogenes Aureus Develops

If we receive a wound penetrating through the skin layer, micrococci either present on the skin or on the agent producing the wound will be introduced into the broken tissue. Also if micrococci penetrate sweat glands, hair follicles, etc., they may produce an infection.

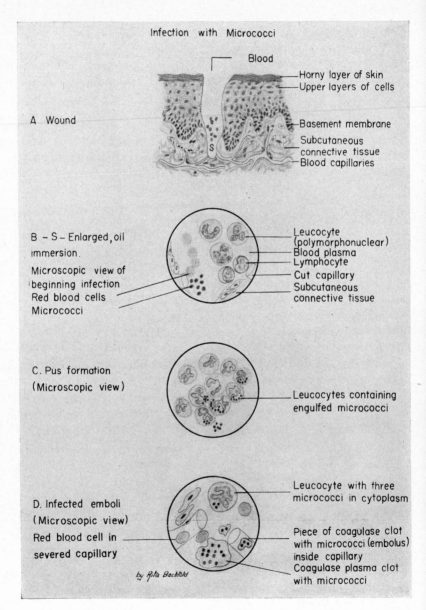

Fig. 25.—*A*, Wound in skin.   *B-S*, Beginning sepsis with micrococci.   *C*, Pus formation.
D, Infected emboli (see text for description of infecting process).

Fig. 25,*A* represents an injury where the skin is broken, *S* representing the area where the micrococci are deposited to initiate a septic infection. The severed capillaries ooze blood into the cut area, and soon there is a blood clot. Into this clot are deposited white blood cells, red blood cells, and antibodies. If the micrococci, which are introduced into the wound, are quickly dispatched by the body defense mechanisms (leucocytes, antibodies, etc.) and are aided by proper cleansing and disinfection, then there will be no infection and the injury will promptly heal (Fig. 25,*B*).

If the micrococci begin to multiply and get the upper hand, temporarily at least, many leucocytes will be killed by the bacterial leucocidin. The defense mechanisms are then pressed into jet activity and the area literally swarms with the leucocytes and antibody from the blood (Fig. 25,*C*). The antibody coats the bacteria, thus favoring clumping or agglutination of the bacteria. The leucocytes are able to travel by ameboid movement, extending their protoplasm much as an amoeba. Their ability to capture and engulf single bacteria is limited. However, when the bacteria are herded together in clumps or groups (agglutination) the work of the leucocyte is greatly decreased and efficiency is improved. Several bacteria are therefore phagocytized at a time. During this process of defending itself, the body activates several mechanisms. The capillaries enlarge, affording greater blood supply to the part, thus a reddening; an increased amount of fluid collects, thus causing a slight bulging of the tissues, and with this there is slight pain and impaired function. This part of the defense mechanism is called inflammation. As increased numbers of leucocytes are killed, there is an increased amount of pus (dead leucocytes, bacteria, tissue debris, etc.) collecting, causing a bulging of the tissue, the center of which is generally a cream color. If the pus causes sufficient pressure, or the surface area is incised, the pus will be released and ooze to the surface of the skin as a thick, creamy mass. After almost complete drainage of the pus, and all bacteria are killed, the area begins to heal. The lymphocytes and monocytes then take over the task of cleaning up the debris in the wound. They phagocytize dead tissue, dead leucocytes, dead bacteria, and by their enzyme action digest this material. The majority of local infections caused by *M. pyogenes* var. *aureus* remain localized. Some may become chronic, however, as in osteomyelitis. Occasionally micrococcus infections break loose from their localized area and spread to other parts of the body. The exotoxin, coagulase, produced by many strains of this organism causes a fibrin clot, thus surrounding the microorganisms with a gelatinous-like mass. This mass helps prevent entry of sufficient antibody and prevents leucocytes from attacking the bacteria. If the particular strain of micrococcus produces fibrinolysin, and some do, although slowly and in not a high concentration, the fibrinolytic activity will dissolve tiny pieces of the clot from the large parent clot. In these tiny pieces, several viable micrococci are frequently present (see Fig. 25,*D*). If these tiny infected clots (infected emboli) enter a capillary, they will be carried to many parts of the body, such as the liver, lung, kidney, brain, etc., and set up new areas of infections, thus producing multiple abscesses. Vigorous squeezing of a boil or abscess may aid the initiation of such a spread by injuring small capillaries.

Antibiotics have difficulty penetrating these infected emboli, therefore treatment is frequently unsuccessful.

Immunity

Only transitory immunity is produced after recovering from infections due to *M. pyogenes* var. *aureus.* Frequently a draining boil or furuncle will locally infect adjacent sebaceous glands, sweat glands, and hair follicles, producing multiple boils, one crop following another continually for weeks on end if not properly treated.

Prevention

Since infection of minor cuts, scratches, and abrasions is frequent with micrococci, proper care by thoroughly washing the area immediately with warm water and soap and applying a mild, nonirritating germicide, such as mercurochrome, is necessary. Children should be instructed to report all such skin abrasions immediately so that parents can properly care for them. A physician should be consulted in case of serious wounds or in case infection has set in. Tight collars or other tight clothing rubbing against an infected area frequently causes a spread of these microorganisms.

Vaccination against micrococcus infections has not been of value. Prophylactic and therapeutic use of antibiotics and sulfonamides are our most efficient weapons against this microorganism.

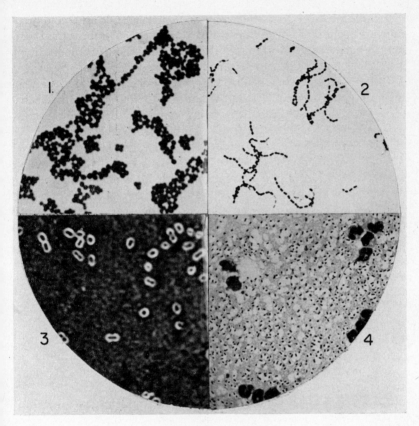

Fig. 26.—*1, Micrococcus pyogenes* var. *aureus. 2, Streptococcus pyogenes. 3, Diplococcus pneumoniae* (showing capsules—Congo red capsule stain). *4, Diplococcus pneumoniae* in the peritoneal exudate of a mouse. Note the fine strands of fibrin and the horseshoe-shaped leucocytes.

STREPTOCOCCI

Streptococci, like micrococci, are found almost everywhere. They were first described in 1881 by Ogston, but it was not until Rosenbach (1884) called attention to the constant presence of these chain-forming cocci in certain infections that attention was centered on these microorganisms as serious disease-producers. He suggested the name *Streptococcus pyogenes*, the name used at present for the virulent form producing human diseases. This genus, *Streptococcus*, comprises many species, several of which are capable of producing disease in both man and animals.

Streptococci, both pathogenic and nonpathogenic are found on the mucous membranes of the nose, throat, mouth, gastrointestinal tract, and frequently in the vagina. They are also frequent inhabitants of the skin. Soil, milk, and water all harbor species of this genus.

Streptococci, when multiplying, tend to constrict and divide in more or less in a straight line. As a result, the round cells of this bacteria form chains of organisms, being connected by a thin strand of protoplasm. They are generally round, but some forms may be slightly flattened against each other and others may be slightly pointed. This is best observed by the electron microscope. They vary in size from about 0.7 to 1.0 μ in diameter with little or no change, using different culture media. They retain the gram-positive stain quite well, although some cultures may show gram-negative forms interdispersed within a chain, indicating the organism probably died before staining.

Many species, particularly pathogens, possess capsules, which have a high content of hyaluronic acid, some polysaccharide and protein material. No known species forms spores and none are motile.

The pathogenic species are more exacting in their food requirements than are the micrococci. Culture media enriched with blood or fresh animal protein extracts and fluids will readily support growth. The majority of the pathogenic species require about thirteen different amino acids along with nicotinic acid, riboflavin, pyridoxine, pantothenic acid, probably biotin and some nucleic acid derivatives. Most of the above are supplied when whole blood is added to nutrient broth or solid media. The saprophytic specie generally grow on less complex media. This latter group includes the *S. faecalis*, *S. lactis*, and others.

The pathogenic group grow more luxuriantly at 35-37°C, while the saprophytic group may grow over a wide temperature range, 10-50°C.

Most pathogenic specie are aerobic or facultative anaerobes, but a few species are strict anaerobes. At times, however, *S. pyogenes*, when isolated from a wound, fails to grow under aerobic conditions and for initial isolation anaerobic media must be used. After isolation they grow well under aerobic conditions. These are frequently referred to as microaerophiles or Meleney types of *S. pyogenes*.

Enzymic activity is not as marked with the streptococci as with the micrococci. Dextrose is fermented with the production of acid but no discernible gas. Ethyl alcohol, acetic and formic acids are formed. Also lactose, mannitol, maltose, and sucrose are fermented.

The pathogenic beta hemolytic streptococci are fairly active and prompt exotoxin producers. They produce a very potent beta hemolysin, which plays an important role in infection. Also a leucocidin, an erythrogenic toxin, hyaluronidase (spreading factor) and a fibrinolysin which is very active. The alpha hemolytic species and the saprophytic species are less active toxin producers.

Some strains produce fairly large quantities of ammonia from peptone.

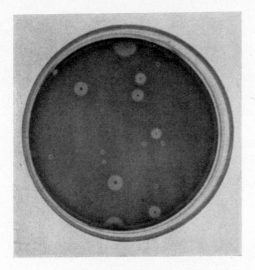

Fig. 27.—The clear zones around the colonies (*Streptococcus pyogenes*) are beta hemolysis on a blood agar plate.

Classification

Rebecca Lancefield[2] discovered that certain substances in the cell wall were antigenically specific for certain types of streptococci, particularly the human pathogens. This substance is called "C" substance or a carbohydrate-like material that will react with specific antisera. This technique has divided the streptococci into fourteen groups, A, B, C, etc., to N. In Group A are placed the beta hemolytic streptococci, primarily human pathogens. Group B, beta hemolytic streptococci producing primarily diseases of animals, particularly mastitis in cattle; Group C, beta hemolytic streptococci causes diseases pri-

[2] J. Exper. Med. **42:**377, 1925; J. Exper. Med. **57:**571, 1933; J. Exper. Med. **84:**449, 1946.

marily in animals, but may also infect man; Group D is found primarily in cheese; Group E is found primarily in milk.   Other lesser important groups are found in the intestinal tract, in water, soil, etc.   From 92 to 100 per cent of all human infections are caused by the beta hemolytic streptococci of Group A.

Those organisms classified as Group A by the specific carbohydrate are further divided into types.   Griffith[3] found that a type specific antigen could be detected by agglutination, and later he extracted these by chemical means from the body of the bacterial cells.   He classified the organisms as types 1, 2, 3, etc., and at present nearly 40 different types of Group A have been found.   The chemical substance determining type specificity are called "M" (nucleoprotein), "T" (proteinlike).   Other groups, "B" and "D" have type specific poly saccharide ("S") substance.

There is no strict correlation between diseases caused and the *type* of streptococci that may be isolated within a specific *group*.

Diseases Caused by Streptococci

Both the micrococci and streptococci are responsible for more infections of man, both mild and serious, than all other bacterial diseases combined.   Thus, streptococcal infections are very frequent

Cultures of infected tonsils will reveal beta hemolytic *Streptococcus pyogenes* in 100 per cent of the cases.   Severely sore septic throats are more frequently caused by this organism than any others.   Sinus infections, middle ear infections likewise harbor both the micrococci and streptococci.   Various wound infections invariably harbor either alpha or beta hemolytic streptococci.   Bronchial pneumonia is frequently caused by streptococci, both the alpha and beta hemolytic types.

Scarlet Fever:—Two research workers, George and Gladys Dick[4] (1923), proved that scarlet fever was caused by beta hemolytic streptococci.   The disease agent gains entrance to the nose and throat of susceptible individuals by direct contact with organism carriers. After 2-5 days the throat becomes red, intensely sore, a temperature of 101-104°F. or more develops along with a headache.   Vomiting is sometimes seen.   The beta hemolytic streptococci gain entrance to

---

[3]Griffith, F.:   The Serological Classification of *Streptococcus pyogenes*, J. Hyg. **34:**542, 1934.

[4]Dick, G. F., and Dick, G. H.:   Scarlet Fever, Chicago, 1938, The Year Book Publishers.

the blood stream, and are deposited, along with toxins (particularly the erythrogenic toxin) in the skin and other organs. A rash becomes visible within a day or two; at first, slight pink dots, then turning to a scarlet. The tongue is slightly swollen and red, resembling a large strawberry. The body is a pinkish color, except around the mouth, which is a pale in comparison, this being called circumoral pallor. The patient is very toxic and ill. Complications may follow, such as middle ear infections, and kidney and heart damage. Scarlatina is frequently used to denote mild cases of scarlet fever, the after effects being less severe than the usual scarlet fever.

Immunity develops against the toxic principles, but immunity to the bacterial cell is not as secure or lasting. Repeated sore throats are not uncommon, but second attacks of scarlet fever with rash are uncommon. The injection of antitoxic serum into the skin of the patient will bleach the rash, this being called the *Schultz-Charlton* test. Also a skin test is available, called the Dick test, to determine antitoxic immunity or susceptibility to the toxin. One-fiftieth of a skin test dose of toxin is injected intradermally, and if immunity is present no reaction results because antitoxic antibodies neutralize the injected toxin. If the individual is not immune a red, slightly swollen area will develop at the site of the toxin injection within 24 to 58 hours. Subclinical scarlet fever, or mild sore throats due to a beta hemolytic streptococcus, which produces small quantites of the erythrogenic toxin, may produce a solid immunity in the individual. Such immunization by having had a mild disease without the rash is a fairly common occurrence in the population.

This disease is a winter and early spring disease, the highest incidence being in February, March, and April, with comparatively few cases in July, August, and September. The disease agent is spread by direct contact with those harboring the organism in the nose and throat, the nasal carriers being perhaps the most important. The streptococci may be contracted through contaminated water (swimming particularly), dishes, bedding, etc. Therefore isolation of the patient as well as sterilization of all articles handled by a patient is important.

The diagnosis of this disease is by clinical signs and symptoms, and isolation of beta hemolytic streptococci from the nose and throat of the patient.

Some physicians advocate prophylactic immunization with toxoid, but this technique has not proved too successful and is not widely used.

Treatment is by the use of sulfonamides and antibiotics, and some recommend the use of antitoxic serum in severe cases.

Rheumatic Fever.—This disease may develop following repeated attacks of septic sore throat caused by beta hemolytic *Streptococcus pyogenes*. There is a suggestion of a sensitization of body tissues resulting in an allergic component not unlike atopic-like hypersensitivity. Swollen tender joints, slight fever, and later heart valve damage are seen in this disease. Whether streptococci are the cause alone, or in conjunction with some other microorganisms, is not known.

Puerperal Fever.—When the female genitals are infected with group A beta hemolytic *Streptococcus pyogenes*, after childbirth, the disease puerperal fever (childbed fever) produced is severe and fulminating. Blood stream invasion, with septicemia and a high fever, is frequent, and a very toxic patient results.

Erysipelas.—This is an infection of the skin due to beta hemolytic streptococci, producing oozing, draining, scaly lesions of the skin, which frequently spread rapidly from one area to another by self infection.

Subacute Bacterial Endocarditis.—This disease, characterized by a low-grade fever and ulcerative lesions on the heart valves, is frequently caused by the alpha (greenish hemolysis) hemolytic streptococci of the viridans group (*Str. salivarius*, *Str. mitis*).

PNEUMOCOCCI

This group of microorganisms was first observed by Pasteur, but it was not until later that Frankel and Weichselbaum proved that they were the cause of lobar pneumonia. *Diplococcus pneumoniae* is the accepted name of this group.

These organisms are gram-positive diplococci occurring frequently in chains, slightly lancet-shaped with the pointed ends outward from their mate. Virulent forms are capsulated. Neither spores nor flagella are formed.

Culturally they are moderately fastidious, multiplying more vigorously on blood- or serum-containing media. Some types will grow in synthetic media containing a reducing agent, certain vitamins (nicotinic acid and pantothenic acid particularly) plus certain critical amino acids and dextrose. Colonies are slightly larger than those of streptococcus, and are alpha hemolytic or green-producing on blood agar. The majority are aerobic, preferring a temperature of 35-37°C. A few species are, however, obligate anaerobes.

A few sugars are fermented with the production of lactic acid and other compounds. They do not produce catalase, so that minute quantites of hydrogen peroxide produced are collective and eventually kill the organisms. They ferment inulin and are soluble in bile. The bile solubility aids in differentiating the pneumococci from the alpha hemolytic streptococci.

The pneumococci are classified by the kind of capsular material they possess. The capsular material is a soluble polysaccharide designated as "SSS" or specific soluble substance. This specific soluble substance is antigenically different and is immunologically specific for each of the 75 or more types known. The body cells or somatic antigen are homologous and are the same immunologically for each type. Immune serum produced against the various specific capsular substances will cause a swelling of the capsule, the *Quellung* reaction. Some types (1, 2, 3, - - - - - - 75 - - - -) are interrelated, but, in general, classification into types may be accomplished by the *Quellung* test.

One unusual characteristic, however, is that we may convert one type into another. If type 1 is grown in broth and then heat killed and a noncapsulated type 3 or other type is grown in this type 1 capsular environment, the previous type 3 will now produce antigenically specific type 1 capsules.

If the capsule is lost (rough colony type) by subculturing or other processes, the organism loses its ability to produce infection, except in huge doses. By animal passage (mice) recapsulation is again evident. Capsules (smooth colony type) then bear a definite relationship to virulence and invasiveness. Ribonucleic acid in a polymerized form is probably responsible for this type change.

Diseases Caused by Diplococcus Pneumoniae.

The most serious disease produced by this group of microorganisms is lobar pneumonia. Types 1, 2, 3, 7, and 8 produce the greatest number of human infections. The disease agent is transmitted directly from carriers or active cases through sputum droplets, coughing, or sneezing.

The first symptoms, after a short incubation period, are cough, fever, and tight feeling in the chest. Shortly, pain in the chest begins and becomes progressively more severe. The cough becomes severe and a rusty, prune juicelike sputum is coughed up. Difficulty in breathing is frequent.

Diplococcus pneumoniae may be isolated by culture and mouse inoculation from the sputum, and in about one-half the cases organisms can be cultured from the blood stream. Identification of types is unnecessary since antiserum is no longer used in therapy. Antibiotics and sulfonamides are the treatments of choice.

One attack usually confers a fairly solid immunity for the immunologic type causing the infection, but not for other types. Vaccines, using particularly SSS, are fairly successful for prevention of the disease.

Infections other than lobar pneumonia may be caused by the pneumococcus group of microorganisms. Eye infections, meningeal infections, kidney infections, middle ear and sinus infections are some of these. Prompt treatment of the above types usually effects a rapid cure.

Pneumococcus infections, particularly lobar pneumonia, are more frequent in the winter and spring months. The disease is spread primarily by direct contact, coughing, sneezing, etc., from carriers and acute cases of the disease.

Meningitis and Gonorrhea

THE GRAM-NEGATIVE COCCI

NEISSERIA

The generic name, *Neisseria*, of this group is derived from the first person to describe a species of this group. Neisser, in 1879, repeatedly found kidney-shaped gram-negative cocci in gonorrhoeal pus. A similar organism was found in pus from a meningeal inflammatory disease in 1884 by Marchiafava and Celli, but was later identified as a separate species.

Two species of the genus *Neisseria* are highly pathogenic, *N. gonorrhoeae*, and *N. meningitidis* (*N. intracellularis*) while the others are only mildly pathogenic, if at all. *N. catarrhalis* and *N. sicca* are frequent normal habitues of the nose and throat, and probably play no or very little role in infectious processes. Several species may produce pigment, and may at times produce infections (*N. parflava, N. flava, N. subflava*, and *N. flavescens*).

Structure

All of the *Neisseria* are gram-negative, bean- or biscuit-shaped cocci, flattened in diplococcal fashion against their neighbor. In pus they are frequently seen in the cytoplasm of leucocytes as well as extracellular. Some strains may be capsulated. They are nonmotile.

Cultivation

The *Neisseria* are fastidious in their growth requirement, growing best on blood-enriched media which has been heated to liberate hemo–globin. A variety of synthetic media have been devised for culture of this organism containing particularly glutathione, casein digest, cystine, etc. A temperature of 35-37°C. is necessary as well as a water-saturated atmosphere. For initial isolation carbon dioxide in about a 10 per cent concentration is required, although they are strictly aerobic. After several culture transplants removed from a human source, they will grow without carbon dioxide. Adequate growth is obtained in 24-48 hours.

Colonies are about the size of pneumococci, are glistening, and frequently have a slight blue gray appearance. Blood is not hemolyzed.

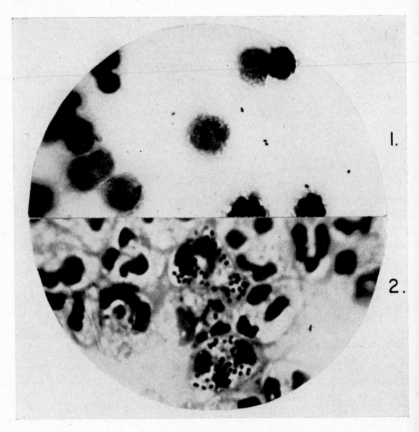

Fig. 28.—*1, Neisseria meningitidis* in spinal fluid. Note the large number of lymphocytes. *2, Neisseria gonorrhoeae* in pus. Note the large numbers of microorganisms in the cytoplasm of the leucocytes.

These organisms are not active biochemically; *N. gonorrhoeae* ferments only glucose with the production of lactic acid; *N. meningitidis* ferments maltose and glucose. Catalase and indophenol oxidase are produced. The presence of indophenoloxidase may be detected by the addition of tetramethyl-p-phenylenediamine (1 per cent aqueous) to the colonies. A group of color reactions take place, at first a light rose, then deep blue rose, an finally a deep purple.

Although colonies of other genera of microorganisms may produce this color reaction, the gram-negative diplococci may be determined by the Gram stain. This test, therefore, aids in picking colonies of the genus *Neisseria* from a plate on which several different genera of microorganisms may be growing.

Diseases Produced by N. Gonorrhoeae

This microorganism is the cause of the venereal disease of humans, gonorrhea. It is usually transmitted by direct sexual contact, but may be transmitted by indirect methods, as contaminated fingers, clothes, instruments, etc. A fairly constant discharge of pus is characteristic in infected persons. Not all infections producing pus discharges from the genitals is gonorrhea, and a proper diagnosis is necessary.

In males the disease agent affects nearly all parts of the genital tract. In females the cervix uteri and the oviducts are the principal areas for localization of this microorganism. Sterility may be produced in both sexes by this infection. Immunity following infection is of a low order and probably lasts only a short time.

Other infections may be caused such as joint infections, heart valve infections, etc., which are usually secondary to the initial genital involvement. Ophthalmia neonatorum an infection of the eyes of newborns may be caused during birth, contracted from the infected mother. This disease may cause blindness. Eye instillation of silver salts or penicillin usually prevents such infections.

Diseases Produced by N. Meningitidis

The most serious disease produced by this microorganism is a meningitis or an inflammation of the coverings of the brain and spinal cord. When this disease attacks many individuals in a given area it is called *epidemic* cerebrospinal meningitis.

The microorganism is spread by direct contact, by nasopharyngeal droplets from carriers of the microorganism. They lodge in the nasopharynx, multiply, and produce a sore throat. The organisms then gain entrance to the blood stream, then to the meninges of the nervous system. Temperature may vary from 101° to 105°F. with stiffness of the neck and back muscles and vomiting. Headache is usually present. Frequently a pink rash develops over the body from which microorganisms may be isolated.

Spinal fluid from these cases frequently shows great numbers of pus cells producing a ground-glass appearance, instead of the water clear spinal fluid of normal persons. Microorganisms are seen, in stained specimens, both intracellularly and extracellularly in these fluids. Prompt diagnosis is necessary for prompt treatment to spare the patient's life.

Sulfonamide drugs and penicillin is the treatment of choice.

The disease, epidemic meningitis, is more frequent in the winter and spring months, the highest incidence being from January through May. Individuals may become carriers of this microorganism in their noses and throats but show no symptoms. Usually a carrier rate of 20 to 30 per cent is required before many active cases of the disease develop. When the carrier rate of the population reaches 80 to 90 per cent, then an epidemic usually breaks out.

Prevention of the disease is by restricting too close sleeping of groups (e.g., in army barracks, etc.) by alternating head and foot of bed. Isolation if possible of carriers and cases and the prophylatic use of sulfadiazine is employed. Closing of schools and other places of gathering is probably not necessary unless a severe epidemic is evident. The disease agent usually attacks children and young adults but may attack adults.

Immunity following an attack is fairly solid, although second attacks may occur, probably with an antigenically different species. Subclinical or inapparent infections probably account for the immunity or resistance in adults. Vaccination is of doubtful value.

Selected Reading References

Aycock, W. L., and Mueller, J. H.: Meningococcus Carrier Rates and Meningitis Incidence, Bact. Rev. **14**:115, 1950.

Hedrich, A. W.: Recent Trends in Meningococcal Disease, Pub. Health Rep. **67**:411, 1952.

Hodges, R. G., MacLeod, C. M., and Bernhard, W. G.: Epidemic Pneumococcal Pneumonia. III. Pneumococcal Carrier Studies, Am. J. Hyg. **44**:207, 1946.

Hotchkiss, R. D.: Transfer of Penicillin Resistance in Pneumococci by Desoxyribonucleate Fractions From Resistant Cultures, Cold Spring Harbor Symposia on Quant. Biol. **16**:457, 1951.

Kligler, B., and Blair, J. E.: Correlation Between Clinical and Experimental Finding in Cases Showing Invasion of the Blood Stream by Staphylococci, Surg., Gynec. & Obst. **71**:770, 1940.

Powers, G. F., and Boisvert, P. L.: Age as a Factor in Streptococcosis, J. Pediat. **25**:481, 1944.

Rantz, L. A.: The Prevention of Rheumatic Fever, Springfield, Ill., 1952, Charles C Thomas, Publisher.

Rogers, D. E., and Tompsett, R.: The Survival of Staphylococci Within Human Leucocytes, J. Exper. Med. **95**:209, 1952.

CHAPTER 22

BACTERIAL DIARRHEAS AND DYSENTERY

The Enteric Bacteria and Cholera

At least eight genera comprise this group of microorganisms that frequent the gastrointestinal tract of man and animals. Some are not harmful, others are capable of producing disease. All species are gram-negative and species of seven of these genera are motile; some form capsules and some do not. None of this group form spores. All are rod forms except one, this being a vibrio or curved rod. The most important types of diseases caused by this group are, diarrhea and dysentery, and bacteremia or blood stream invasion. All species are aerobic and will grow readily on ordinary culture media over a rather wide temperature range (15-40°C), the optimum being 35-37°C.

ALCALIGENES

This genus comprises several species frequently found in the human intestinal tract, milk, and food. Only one specie, *A. fecalis*, produces mild infections of the genitourinary system and occasionally a very mild gastroenteritis. Morphologically and culturally it may be confused initially with the disease agent of typhoid. *A. fecalis* is motile and produces an alkaline reaction in milk. It has no fermentative ability on carbohydrates.

PROTEUS

There are only four species in this genus, *Pr. vulgaris* and *Pr. morganii* being the most important. These species of microorganisms are frequent inhabitants of the normal flora of the gastrointestinal tract. They may cause mild chronic infections of the gastrointestinal tract, kidneys, ulcers of the skin and are fairly frequent secondary contaminants of other types of infections, acting as synergists.

They possess active enzymic systems, acid and gas being produced from glucose; gelatin being liquefied and milk acidified at first, then alkalinized and curdled and later peptonized. Gelatin is rapidly liquefied and urea is hydrolyzed.

A rather important characteristic of this organism is its ability to spread from a few colonies, completely covering a culture plate as a fine mistlike growth. The active flagella (H) are responsible for this spread because of their great motility.

Another important characteristic is the fact that the Proteus X strains contain antigenic substances that are similar to antigenic components of rickettsia. Antibodies produced by certain rickettsial infections will agglutinate these proteus X strains, this test being called the *Weil-Felix* reaction.

PSEUDOMONAS

Several species comprise this genus, only one of which is mildly pathogenic. Most of the species are found in water, milk, soil, and other organic material.

P. pyocyanea is a pigment producer, showing a blue green coloration on the media. The blue pigment *pyocyanin* is predominant, while the yellowish fluorescent pigment is not readily observed. These organisms produce a potent pyrogen.

This organism is a motile gram-negative rod. It grows readily on simple cultural media at a rather wide temperature range (10-40°C.). It ferments dextrose with acid, but no gas is produced. It usually produces a mild alkalinity in media containing an adequate source of nitrogen.

This organism may produce a variety of mild infections such as infant diarrhea, ear infections, and occasionally genitourinary infections. It may also produce a "blue pus" beneath plaster of Paris casts.

AEROBACTER

Two species, *A. aerogenes* and *A. cloacae*, are closely related to *Esch. coli* and are frequently found in the gastrointestinal tract, water, food, milk, soil and, on occasions, in infections.

Aerobacter aerogenes is a short, plump gram-negative rod frequently capsulated. It is usually nonmotile.

Acid and gas is produced from a number of sugars including lactose, glucose and maltose. It does not produce marked acidity; therefore the methyl red (MR) test is negative. Acetyl methyl carbinol is not produced, therefore the Voges-Proskauer (V.P.) test is negative. Sodium citrate can be utilized as a source of carbon. Indol is not produced.

ESCHERICHIA

This genus was first described in 1886 by Escherich and contains several species; only one, however, *Esch. coli*, will be discussed. This organism is found universally in the intestinal tract of humans and lower animals. Any water, soil, food, etc., contaminated with fecal discharges will contain *Esch. coli*.

Esch. coli is a motile, gram-negative rod, without capsules or spores. It tends to exhibit pleomorphism and may vary from 0.4 to 0.8 micron wide and 1.5 to 4 or more microns long.

The organism grows readily on simple laboratory media and will also grow quite well on synthetic media. A rather wide temperature range will support growth, from 15-40°C. It is aerobic but will grow moderately well under strict anaerobic conditions.

Esch. coli has a well-developed group of enzymes and is capable of fermenting, with the production of acid and gas, several carbohydrates, including lactose, dextrose, and maltose. The *IMViC* test is applied (methyl red test, production indol, acetyl methyl carbinol production, and citrate utilization), which separates *Esch. coli* from *A. aerogenes* (see Chapter 13 on water bacteriology).

There are at least three specific antigenic components in the coliform microorganisms. (1) A heat stable somatic O antigen; (2) cell wall antigen, or K antigen, of which there are three components: L, a thermolabile antigen; A, a thermostable antigen; and B, a thermolabile antigen differing from the L antigen. The K group tends to block agglutination of the somatic (O) antigen with specific antisera; (3) flagellar of *H* antigen. Another organism of this group, frequently referred to as a *paracolon*, is similar to *Esch. coli*, but the enzyme systems are not as well developed, and lactose fermentation is frequently delayed several days or even two weeks or more.

Esch. coli is useful as a tracer organism, and when found in water, food, etc., is an indication of fecal contamination.

Although normally present in the intestinal tract of man and animals, it is at best a mild disease producer. Chronic and acute genitourinary tract infections are probably the most common.   Occasionally appendicitis and bowel abscesses may be caused by this microorganism.   It is also found, on occasions, associated with staphylococcal and streptococcal infections and acts symbiotically with these types of microorganisms, aiding in producing a more severe type of infection.

KLEBSIELLA

This genus was first described by Friedländer and frequently goes by the discoverer's name, Friedländer's bacillus.   However, *Klebsiella pneumoniae* is the accepted name.   Several species have previously been assigned to this genus, but recent studies show that they are a different antigenic type of the same species.

K. pneumoniae is a capsulated, nonmotile gram-negative rod, slightly more plump than *Esch. coli.*

The organism grows readily on simple laboratory media, the colonies being rather large, mucoid, and frequently resemble large drops of rain water.   It is aerobic and prefers a temperature of 35-37°C. but will grow over a temperature range similar to *Esch. coli.* Acid and gas are produced by most strains from several carbohydrates, but the biochemical activities are so variable that indentification by this means is not accurate.   Most strains ferment lactose with acid and gas.   The V-P test is negative and the methyl red and citrate tests are positive.

Three antigenic types have been described, types A, B, and C, type A being the principal one infecting humans and type B found in animal infections.   An unclassified group X comprises about 15 per cent of the strains isolated.

This organism produces infections primarily of the respiratory tract and sinuses.   Pneumonia produced by this disease may be fatal unless it is promptly treated.   Thick mucoid material produced by the capsular substance is difficult to dislodge.

SALMONELLA

In this genus there are over one hundred ninety-six serological varieties, most of which are pathogenic for man.   Microorganisms of this genus are the most frequent causes of food poisoning and also produce epidemics due to fecal contamination of water supplies.

The first species of this genus was discovered by Eberth in 1880, and was, until 1948 named *Eberthella typhosa*, the causative agent of typhoid fever.   The generic name is now *Salmonella*.

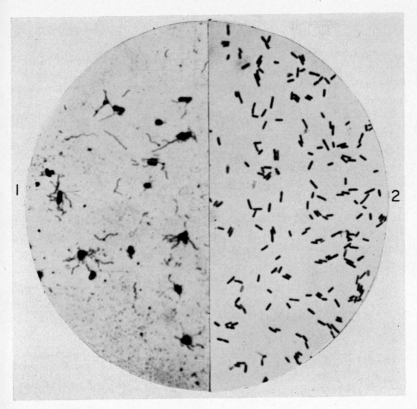

Fig. 29.—*1, Salmonella typhosa* showing flagella.   *2, Shigella dysenteriae.*

All species of this genus are gram-negative noncapsulated rods. All are motile except two species, (*Sal. gallinarium* and *Sal. pullorum*) these being pathogenic for animals, but rarely pathogenic for man. None of the species ferment lactose.   This is an important differential in distinguishing the *Salmonella* from the *Esch. coli*.   Indol is not produced.   Several carbohydrates and higher alcohols are fermented, many species producing both acid and gas.

Typhoid Fever

Typhoid fever is frequently a seasonal disease or "vacation disease," the greater number of cases being seen more frequently in the summer and early fall months. The disease organism, *Salmonella typhosa*, is spread by contaminated water and food. A small per cent of the people who recover from the disease may be chronic carriers of the organism. The organisms may grow in the gall bladder, liver, bile ducts, kidneys, or intestinal tract in these convalescents and their fecal material may contain virulent microorganisms for the life of the individual.

After ingesting *Salmonella typhosa* organisms the patient is without symptoms until six or seven days, when there is generally vague abdominal pains and a mild fever. The fever gradually increases and the patient becomes acutely ill. The dangers of this disease are the damage done to the intestinal tract, with frequent ulcerations with bowel rupture.

Organisms may be isolated during the first week of the disease from the feces, blood stream, and sometimes from the urine. An agglutination test, called the Widal test, using the patient's serum and known typhoid organisms is generally positive after the second week. The Vi or virulent phase organism must be heated before agglutination is obtained.

Prevention is by vaccination and maintaining sanitary water and food supplies.

Other Organisms of the Paratyphoid Fevers and Food Poisoning Group

This group may present a disease process resembling typhoid fever and others resemble dysentery-like diseases.

The most frequent and probably the most serious diseases of this group are caused by *Sal. paratyphi A*, *Sal. paratyphi B*, *Sal. enteritidis*, and *Sal. typhimurium*. Other species are less frequently encountered. These are all actively motile. Acid and gas are produced from several sugars and alcohols by most species.

Like typhoid fever, the paratyphoid fevers are summer and early fall diseases being charaterized by vomiting and diarrhea. Water and food contaminated by fecal material is the most frequent method of spread. The organisms may penetrate the gut via the lymphatics and enter the blood stream like *Sal. typhosa*, but more frequently they remain localized in the intestinal tract.   The bodies of these organisms are toxic (endotoxin) and even when killed are able to produce gastroenteritis.

Classification of the Salmonella

The microorganisms of this genus possess a multiplicity of different antigenic components of both the body (somatic) and flagella.   These are designated O (somatic) and H (flagellar) antigens.   Several different specific antigens may be found in any one given bacterial body or flagella.   Likewise many species of bacteria of this genus have antigens in common and as a result are closely related immunologically.   Somatic (body) antigens are designated by Roman numerals; flagellar antigens are designated by arabic letters and numbers.

For example the antigenic formula of *Sal. typhosa* is IX, XII, [Vi]: d:—.   The antigenic formula of *Sal. paratyphi* B is: [I] IV, [V], XII: b: 1, 2.   Antigen I and V may be absent in some strains. The application of immunological methods, using the agglutination technique and agglutination adsorption is necessary for exact classification.

SHIGELLA

The genus *Shigella* is comprised of many species, only a few of which will be briefly discussed.

The microorganisms of this genera are all gram-negative, nonmotile and all are capable of producing disease.   Gas is not produced on carbohydrates or higher alcohols, but many of these chemicals are attacked by some of the enzymes of the bacteria, producing an acid reaction in the media.   They are divided into two primary group termed mannitol fermenters or nonmannitol fermenters.

MANNITOL FERMENTERS	NONMANNITOL FERMENTERS
Sh. paradysenteriae	Sh. dysenteriae
Sh. sonnei	Sh. parashiga
Sh. alkalescens	Sh. ambigua
Sh. dispar	

Members of this genus produce a rather large number of our usual "summer diarrhea" cases.   These microorganisms are frequently spread by contaminated water and food and are also frequently spread by flies and rodents.   The disease is characterized by mild to severe diarrhea, with only mild fever.   Some may produce bloody diarrhea.

One member of this genus, *Sh. dysenteriae*, produces a potent exotoxin, thus producing a toxic condition in the patient. Only rarely do the bacteria of this genus invade the blood stream, but remain in the intestinal tract.

TABLE XV

Seed Differential Media
↓
Fish colorless colony
↓
Seed to Russell's or Kligler's iron (Stab and Slant)

Acid and gas in butt
Slant alkaline
↓
Gas-producing *Salmonella*

Acid in butt; alkaline slant
↓
Sal. typhosa or *Shigella*
(No gas produced)

Seed to Biochemical Media

SPECIE	LACTOSE	MALTOSE	XYLOSE	ARABINOSE	MANNITOL	SUCROSE	TREHALOSE	INDOLE	ADONITOL	GROUP	ANTIGENIC TYPES
Sal. paratyphi A	O*	AG‡	O	AG	AG	O	AG	O		A	1
Sal. paratyphi B (Schottmuelleri)	O	AG	AG	AG	AG	O	AG	O		B	31
Sal. typhimurium	O	AG	AG	AG	AG	O	AG	O			
Sal. cholera suis	O	AG	AG	O	AG	O	O	O		C	50
Sal. enteritidis	O	AG	AG	AG	AG	O	AG	O		D	23
Sal. typhosa	O	A	A	O	A	O	A	O			
Sh. dysenteriae	O	O	O	O	O	O		O	A		1
Sh. ambigua	O	O	O	O	O	O		P			1
Sh. paradysenteriae	O	O	O	O	A	O		O			8 Flexner types 6 Boyd types
Sh. sonnei	A†	A	O	A	A	A		O	O		2

O* = no reaction.

A† = acid.

AG‡ = acid and gas.

Identification of the Salmonella and Shigella

Stool and blood specimens are samples of choice in *Salmonella* infections and stool specimens are more suitable for *Shigella* infections. Urine samples may also be taken. Blood samples are seeded directly to enriched broth (Selinite F, tetrathionate, and others). Stool and urine samples are seeded on agar plates containing a special selective media (S.S.; E.M.B.; bismuth sulfite, MacConkey agar, and others). Isolated colorless colonies are fished and seeded into various biochemical media (see Table XV).

Final identification of most of the *Salmonella* sp. and *Shigella* sp. is determined by the agglutination technique. Several *Salmonella* typing centers are equipped to do the more complicated antigenic analyses, a few of these centers being: Kentucky Experiment Station; U. S. Public Health Service Laboratory, Alabama; New York Typing Center and the Serum Institute, Copenhagen, Denmark.

VIBRIO

Although the genus *Vibrio* is generally not classified with the enteric group of microorganisms, it will be briefly discussed in this chapter because the species of this genus produce severe diarrhea.

Vibrio cholerae is a gram-negative, slightly curved rod, highly motile and has a tendency to multiply so as to produce several organisms end-to-end, the comma being reversed with each organism, the net result being that the group resembles a spirillum.

This organism prefers an alkaline medium (pH 8) and an abundance of oxygen. It will grow over a wide temperature range (16-42°C.); therefore, in relatively warm water with high contamination this organism will undergo some multiplication.

Asiatic cholera, the disease caused by this microorganism, is transmitted to people by water and food. This disease is fairly common in many parts of Asia and Asia Minor, but a few epidemics have occurred in the United States.

A very severe diarrhea is produced in cholera patients, with flecks of intestinal epithelium in the stool. These stools are frequently called rice water stools. Great amounts of fluid are lost during the severe diarrhea, resulting in sunken eyes and leathery wrinkled skin of the patient. Immunity on recovery lasts several years.

Control of this disease is by vaccination and by sanitary control measures of culinary water supplies and adequate sanitary measures of handling food.

Selected Reading References

Kauffmann, F.: The Diagnosis of Salmonella Types, American Lecture Series No. 62, Springfield, Ill., 1950, Charles C Thomas, Publisher.

Linton, R. W.: The Chemistry and Serology of the Vibrios, Bact. Rev. **4:**261-319, 1940.

Report of the Scientific Advisory Board, Indian Research Fund Association, New Delhi, India, 4-21, 1944.

Weil, A. J.: Medical and Epidemiological Aspects of Enteric Infection, Ann. Rev. Microbiol. **1:**309-332, 1947.   C. E. Clifton, Ed., George Banta Publishing Company, Menasha, Wis.

CHAPTER 23

PLAGUE, WHOOPING COUGH, AND BRUCELLOSIS

1. PLAGUE AND TULAREMIA

The microorganisms comprising this group are all nonmotile gram-negative rods and moderately to strictly fastidious in their growth requirements. All are aerobic, and grow best at 30-37°C. The organisms of this group have probably been responsible, in the past, for more severe illness and death in man and animals than any other group of pathogenic bacteria.

Plague.—Caused by *Pasteurella pestis.*

Bubonic plague or Black Death presents a horrible and terrifying example of the lack of public health work and of medical knowledge in the fourteenth century and up to the nineteenth century. In one year (1348-1349) over 25 million Europeans perished. In London in 1665 over 70,000 perished of this disease. In 1894 a moderately severe epidemic raged in Hong Kong, which spread by 1900 via infected fleas on rats on board ship to Scotland, Australia, and United States. Over 500 cases have appeared in the continental United States since 1900. Only an occasional case appears in the United States at the present time, but in India and China the disease is endemic with many cases appearing annually.

Pasteurella pestis.—This organism is a short, plump coccobacillary rod, possessing a capsule and is nonmotile. It is gram-negative and by special methods exhibits polar staining. Irregular and vacuolated forms are seen frequently.

P. pestis grows quite well on nutrient broth and agar, but enriched media enables the organism to multiply more profusely. Temperatures slightly above room temperature (25-30°C.) are more favorable than body temperature (37°C.) (98.6°F.). Biochemical reactions on sugars are undependable as a means of absolute identification.

Identification by agglutination with specific antiserum and specific bacteriophage lysis is more reliable.

The disease agent, *P. pestis*, is transmitted to man primarily by the bite of the rat flea *Xenopsylla cheopis*, and *Ceratophyllus fasciatus*. Fleas bite an infected rat, the organism then reproduces in the gut of the flea with no great damage to this insect. The flea now bulging with *P. pestis* organisms bites man and the regurgitation of bacteria from the intestinal tract of the flea infects man. Rubbing the organisms on the shaved or scarified skin of rats or guinea pigs will produce infection. Sylvatic plague, an endemic and epidemic disease of rodents (squirrels), has been found in the Western part of the United States since 1908, although in Asia it was known previously to this. Many areas have been decimated of ground squirrel populations because of sylvatic plague.

The disease, bubonic plague, is characterized by swollen lymph nodes which sometimes ulcerate, high fever, and generalized capillary hemorrhages. The blood hemorrhages into the skin and internal organs which turn dark or even black, accounting for the name Black Death. The pneumonic type is frequently transmitted from a sick person to a normal person by sputum droplets. On recovery, a patient enjoys complete immunity for life. The death rate of untreated bubonic plague is about 75 per cent, and of pneumonic type it approaches 100 per cent.

Tularemia.—Caused by *Pasteurella tularensis*.

Tularemia is a naturally occurring disease of animals, primarily rodents, rabbits, deer, etc., and man is only an accidental host. The disease is frequently referred to as "rabbit fever" and "deer fly fever." It was first described in man by a Utah physician in 1911 and the organism *Past. tularensis* was first isolated from a ground squirrel in Tulare County, California, in 1911.

Pasteurella tularensis.—This organism is a tiny, gram-negative coccobacillus, with a tendency to pleomorphism, rod forms being predominant in old laboratory cultures. Capsules are present in freshly isolated organisms. These organisms do not possess flagella.

This species of organism requires special cultural media, blood-cystine-dextrose agar being the most suitable. It may be cultured in casein hydrolysate broth, to which has been added liver extract, biotin, and red blood cell extract. Shaking during broth cultivation enhances growth. The organisms are aerobic, but will grow feebly under microaerophilic conditions. Several carbohydrates are fermented, dextrose, mannose, and glucose showing acid but no gas.

Identification is more dependable by the agglutination test and mouse inoculation tests for isolation of the microorganism.

This organism is one of the most invasive known, a single organism being capable of producing a fatal infection in mice. Rubbing the organism on the shaved skin will produce infection in man or laboratory animals.

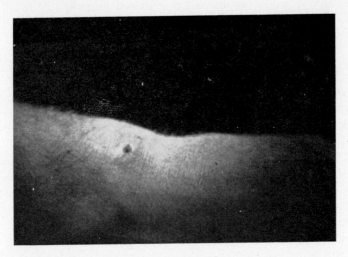

Fig. 30.—Tularemic ulcer near the wrist. Note the dark center.

The disease is frequently contracted by man as a result of handling infected rabbits; thus hunters and those dressing wild rabbits are the most common victims. The disease also may be transmitted by deer flies (*Chrysops discalis*), wood ticks (*D. andersoni*), and several other insects. Certain mosquitoes may transmit the disease by biting.

It has been observed that when a rabbit infected with this bacterium becomes moribund, if deer flies are in the area, these flies swarm by the hundreds on the helpless animal. As soon as the rabbit's heart stops beating, the flies leave as if by a signal. Wild rabbit populations in the Western United States have been greatly decreased by this disease. Epidemics in beavers and muskrats have been reported in the Rocky Mountain area of the United States.

Prevention of this disease is primarily by avoiding handling wild animals that have been found dead, or dressing freshly killed animals that have appeared ill before they were killed.

Vaccination offers some protection, but the immunity produced is undependable. Two means of immunization have possibilities: (1) Foshay's whole organism vaccine; (2) Larson's extract vaccine.

2. THE HEMOPHILIC GROUP

This genus, *Hemophilus*, comprises several species, most of which are capable of producing diseases in man and animals. Although the name implies "blood loving" most of the species will grow without blood if the proper culture medium is utilized.

Hemophilus Influenzae

This organism, *H. influenzae*, was first isolated with such regularity from cases of influenza that it was assigned as the cause of this disease. True influenza is caused, however, by a virus, and many bacteria, including *H. influenzae*, are secondary invaders.

This organism is a gram-negative rod, rarely more than 0.3 micron wide, but varies in length from 0.8 to 2 or more micra when seen in direct smears in fluids from infected patients. Colonies are tiny (0.5 to 1.5 mm. diameter) gray white, and produce a characteristic mustylike odor. A capsule is frequently present on freshly isolated strains of Types A and B. Growth requirements are exacting, blood generally being necessary for adequate growth. Two growth factors X (hematin-like) and V (coenzyme-like) are both found in blood. Fresh rabbit blood added to a suitable base medium and heated to 90° C. for 3-4 minutes to release the hemoglobin, produces a "chocolate colored" agar when hardened, which supports good growth. Some strains are hemolytic.

The organism is only mildly active biochemically; some strains produce indol and some ferment dextrose.

Six antigenic types are known, Types A through F, a specific capsular polysaccharide conferring specificity. Capsular-swelling antiserum (anticapsular antibody) will produce a swelling of the capsule similar to that seen with pneumococcus and its specific capsular antiserum. Capsular material in *H. influenzae* is quite soluble and is often not detected after 6-10 hours of cultivation. Fresh animal or human source organisms possess adequate capsules which may be preserved with formaldehyde for future capsular-swelling tests. The smooth forms are the virulent forms possessing capsules, while rough forms are acapsular and avirulent.

A number of disease entities may be produced by this organism, including eye infections, pink eye, (Koch-Weeks bacillus), but the most serious is a meningitis. *H. influenzae* meningitis approaches 100 per cent fatality without treatment. Present-day therapy (sulfa and streptomycin), when instituted promptly, affects cures in 95 per cent or more of the cases.

Other organisms closely related to *H. influenzae* are *H. parainfluenzae*, *H. suis* (associated with swine influenza), *Moraxella lacuanata*, and *H. ducreyi* (cause of soft chancre or chancroid, which may be confused with primary syphilitic chancre).

Hemophilus Pertussis

Whooping cough or pertussis is perhaps the most serious disease that infants under one year of age can contract. This disease is caused by *Hemophilus pertussis*.

This organism is a gram-negative, coccobacillus, frequently capsulated, growing more luxuriantly on blood-containing media. Bordet-Gengou medium is the medium of choice for original isolation, although this organism will grow on artificial medium suggested by Cohen and Wheeler.[1] The organism is hemolytic: colonies on B-G medium are pearl-like, varying from 1-4 mm. in diameter, and are sticky when a loop needle is touched to the colony. Fermentation reactions are variable, but dextrose and lactose are frequently fermented, with acid but no gas formation. The agglutination and complement fixation tests are the most dependable diagnostic tests.

Four antigenic types are recognized, referred to as Phases 1, 2, 3, and 4. Phase 1 is smooth and virulent. Thus, only Phase 1 organisms are used in the preparation of vaccines, since phases 2-4 are devoid of capsular material.

This organism is highly communicable and a single exposure of nonimmunes is sufficient to produce an infection. The incubation period is about two weeks after exposure, the first symptoms being like those of a slight "common cold" with a nocturnal cough. As the disease progresses, the small air passages in the lungs become plugged with the "sticky" capsular material. Coughing becomes more severe, with "spasmodic" or interval coughing spells with a distinctive whoop or crowing sound on inspiration. At times the "lack of oxygen" may produce damage to the nervous tissue. Vomiting may occur. The death rate in infants below one year of age may approach 25 per cent.

---

[1]Cohen, S. M., and Wheeler, M. W.: Am. J. Pub. Health **36**:371, 1946.

The disease is fairly easily prevented by specific vaccination. Vaccination should be started by the second or third month of life, particularly during the whooping cough season, early winter through the spring months. Vaccination may not prevent the disease in all cases, but those vaccinated individuals who do contract the disease usually have a very mild infection.

3. THE BRUCELLOSIS GROUP

Brucellosis.—Caused by *Br. melitensis, Br. abortus,* and *Br. suis.*

Although the disease we now know as brucellosis was first reported in 1859, the causative bacterium was not described until 1887 by Bruce. The disease was known at that time as Malta fever and undulant fever. The *Br. abortus* species was isolated in Denmark by Bang in 1897 from abortion material in cattle. The disease in cattle is frequently referred to as Bang's disease. The species common to swine was isolated in 1914. Organisms of the genus *Brucella* are gram-negative coccobacilli.

These organisms are moderately fastidious when first isolated from a sick person or animal, and require special types of media, trypticase-soy agar, or milk-tryptose-crystal violet medium being superior to simple liver infusion media. Synthetic media must contain at least 18 amino acids plus added salts and vitamins. Carbon dioxide is required by most strains of *Br. abortus* but the other two strains propagate well in atmospheric oxygen. Colonies on enriched agar media are about 1 to 3 mm. in diameter, slightly convex with an amorphous smooth appearance. Growth is optimum at temperatures of 35-36°C.

Enzyme systems are not well developed in members of this genus. Only a few carbohydrates are attacked, with acid but no gas being produced. Nitrates are reduced and hydrogen sulfide is produced in variable amounts. Catalase is produced by all three species.

Identification is primarily dependent on dye tolerance and immunological tests. Thionin 1:800 inhibits the growth of *Br. abortus;* neither of the other two species are affected by this dye concentration. Basic fuchsin 1:200, pyronin 1:800, or methyl violet 1:400 inhibit growth of *Br. suis,* but have no effect on the other two species. Agglutination absorption tests will differentiate *Br. melitensis* from the *abortus-suis* group. Three antigens appear to be present—somatic A and M and a virulent antigen L. Brucellergen, an extract of the organisms, when injected intradermally will produce a local reddening and swelling if the patient has the disease.

The disease agent is spread in several ways, the most common being direct contact with a sick animal or its body discharges (aborted placenta, feces, urine, etc.). Accidental laboratory infections are sufficiently common to make the worker take every possible precaution. The organism may penetrate small abrasions of the skin. Ingestion of contaminated milk (goats, cows, etc.) which is unpasteurized accounts for many cases of this disease in man each year. Eating cheese and other dairy products made from contaminated unpasteurized milk and cream is another way in which the organisms are spread. Rats are also considered as carriers, particularly of the *suis* species.

The disease is rather slow in developing, frequently appearing 2 to 4 weeks or longer after infection. The disease at first resembles a common cold with influenzal-like symptoms, low fever, headache, chills, sweats, and a general feeling of weakness. Constipation is frequent. The disease is often undulating, with a debilitating period of several weeks or longer, then remissions, then a second attack. The disease may continue off and on for several months to even years. This disease therefore is of an acute and chronic nature.

Prevention of the disease is by means of the eradication of infected cattle, sheep, and goats. Testing programs (agglutination and other tests) to locate infected animals is an essential public health method for the detection and eradication of the disease. Pasteurization of all milk and milk products is an essential preventive. Education of handlers of livestock, hides, offal, etc., also aids in decreasing this disease, since it is an occupational disease in packing house workers, farmers, hide dealers, etc. Vaccination of calves has decreased, to a certain extent, the spread of this disease within herds.

The purchaser of dairy cattle or other livestock should be certain of a veterinarian's certificate that the animals purchased are free of this disease. The *Br. abortus* not only affects man, but produces great loss in the dairy industry in causing abortion and loss of milk production. Infected bulls may pass this organism via the semen.

Selected Reading References

Brucellosis, A Symposium, AAAS, 1950, Washington, D. C.

Gordon, J. E., and Kines, P. T.: Flea Versus Rat Control in Human Plague, Am. J. M. Sc. **213**:362, 1947.

Huddelson, I. F.: Brucellosis in Man and Animals, New York, 1943, The Commonwealth Fund.

Larson, C.: Immunization of Guinea Pigs With a Soluble Antigen Obtained from *Br. abortus*, J. Immunol. **63**:471, 1949.

Zinsser, H.: Rats, Lice and History, Boston, 1935, Little, Brown and Company.

CHAPTER 24

THE MOST RESISTANT BACTERIA

AEROBIC AND ANAEROBIC SPOREFORMERS

1. AEROBIC SPOREFORMERS—GENUS BACILLUS

The sporeforming aerobic bacteria are classified under the generic name *Bacillus*. Unfortunately, many rod forms of bacteria are frequently referred to as "bacillus"; this term does not necessarily mean the true genus *Bacillus* which are sporeformers and prefer oxygen tension equivalent to that of atmospheric oxygen.

A number of species (about 33) comprise this genus, all except one of which are harmless to man and animals. The majority are soil and water organisms which aid to a certain extent in soil fertilization.

All of the members of the genus *Bacillus* are large gram-positive rod-shaped bacteria, possessing endospores. These endospores will withstand boiling for from several minutes to several hours. Many will remain viable in the soil and in water for many years. They are many times more resistant to chemicals than their parents, the vegetative cells. Nearly all are motile except the pathogenic species *B. anthracis*.

Perhaps the most written about species of this genus is the *Bacillus subtilis*, the one which plagued Pasteur in many of his experiments. It is widely disseminated and found universally in hay, therefore, frequently referred to as the hay bacillus. The organism is a common contaminant of cultures in the laboratory and fairly frequently is found on the skin of humans, thus sometimes interferring in diagnostic bacteriology. The endospores are resistant to heat and chemicals and sunlight, thus their constant presence on dust particles in the air, as well as in other areas.

Bacillus subtilis grows readily on practically all laboratory media· The colonies on solid media are large, fringed-edged, irregular to round and are frequently confluent. They grow over a wide temperature range, thus great numbers are found in brakish water, soil, etc.

The organism possesses moderate enzymic activity, ferments some carbohydrates, hydrolyzes starch, reduces nitrates, and produces acetylmethyl carbinol.

The organism is gram-positive, possessing elliptical to cylindrical endospores, usually central, but some may be terminal. Special stains are required to stain these spores effectively, but an ordinary Gram stain will enable one to see "vacant" or hollow areas in the vegetative cell, which are the unstained endospores. Spores may be discharged from the bacterial cell, and when specially stained appear as tiny round to ovoid bodies. The organism is large—0.7-0.9 microns wide by 4-8 microns long, frequently seen in chain formation.

This organism is not pathogenic, but is useful as a sensitivity test organism for standardizing penicillin.

Other saprophytic members of this genus frequently encountered are *B. megatherium*, *B. mycoides*, *B. firmus* and others.

The only pathogenic species of this genus is *B. anthracis*, the causative agent of the disease anthrax. Pasteur produced one of the first types of bacterial vaccines with this organism, initiating the concept of preventive vaccination against a specific bacterial disease.

This microorganism is a large gram-positive rod, nonmotile, frequently appearing in long chains resembling a jointed pole. Endospores are formed. In freshly isolated cultures from animals or man the organism is capsulated.

Growth on a variety of culture media is easily obtained. The colonies are large and grizzly looking. Unlike most other bacteria, organisms from smooth type colonies are generally less virulent, while rough colonies more frequently represent highly virulent microorganisms. The organism grows readily at room temperature, but very slowly below 10°C. Moderate growth is obtained at 41-42°C. or 4 to 5°C. above body temperature. An adequate oxygen supply is essential and poor growth results when the oxygen tension is reduced only slightly. Dextrose and trehalose are fermented within a few hours with acid production, but no gas. Lactose, mannitol, and several other sugars are not fermented. Milk is curdled and peptonized. No exotoxin is produced.

The organism, or tissue infected with the organism, when boiled produces a soluble antigen which will react to specific antiserum, the Ascoli test.

The disease is a natural infection of certain animals (sheep, cows, swine, etc.) but may accidentally be transmitted to man.   By selective breeding, certain species of sheep have developed a rather high natural resistance to the disease.   Humans handling livestock, hides, or wool have the highest morbidity rate.   Veterinarians may contract the disease, most frequently from injuries of the hands or arms during post-mortem examinations of diseased cattle, sheep, etc.   The infection in man is more frequently seen on the skin, but pulmonary and gastrointestinal cases do occur.   In watching a skin lesion develop in a human, it resembles a mosquito bite for a day or so.   The area then reddens, with a serum ooze, then there is a gradual color change to black.   Blood stream invasion is moderately frequent, and unless promptly treated, death may result.

The disease may be prevented by adequate immunization of livestock—and care in handling livestock suspected of having the disease. Disposal of dead livestock by burning prevents ground contamination, but pastures which become contaminated may remain dangerous to livestock for several years.   Decontamination of wool, animal products, bristles of hogs, etc., from infected animals, is essential before they can be processed.

2. ANAEROBIC SPOREFORMERS—GENUS CLOSTRIDIUM

This group of microorganisms is classified in the genus *Clostridium* and all species are anaerobic or microaerophilic, form endospores, and are gram-positive rods.   Many species inhabit the soil as saprophytes and are of benefit in soil fertilization.   Several species which also may be found in the soil are highly dangerous, producing disease in man and animals.   The endospores of this group are extremely resistant to heat; some of them may withstand boiling water for two to four hours.   All of the disease-causing species produce potent specific exotoxins.

Atmospheric oxygen inhibits the growth of all the Clostridia, and cultural methods excluding atmospheric oxygen must be used. Reducing agents added to adequately prepared culture media, such as sodium thioglycollate, cysteine, etc., will enable these strict anaerobes to propagate.   Microorganisms of this genus lack catalase, peroxidase, cytochrome, and cytochrome oxidase enzyme systems.

The clostridia may be classified into three general groups according to the type of disease produced in man: (1) *Cl. tetani*, the lockjaw or tetanus organism (10 types); (2) *Cl. botulinum*, the food poisoning or intoxication group (5 types); (3) *Cl. perfringens*, the gas gangrene group (4 types of *Cl. perfringens* and 15 associated species producing mild to severe gangrene). In cultures all produce a foul, nauseating odor.

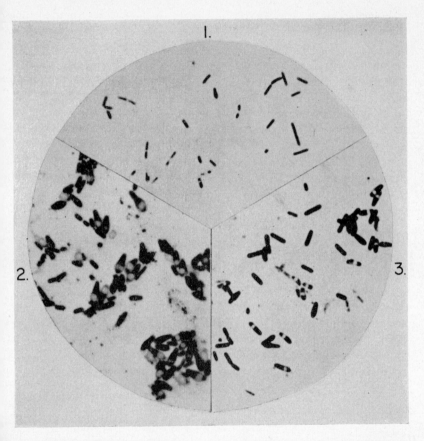

Fig. 31.—*1, Clostridium perfringens.   2, Clostridium tetani.   3, Clostridium botulinum.* Note the light staining areas in the bacterial cells which are endospores.   Note the marked bulging of one end of the *Cl. tetani* organism, produced by the large terminal endospores.

Clostridium tetani is a large, gram-positive motile rod, with a terminal spore giving a clubbed-end effect. It fails to ferment carbohydrates. Colonies on blood agar show *beta* hemolysis, are granular, irregular, and spread rhizoid-like a short distance from the mother colony.

A potent exotoxin is produced, causing hemolysis (tetanolysin) and also a severe toxic action on nerve cells (tetanospasmin). Purified toxin[1] contains over 6 million lethal doses per milligram for mice.

Despite several somatic antigenic types the toxin from all types is homogenous and a specific antitoxin against one will neutralize toxin from all other strains of the species.

Tetanus or lockjaw is produced by the introduction of *Cl. tetani* organisms or spore into the tissue, more frequently in deep penetrating wounds, but, at times, superficial wounds promote growth of this microorganism. Knife wounds, gunshot wounds, stepping on nails, severe burns etc., may introduce this organism and cause tetanus. Firecracker burns used to be a frequent cause of tetanus. The organism is localized but the toxin is spread throughout the body in the blood stream. The toxin attacks nerve cells and produces at first a tightening of the muscles of the back and neck, then the jaw. Soon thereafter muscle spasm develops and the jaw muscles set tight. Immediate treatment is necessary to prevent death; without treatment the death rate approaches 90 per cent.

Prevention is primarily achieved by vaccination with tetanus toxoid. Prophylactic antitoxin at the time of injury is also an important factor to suppress a possible infection. Since spores are found in the soil, in manure, etc., any injury in which *dirt* may be introduced should be considered potentially dangerous and a physician should be consulted immediately.

Clostridium botulinum is a short, plump gram-positive motile rod, with a subterminal spore causing a slight cell wall bulging. This organism was first isolated by Van Ermengen in 1896 from spoiled ham.

The organism is not fastidious, growing on a variety of cultural media, alkaline vegetables and meats, so long as anaerobic conditions exist. Milk is peptonized and hydrogen sulfide is produced. It is hemolytic on blood agar and the colonies (1-4 mm. diameter) are mostly smooth, but irregular serrate margins may be seen. Some types are proteolytic and some are not. A great amount of gas is produced when sugars are fermented and proteins broken down.

This organism produces the most potent exotoxin known, one milligram being sufficient to kill 20 million mice. Five specific types are known, A-E, each producing a specific toxin, and each type toxin being neutralized only by its specific antitoxin.

---

[1]Picket, M. J., et al.: J. Bact. **49**:515, 1945.

The disease "botulism" is an intoxication and only rarely an infection. Preformed toxin is ingested, more frequently found in inadequately home processed vegetables. Rarely are commercially canned foods contaminated. Heating opened canned foods to boiling for 15 minutes destroys the toxin, and the bacterial spores and vegetative cells can be ingested with impunity. Symptoms in man may begin from 12-18 hours after ingestion of the spoiled food. More frequently, however, 24-36 hours elapse before any effect is noted. The legs feel heavy. There is no fever. Only mild gastrointestinal distress is noted.

A change in vocalization of words, to a guttural, thick-tongued word pronunciation is one of the early symptoms. Double vision is frequent at the early stage. Loss of equilibrium is frequent in advanced stages of the disease. The death rate is high without specific antitoxic treatment. Types A and B are the most frequent types causing the disease in humans. Type A is the most common on the Pacific Coast and in the Rocky Mountain region.

Prevention is primarily abstinence from eating possibly spoiled food—particularly food from bulged cans or bottled food that is under pressure when opened. Occasionally a cheesy smell may be detected, but frequently no off odor is noted. In the analysis of 10 samples in our laboratory of food that proved fatal to humans, only two were found to be sufficiently off odor to be suspicious. Always boil home bottled or canned vegetables that may present the slightest suspicion of being spoiled. Home canning of peas, corn, beans, spinach, swiss chard, etc., should be done in a pressure cooker.

A disease occurring naturally in wild ducks in the marshes frequently kills from 5,000 to 20,000 ducks in a single epidemic. Type C *Cl. botulinum* is the causative organism. In slow flowing or stagnant shallow marsh waters, the *Cl. botulinum* is deposited in the mud and debris. *Ps. pyocyanea* growing in the same tiny pocket furnishes an alkaline substrate, and reduces the oxygen content. Sufficient foodstuff is available for slow propagation of the *Cl. botulinum*, particularly during hot summer days. Ducks feeding in the mud absorb sufficient toxin to produce severe illness and death. In the summer of 1952 it is estimated that between 10,000 and 20,000 wild ducks developed the disease in the marshes north of Salt Lake City, Utah. They were burned by the truckload. Prevention is by adequate flow of fresh water through the marshes. Type C apparently has not caused the disease in man.

Clostridium perfringens, frequently referred to as the "Welch bacillus" (*Cl. welchii*), is a moderately short, plump, gram-positive, nonmotile encapsulated rod. The spore is generally centrally placed and does not produce bulging of the bacterial wall. This organism was first isolated by Welch and Nuttall in 1892 from a gangrenous wound. This organism produces large amounts of gas from certain carbohydrates (lactose, maltose, and sucrose); mannitol is not fermented. Milk is coagulated and the lactose fermentation is so vigorous that the clot is blown to bits—the so-called "stormy fermentation" of gas gangrene cultures.

Four antigenic types are known, A-D, all of which produce some exotoxins and enzymes. The following toxins may be produced by one or more types, but not all types yield all of the exotoxins and enzymes: hemolysin, hyaluronidase, fibrinolysin, necrotoxin, collagenase, and lecithinase. Specific antitoxin can be produced against the various exotoxins. The greater part of the lethal activity of the toxic principle is probably due to lecithinase. This toxin is assayed by determining the lecithinase activity by the application of the *Nagler reaction*. Lecithin of egg yolk is split into phosphocholine and diglyceride.

All of the other members of the gas gangrene group, *Cl. novyi*, *Cl. histolyticum*, *C. septicum*, etc., are capable of producing a mild form of gas gangrene in man.

The disease, gas gangrene, is a disease following severe wounds, including compound fractures, in which soil or contaminated material is introduced into the wound. As the *Cl. perfringens* grows, much gas and toxic material is produced—thus gas in the tissues, and beginning death of the latter. Tissue death may become so extensive, if treatment is not instituted, that a part must be amputated. Severe wounds should be given attention immediately by a physician, specific antitoxin and other measures being necessary.

The prevention of this disease is primarily by immediate treatment of a severe wound by a physician. Vaccination with toxoids has not yet proved effective in man, although experimentally, in animals, it has proved moderately effective. The *Cl. perfringens* spores are found universally in the soil, throughout the world, thus injuries where soil is introduced into the wound are the most frequent types producing gas gangrene.

NONSPOREFORMING ANAEROBES

Although this group of organisms do not technically belong with the sporeformers, their anaerobic nature may allow them to be studied as a group following the gram-positive, spore-bearing anaerobes.

The nonsporeforming rod form of anaerobic microorganisms are classed in the genus *Bacteroides*. They are gram-negative, nonsporeforming, filamentous rods that are nonmotile. Most species are strict anaerobes; a few are microaerophilic. One species (*Bact. melaninogenicum*, newer terminology—*Bacteroides nigrescens*) is so fastidious that it is difficult to obtain in pure culture on blood or chocolate media. This species produces black colonies. Other species (*B. fragilis, B. funduliformis*, etc.) are easily grown on laboratory media.

Organisms of the genus *Bacteroides* comprise a great part of the intestinal bacterial flora of man. They may also be isolated from indolent skin abscesses, pulmonary abscesses, etc. Occasionally they may produce septicemia.

Selected Reading References

Dack, G. M.: Food Poisoning, Chicago, 1943, University of Chicago Press.

McLennan, J. D.: Anaerobic Infections of War Wounds in the Middle East, Lancet **2**:94, 1943.

Meyer, K. F., and Dubovsky, B. J.: The Distribution of the Spores of *B. botulinus* in the United States. IV. J. Infect. Dis. **31**:559, 1922.

Tytell, A. A., Logan, M. A., Tytell, A. G., and Tepper, J.: Immunization of Humans and Animals With Gas Gangrene Toxoids, J. Immunol. **55**:233, 1947.

CHAPTER 25

DIPHTHERIA

The disease diphtheria was named by Bretonneau in 1826,[1] "la diphtherite," the English version, diphtheria, meaning in Greek "prepared hide or leather." The disease undoubtedly existed many centuries previous to the initial description given by Bretonneau. The causative microorganism was first described morphologically by Klebs in 1883 and cultured by Loeffler in 1884. The organism is to this day frequently referred to as the K-L organism.

The microorganism causing the disease is *Corynebacterium diphtheriae*. It is a moderately thin aerobic gram-positive rod, without spores, capsules, or flagella. It is frequently club-shaped, with internal granules called Babs-Ernst granules. Arrangement on stained microslides is fairly bizarre, some arranged parallel in picket or palisade-like arrangement; some form bizarre shaped letters of the alphabet such as X, Y, H, etc., this being due to the snapping of the organism during binary fission. Special staining is required to reveal the internal structure, Loeffler's methylene blue, Neisser's or Albert's stain being the most commonly used.

C. diphtheriae grows poorly on plain nutrient agar, but grows vigorously on agar enriched with blood or serum. The addition of pantothenic and nicotinic acids aids the growth of most strains. Coagulated serum media as advocated by Loeffler is probably the most commonly used medium, but blood agar is equally suitable. Loeffler medium may tend to inhibit some of the many microorganisms associated with the disease process in man. Colonies on Loeffler's media are small (1-3 mm.) and granular, while on blood agar they are considerably larger. On media containing potassium tellurite the colonies are dark gray or black. An alkaline medium (pH 7.5-8.0) and an abundance of

---

[1]Bretonneau, P. F.: Des inflamations speciales du tissu mugneux et eu particulier de la diphtherite, ou inflammation peculiare, connu sous le nom de croup, d' angine maligne, d'angiene gangreneuse, etc. Paris, Crevot. 1826.

oxygen at a temperature of 34-36°C. favors optimum growth and toxin production.

Three types are known, *gravis, intermedius,* and *mitis.* All three types produce a potent exotoxin, each of which is homologous. The *gravis* produces the greatest quantity, *mitis* the least. All three types

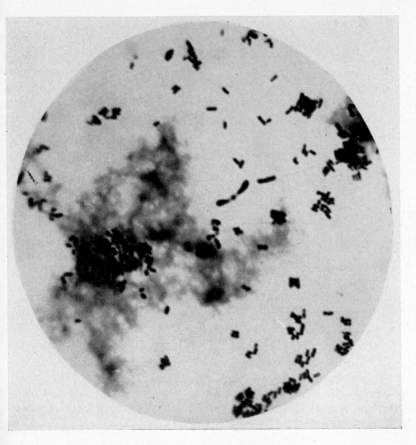

Fig. 32.—*Corynebacterium diphtheriae.*

ferment maltose and dextrose with acid production, but no gas. *Intermedius* and *mitis* types fail to ferment dextrin and starch, while the *gravis* types regularly ferment these two chemicals. The *mitis* type strains usually produce hemolysis on blood agar; the other two generally do not.

The disease is a local infection with a generalized intoxication. From two to four days after exposure (from a carrier or a person actually ill with the disease) the patient shows a slight fever and sore throat. The organisms grow on the mucous membranes of the throat, tonsil, nose, trachea, and bronchi. By their local growth and toxin production they produce a gray, sometimes cream-colored leathery membrane. Toxin is liberated locally and passes into the blood capillaries which extend into this membrane. The toxin produces injurious effects on nerve cells, periphereal nerve fibers, heart muscle, and certain glandular tissues. Paralysis of the muscles of swallowing may ensue, causing an inability to swallow, imbibed fluid being regurgitated through the nose. Immediate antitoxic treatment along with penicillin is life saving. Recovery produces a fairly lasting immunity in most individuals.

After recovery, the patient should have two negative throat cultures before being allowed to associate with his fellows. A virulence test is made on the organism, using two guinea pigs, one injected with 500 units of antitoxin and the other without toxin. Intracutaneous injection of a culture of the organisms produces a reddening within 24-48 hours in the unprotected animal. The *mitis* strain may take 72-96 hours to produce local reddening and edema in this test.

Antitoxin is produced in horses by giving increasingly large doses of toxin over long periods of time. After adequate immunization, by preliminary tests of the antitoxic qualities of the serum, the horse is bled (about 4 liters), the serum separated, purified, and standardized. One unit will neutralize about 100 guinea pig lethal doses of the toxin.

Prevention of the disease is by vaccination. A variety of detoxified toxins are available, toxin-antitoxin mixtures, formaldehyde-treated toxin (anatoxin), and alum-precipitated toxoid. The latter is the most commonly used immunizing agent. Immunization is started early in life (2 to 6 months of age) and after successful immunization as indicated by the Schick test (1/50 guinea pig lethal dose of toxin injected intradermally), yearly single booster immunizations are given for a period of three or four years. This artificial immunization may wear off in later years. If the child is exposed to diphtheria a booster injection of toxoid may aid in the prevention of the disease. Prevention of the disease, if an unvaccinated child has been exposed to the organism, may be accomplished by prophylactic use of antitoxin.

The disease, diphtheria, is spread from person to person as the most common method. Some persons may be "carriers" or they may have in their nose and throat a virulent *C. diphtheriae* organism, they themselves being immune. Coughing, sneezing, etc., of a person harboring diphtheria organisms may distribute this organism widely in a room, thus exposing many, if it is a schoolroom or other place currently densely populated. All nonimmune age groups are susceptible, age being no barrier to this microorganism. Contaminated milk, food, bedding, toys, and other articles may aid in the spread of this disease.

The most susceptible groups are from 3 months to 2 years of age. Exposure to a mild organism, or one that fails to produce sufficient toxin to produce the disease, has an immunizing effect. By the age of 20 years, even without artificial vaccination, about 70 to 80 per cent of those individuals are immune. During this process, however, without vaccination, many cases of diphtheria will develop in the children, with numerous deaths, as has happened in the past.

Diphtheria is endemic throughout the year, but there is a gradual increase in the number of cases beginning in August, with the greater number of cases occurring in October, November, and December in temperate climates.

NONPATHOGENIC SPECIES OF CORYNEBACTERIUM

One of these nonpathogenic forms, *C. hoffmanii*, may be not infrequently confused with true *C. diphtheriae*. Its morphology is similar to the true diphtheria organism. However, bulging of the subterminal portion of *C. hoffmanii* is not present. The granules within the cell are usually smaller and fewer in number. It is inclined to be a little more plump than the true diphtheria organism. It is less active biochemically, failing to produce exotoxin and failing to ferment any of the carbohydrates or higher alcohols. It is fairly frequently found in the normal throat and associated as a saprophyte in skin infections.

The other one of this group, *C. xerosis*, is occasionally isolated from normal throats, conjunctivae, and as a saprophyte in skin wounds. It is short and plump, often presenting a bipolar type staining. No extoxin is produced. Both dextrose and sucrose are fermented with the production of acid only.

Selected Reading References

McLeod, J. W.: The Types Mitis, Intermedius and Gravis of Corynebacterium Diphtheriae, Bact. Rev. **7**:1, 1943.

Pappenheimer, J., Jr., and Lawrence, H. S.: Immunization of Adults With Diphtheria Toxoid. III. Highly Purified Toxoid as an Immunizing Agent, Am. J. Hyg. **47**:241, 1948.

CHAPTER 26

TUBERCULOSIS AND LEPROSY

THE ACID-FAST BACTERIA

TUBERCULOSIS

Tuberculosis and leprosy have probably plagued man for over 5000 years.[1] Repeated reference to leprosy is made in the Bible. Aristotle described the disease fairly accurately in 345 B.C. The disease tuberculosis has been known by a variety of names, miners' tuberculosis, consumption, phthisis, etc. Tuberculosis is widespread throughout the world, the greatest morbidity being in the American Indians living on reservations.

The causative tuberculosis organism was discovered by Robert Koch in 1882. The tubercle bacterium (*Mycobacterium tuberculosis*) is a short, plump, gram-positive, acid-fast (Ziehl-Neelsen stain) nonmotile, acapsular, and nonsporeforming rod, that is frequently curved slightly and possesses granular material called Much's granules. On continued artificial cultivation the organisms may become somewhat more slender and elongated. The organisms contain an acid-fast wax and phospholipid (saturated and unsaturated fatty acids).

, Two main types pathogenic for man are known, *M. tuberculosis* var *hominis* and *M. tuberculosis* var. *bovis*. Culturally *M. tuberculosis* organisms are "slow growers" frequently requiring two or more weeks for visible growth to appear. On special types of media, however (Dubos media), visible growth may be obtained in 18-24 hours. The organisms are moderately fastidious in their growth requirements and initial isolation requires specialized types of media containing egg, potato, serum or serum albumen fraction, and certain fatty acids (particularly the water-soluble esters of fatty acids, Tween 80). Colonies of the human type are slightly wrinkled, dry looking and brittle, varying in color from a light brown to a cream gray (eugonic colonies).

---

[1]Webb, G. B.: Tuberculosis, New York, 1936, Paul B. Hoeber, Inc.

Frequent passage on different types of media may alter colonial structure. Bovine type grows poorly on egg type media (dysgonic). The optimum temperature for growth is about 37°C., but this bacterium will grow at slightly higher or lower temperatures. Carbohydrates are of no value in identification. The human type utilizes glycerol, while the bovis type does so only slightly. A rather pleasant fruity odor is produced in cultures of *Mycobacterium*.

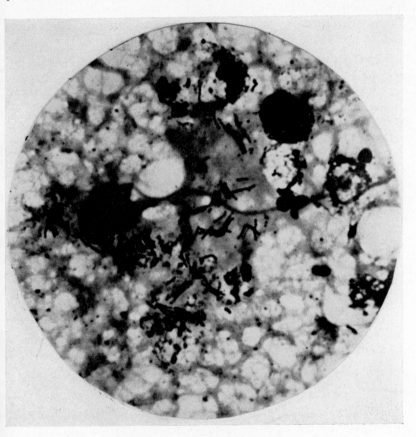

Fig. 33.—*Mycobacterium tuberculosis* var. *hominis* in sputum.

Serologically the human and bovine variety are indistinguishable, but by agglutination adsorption tests four antigenic types can be determined—the saprophytic type, cold blooded animal type, avian type, and the mammalian type.

This group of microorganisms is highly resistant to most physical and chemical agents—3-4 per cent sodium hydroxide for several hours; 2 per cent mineral acids for several hours, chlorine compounds for several hours and 11 per cent trisodium phosphate for many hours. Advantage is taken of this resistance in separating and purifying the tubercle bacillus in sputum and body fluids in which mixtures, many contaminating bacteria are present. Temperatures of 65°C. for a half hour (143-145°F.) are sufficient to kill these microbes and temperatures used in milk pasteurization are adequate for killing these pathogenic microorganisms.

The pathogenic mammalian varieties will attack practically all mammals and some birds. No tissue structure may be spared. Tuberculosis of the bones, skin, kidney, bladder, brain, etc., are fairly common, but pulmonary tuberculosis leads the infection sites in man. Man and primates are highly susceptible to both the human and bovine varieties. The avian variety will infect not only many fowl species, but also a few mammal species, swine, rabbit, and horse.

The disease in man is insidious and usually chronic, the infected person being unaware of the disease until it has developed rather extensively. A gradual decrease in physical vitality, moderately decreased appetite with slow weight loss, and a slight afternoon fever are common initial complaints. When cavity formation in the lungs is evident, coughing paroxysms are frequent, particularly in the early morning as the patient awakens. A continued hacking cough or evidence of flecks of blood in the sputum is sufficient evidence for a thorough medical checkup. In pulmonary tuberculosis the bacteria are found in the sputum and in gastric washings, since many people, on coughing, swallow small amounts of sputum. Dihydrostreptomycin and isonicotinic acid are fairly active against this organism.

Tuberculosis of the bones, particularly of the spine, may produce a deformed or "humped" back.

Tuberculosis is a highly contagious disease being spread from man to man by coughed or sneezed droplets. Hundreds of thousands of the tubercle bacilli are discharged into the air by a pulmonary patient who coughs. People living in poorly ventilated quarters or several persons living in inadequate space are frequently infected in this manner. Tuberculosis is not a hereditary disease, but handling of a child by a tubercular mother or father frequently (almost invariably) spreads the disease to the child.

The disease may also be spread by other contaminated body wastes such as urine, pus from tubercular ulcers, etc. Also milk from an infected cow can spread the disease unless the milk is properly pasteurized.

Immunity or resistance in this disease seems to bear a close relationship to the general health of the patient. Mild infections tend to produce a resistance to the more severe infections if the patient is generally well nourished and physically in good condition.

Control of the disease is isolation and treatment of infected persons; control of dairy herds and proper pasteurization of milk. The tuberculin test (extracts of the tubercle bacteria—PPD—purified protein derivative) which is an intradermal injection of the tuberculin merely tests whether the individual has had the disease, or is in the progress of the disease or has not had the disease. Proper medical and x-ray examinations plus bacteriological examinations are necessary to determine the presence of active infection. Vaccination with an avirulent strain (H37 Ra) using the BCG technique seems to have merit. The World Health Organization has, with the assistance of the international tuberculosis campaign, tuberculin-tested nearly 40,000,000 persons and nearly 17,000,000 have been vaccinated with BCG. A rather high percentage of adults in the United States are tuberculin positive, but of these many are "healed" tuberculosis. Recent campaigns by mobile x-ray units throughout the United States have discovered a great number of hitherto unknown cases of tuberculosis in men, women, and children. Continued efforts and x-ray surveys and other methods will be required to remove this disease from near the top of the list as killers of man. More concerted effort must be made among the lower economic classes in both general education and health education and medical care, since the morbidity and death rates are very much higher in this group than in the middle or higher economic classes of our population.

LEPROSY

The word "leprosy" frequently frightens people, but the disease is neither as serious nor as prevalent as tuberculosis. There are about three million sufferers from the disease in the world, with about 750 of these in the continental United States. The principal focus is the Orient, with the Pacific Islands and South America being the next most frequent areas in which the disease is present. The probable cause is *Mycobacterium leprae*, an acid-fast rod resembling the tubercle bacillus.   The leprae bacillus was discovered in 1878 by Hansen in tissues from a patient.   The disease is now widely designated as Hansen's disease.

The organism purported to be the cause of human leprosy has never fulfilled Koch's postulates.   It has not been successfully cultivated and no experimental animal has been found susceptible.   A type of leprae bacilli, causing rat leprosy, has been cultivated and the disease in rats has been studied to some extent.

The disease in man is transmitted only with very close association for long periods of time of infected with uninfected individuals. The incubation period may be months in some cases, years in others. Two forms are known—the nodular type producing thickened skin areas and nodules and the nerve type, in which the peripheral nerves are destroyed (anesthetic type).

The disease is not hereditary, but children are more frequently infected than adults.   The disease progresses slowly and most patients live a normal life span.

Prevention is by segregation and isolation.

Several leprosaria are in existence which have been financed by the United States government: one in the Philippine Islands, one in the Hawaiian Islands, one at Carville, Louisiana, and one in Puerto Rico. Many leprosaria in other parts of the world are supported by the American Mission to Lepers.

SAPROPHYTIC MYCOBACTERIUM

Probably the most common of this group is *Mycobacterium phlei*, found in soil, hay, etc.   It produces a yellowish to orange pigment, grows readily on ordinary laboratory media and produces good growth within two to three days.   The *M. butyricum* is frequently associated with butter.   *M. smegmatis* is a saprophyte found frequently around the geinitals of the male and female.   All of the above organisms are nonpathogenic, acid-fast rods, morphologically similar to the true tubercle bacillus.   These may be confused with the true tubercle bacillus, particularly in simple stained smears, using the acid-fast technique.

Selected Reading References

Aronson, J. P.: Protective Vaccination Against Tuberculosis With Special References to BCG Vaccination, Am. Rev. Tuberc. **58**:255, 1948.

Guralnick, L., and Glaser, S.: Tuberculosis Mortality in the United States, Pub. Health Rep. **65**:468, 1950.

Johnson, F. A., et. al.: Promacetin in Treatment of Leprosy, Pub. Health Rep. **65**:195, 1950.

Lurie, M. B.: Native and Acquired Resistance to Tuberculosis, Am. J. Med. **9**:591, 1950.

Wilmer, H. A.: Huber the Tuber, New York, 1943, National Tuberculosis Association.

Wilmer, H. A.: This is Your World, Springfield, Ill., 1952, Charles C Thomas Publisher.

CHAPTER 27

SYPHILIS AND TRENCH MOUTH GROUP

The Spirochetes

A rather large group of genera of microorganisms belong to the order *Spirochaetales*, many of which are disease producers in both man and animals. Probably the most important organism of this group was brought from America when Columbus returned to Europe before the beginning of the sixteenth century. At least the disease syphilis began to be seen in epidemic form about this time.

GENUS TREPONEMA

Several species of microorganism comprise the genus *Treponema*.

Treponema Pallidum

T. pallidum, the causative agent of syphilis, is the most common spirochete in this group which produces disease in man. This microorganism was first discovered by Schaudinn and Hoffman in 1905. The name syphilis was, however, assigned this disease by the Italian poet-physician Fracostoro in the early 1500's, when he wrote a poem about a shepherd by the name of Syphilis who had the venereal disease we now call syphilis.

The *Treponema pallidum* are spiral organisms resembling the teeth of a 10 point saw. They appear at times to resemble a loosely coiled spring, from 5 to 15 or more microns in length and very thin, rarely over 0.25 to 0.35 microns in thickness. They are gracefully motile, by means of tufts of flagella, undulating and diving slowly in and out of the field of vision in a dark-field microscopic preparation. Their progression across the field of vision is slow, but deliberate. Their whole motion resembles a dignified hula dancer, with much graceful movement in various planes, but slow progression from one point to another. Staining is not accomplished by the ordinary types of stains used for bacteria, but special stains, using silver, gold, etc., are used. These organisms are anaerobic, but are not cultivatable in ordinary laboratory media or special media, including tissue culture techniques.

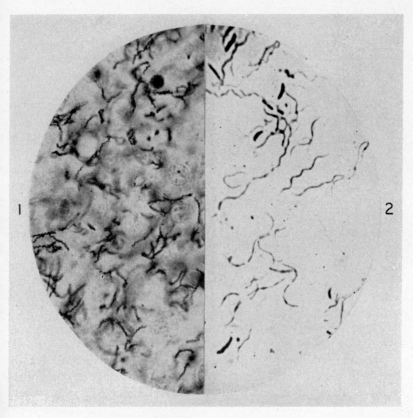

Fig. 34.—*1, Treponema pallidum. 2, Borrelia vincenti*; note the large fusiform bacteria in the upper part of the photograph.

The *T. pallidum* are sensitive to heat and drying, being inactivated in a short period of time (few hours to a few minutes). Cold has a preserving effect, solid carbon dioxide (-70 to $80°C$) will preserve the organisms for several years. In our laboratory, using "full" capillaries, we have preserved these organisms from human cases for over seven months in the deep freeze ($-25°$ C.), and on thawing they were motile.

Syphilis

Syphilis is primarily a disease of man, but may be experimentally produced in rabbits or higher apes. The disease in man is a venereal disease, being transmitted most frequently sexually, but the organisms may infect other areas if there are abrasions (fingers, lips, mucosa of the mouth). Contaminated, unsterilized instruments may also transmit the disease.

The disease in man usually progresses in three stages, but variations of each may be seen. The first or "primary" stage is the formation of a firm, round, buttonlike smooth nodule on the genitals. No pain is involved. This is seen from a few days to a week or two after exposure. This hard chancre contains large numbers of *T. pallidum* and they can be found in the exuded serum, by use of the dark-field microscope.

The "secondary" stage may evolve slowly, taking months or years before one is aware of its presence. The spirochete has now invaded numerous tissues of the body including the blood vessels and peripheral nerves, etc., producing many symptoms. A rash with fever is frequently seen; white patches in the mouth and other mucus membranes, severe itching, shooting pains are all frequent symptoms. A blood test (STS—serological test for syphilis) is positive at this time. A complement fixation test (Wassermann test) or a flocculation test (Kahn, Hinton, etc.) are the tests used. This serological test is reversed (becomes negative) after adequate treatment.

The "tertiary" stage of syphilis is the stage in which nervous tissue is damaged, and usually other organs are also involved.

Damage to parts of the spinal cord makes it difficult for the patient to walk if he does not watch his feet. Damage to parts of the brain frequently produces insanity. This latter stage may not be seen until many years after the primary infection.

If we could assign intellectual aptitudes to pathogenic microorganisms, the *T. pallidum* would come under the genius group. These organisms not only fail to kill their host quickly, but frequently produce such minor symptoms that the patient fails to seek medical aid in order to eradicate these microbes. A syphilitic patient may enjoy a normal life expectancy. The latter part may, however, be in a dream world apart from reality. Some of our fine literature has been written by syphilitics in the beginning stages of neurosyphilis.

Serological tests for syphilis are used in the third stage, using spinal fluid as the test fluid, for a diagnosis.

The disease is not inherited, but may be contracted in utero from a syphilitic mother. Adequate treatment during early pregnancy will prevent in utero infection. Many congenital syphilitics are born dead. Those that survive frequently have a variety of deformities and abnormalities. Hutchinson's teeth, saddle nose, saber shins, deafness, cardiovascular disease, and blindness may result.

The disease is widespread throughout the world, and it has been estimated that about 10 per cent of the population is infected. It is more frequently seen in the lower income groups, in criminal tendency groups, and in those who are socially degraded by alcohol, drugs, and mental deficients. Promiscuous sexual relations and perversions are the primary causes of widespread continuance of this disease. Prevention is by adequate and proper education, elimination of houses of prostitution, and a more sound social and psychological adjustment of the hypersexual individual. Adequate and early treatment (penicillin is the drug of choice) is also essential for the prevention of spread.

Other Treponema

Two other species of *Treponema*, indistinguishable morphologically from *T. pallidum*, produce diseases dissimilar to syphilis. *T. pertenue* produces the disease "Yaws", a disease primarily found in the tropics. This disease is probably spread by insects and also by direct contact of an infected person with open lesions. Skin lesions predominate. It is not a venereal disease and is predominately seen in children up to the age of 15. Adults may also contract the disease.

The other species, *T. carateum*, likewise morphologically similar to *T. pallidum*, produces a disease called "Pinta." The disease is non-venereal and all ages are susceptible. It is spread primarily by direct contact. The disease is seen primarily in the tropics (Mexico, Central and South America, South Pacific Islands, etc.) and is characterized by mottling and depigmentation of the ankles, feet, hands, wrists, and scalp.

The serological tests for syphilis are positive for both Yaws and Pinta.

Two nonpathogenic species of *Treponema*, *T. microdentium* and *T. macrodentium*, are frequently found in normal mouths. Both are highly motile, a helpful distinguishing characteristic from the pathogenic species of this genus.

GENUS *BORRELIA*

About twenty species comprise the genus *Borrelia*. Most of these species are either tick or louse borne. One, however, *B. vincenti*, is a direct contact disease.

The morphology of different species of these organisms is similar. They are moderately tightly wound, coiled-like organisms, with spirals or coils 2 to 3 microns deep. The organisms are somewhat longer, generally, than *Treponema*, being at times 30 or so microns in length. They are flagellated, rapidly motile organisms. Staining with special analine dyes is easily accomplished.

Trench Mouth

The disease trench mouth (Vincent's angina, ulcerative stomatitis) is caused by *B. vincenti*, associated with a rather plump rod, *Bacillus fusiformis*. The organism (*B. vincenti*) is wavy to spiral, and is anaerobic and difficult to cultivate. The organism is easily stained with ordinary analine dyes. The organism is spread by direct contact (kissing, etc.) and by contaminated eating utensils.

The disease is characterized by a severe sore throat and frequently sore gums and whitish mucous patches on the mucous membranes in the mouth and throat. Some cases are so severe that hospitalization may be necessary.

The disease may appear in epidemic form on college or university campuses, or army barracks. Sporadic cases are seen throughout the year.

Relapsing Fever

A variety of types of recurrent or relapsing fever are caused by different species of the genus *Borrelia*. *B. recurrentis*, *B. barbera*, and *B. carteri* are louse borne. *B. hispanica*, *B. duttoni*, *B. parkeri* and several others are transmitted by *Ornithodoros* ticks.

The disease is characterized by a febrile period of 3-10 days, then an afebrile period, then a repeated febrile period of 3-10 days. This may continue for many weeks or longer. The *Borrelia* are numerous in the blood during the febrile period. Experimentally young white rats are susceptible. Agglutinins for XK strain of *Proteus* are found in a fairly high percentage of human cases. Terramycin and penicillin are effective means of therapy.

The immunity produced on recovery is of short duration. Active immunization with killed *Borrelia* fails to confer a resistant state.

The disease is widespread throughout the world. Many sporadic cases are seen in the United States. Rodents serve as a reservoir for the microorganism and several species of ticks of the genus *Ornithodoros* have been found infected with this microorganism. The Western states, Utah, Colorado, Wyoming, Idaho, Texas, Montana and California are areas in which infected ticks have been found. Other states are also incriminated.

GENUS *LEPTOSPIRA*

Several species of the genus *Leptospira* are pathogenic for man.

The organism is a tight spiral or coiled organism about 5-20 microns long, and very thin, about 0.25 micron in diameter. It has no flagella. One end appears to be hooked or slightly bent. These organisms are fairly easily cultivated on special media at a temperature of about 30°C.

The most common organism of this group is *Leptospira ictero-hemorrhagiae*, the cause of "Weil's disease." This disease produces a moderately severe fever with jaundice in many of the cases. The organism is highly susceptible to penicillin.

It is a natural disease of rats and may be transmitted in urine of infected rats to food and water used by man.

A variety of other species of *Leptospira* cause such diseases as pretibial fever (Fort Bragg fever—formerly thought to be caused by a virus), seven-day fever in Japan, marsh fever in Europe.

Recovery from Leptospirosis infections produces a lasting immunity, the blood usually retaining fairly high agglutinin titers.

Selected Reading References

Gochenour, W. S., and Yager, R. H.: Leptospirosis in North America, Am. J. Trop. Med. and Hyg. **1**:457, 1952.

Parran, Thomas: Shadow on the Land, Syphilis, New York, 1937, Reynal and Hitchcock.

Wilmer, Harry A.: Corky the Killer (A Story of Syphilis), New York, 1950, American Social Hygiene Association.

CHAPTER 28

THE SMALLEST SIZED DISEASE-PRODUCERS

1. VIRUS DISEASES

This group of disease-producing agents are so different in their nature from any of those microorganisms previously discussed that an introductory orientation will better acquaint the student with this specialized field. Many cause primarily human diseases; many produce primarily animal diseases, which may in many instances be transmitted to man; some cause diseases of bacteria and are not transmissible to man; many produce diseases in fish, plants, insects, etc., none of which will infect man.

General Morphology

Most viruses are so small that they cannot be seen by the use of the compound microscope not even by using the best possible oil immersion lenses. However, most of them can be seen by means of the electron microscope and have thus been photographed. They vary in size from about 10 millimicrons[1] to about 450 millimicrons.

Some viruses are raised, round to globoid bodies; some are disc shaped; some are rod forms and some are needlelike in their structure.

When a certain virus attacks a cell, it will frequently produce bodies in the cell, some appearing in the nucelus, others in the cytoplasm. Such so-called inclusion bodies are frequently characteristic and specific for a particular virus. Some of these "inclusion bodies" are clusters or parts of the virus itself and some are probably produced by the cell in response to the presence of the virus. Some of them have specific names, e.g., Negri body of rabies; Guarnieri bodies of vaccinia, etc.

The size of viruses is measured by ultrafiltration techniques, ultracentrifugation and by electronmicroscopy.

-----

[1] A micron is 1/1000 of a millimeter. A millimicron is 1/1000 of a micron.

Cultivation

As previously discussed, most microorganisms of a bacterial type can be cultivated on lifeless foodstuff—meat broth, blood, etc., so long as a utilizable source of carbon, oxygen, nitrogen, certain trace elements, vitamins, water, etc. are present. Viruses, however, will not reproduce on such foodstuff. They require a living metabolizing cell for their energy source.

Susceptible animals are the commonly used "media" for virus studies. Fertilized incubated chicken eggs are also a valuable means of growing viruses. Cultivation of animal tissues in test tubes (tissue culture) also serves as a culture medium for viruses.

Tropism

Most viruses have a predilection for certain tissue, and in many cases even a specific affinity for certain cells of a particular organ. This does not mean that a given virus may not be found in other areas or in body fluids, but simply means that the highest virus concentration and the most extensive destruction of cells is usually found in certain areas, fairly specific for each virus.

For example, mumps virus produces its greatest effect in glandular tissue in humans; encephalitis viruses produce the most damage in nervous tissue; and yellow fever virus attacks the liver cells most vigorously.

Most viruses reproduce within the cells, hence masceration of the tissue is necessary to "free" the virus in order to prepare potent virus-containing suspensions. Some viruses, during the course of infections, escape from the damaged cells and may then be isolated from blood, spinal fluid, or other body fluids.

This tropism has suggested certain generalized classifications of viruses; however, this type of classification leaves much to be desired.

Viruses primarily damaging the nervous tissue are frequently called neurotropic; those damaging the skin dermatotropic; those damaging glands, glandulotropic; those damaging the lungs, pneumotropic, and the ones damaging a variety of tissues, pantropic.

Resistance to Physical and Chemical Agents

Most viruses are quite susceptible to oxidizing agents such as hydrogen peroxide, potassium dichromate, etc. Ultraviolet light acts also quite destructively upon viruses. Heat (55-60°C.), will destroy most of them in about one-half to one hour. Antibiotics and sulfonamides are not effective in destroying viruses. A few of the larger ones are, however, susceptible to those drugs. Drying (quick freezing and drying) has a preserving effect on most viruses. At low temperatures as provided in the dry ice box, deep freeze, etc., many viruses may be kept viable for many years if sealed in glass tubes and stored properly. Glycerol has a preserving effect, probably by inactivating the enzymes of the tissues, which in turn would inactivate the viruses. One virus, psittacosis, is however, inactivated by glycerol within 24 hours.

Immunity Response

Practically all persons recovering from virus diseases have an immunity response lasting from a few years to a lifetime. There are, however, a few exceptions (e.g., herpes). Some people who have poorly developed resistance or immune mechanisms may have the same virus disease more than once. Certain diseases (e.g., influenza, poliomyelitis) can be caused by several antigenically different types of the specific virus—thus immunity to one will not provide immunity to a heterologous type. Artificial active immunity (vaccination) is possible with several viruses (e.g., smallpox, rabies, yellow fever, etc.).

Epidemiology

Many virus diseases are spread by direct contact from one person to another (e.g., measles, mumps, etc.). Others may be spread by an intermediary host, such as ticks, mosquitoes, etc.; others may be spread by means of contaminated water supplies, food, and other indirect contacts.

When we consider and include such diseases as measles, mumps, chicken pox, German measles, poliomyelitis, influenza, etc., the viral infections far outnumber the bacterial infections, such as meningitis, tuberculosis, typhoid fever, scarlet fever, etc.

Many virus infections have a seasonal occurrence; some are more frequent in the summer, some in the fall and winter, and others in the spring.

Plant Viruses

Plants, like humans, are susceptible to specific virus diseases, many of which are devastating in their effects.

TABLE XVI
GENERAL VIRUS DISTRIBUTION IN TISSUES WHERE
THE GREATEST DAMAGE IS MANIFEST

Nervous Tissue

Encephalitides
1. Japanese B
2. Equine encephalomyelitis
    a. Eastern type
    b. Venezuelan type
    c. Western type
3. St. Louis encephalitis
4. Lymphocytic choriomeningitis
5. Poliomyelitis
6. Rabies
7. Encephalomyocarditis

Skin and Mucous Membranes

1. Herpes
    a. Simplex (cold sores, etc.)
    b. Zoster (shingles)
2. Smallpox (variola, vaccinia)
3. Measles (red measles)
4. Varicella (chicken pox)
5. Rubella (German measles)
6. Verruca (warts)
7. Epidemic keratoconjunctivitis
8. Trachoma
9. Foot and mouth disease
10. Molluscum contagiosum

Grandular Tissue

1. Mumps
2. Infectious hepatitis
3. Serum hepatitis
4. Yellow fever
5. Dengue fever
6. Rift Valley fever
7. Mononucleosis
8. Lymphopathia venereum

Pulmonary Tissue

1. Atypical virus pneumonia
2. Influenza
3. Psittacosis (parrot fever)

Variety of Tissues (Pantropic)

1. Coxsackie virus disease
2. Cat scratch fever
3. Viral gastroenteritis
4. Newcastle disease

One of the most remarkable of this group is the tobacco mosaic virus, which has been purified to a state of crystallization.

Other virus diseases of plants are: tomato bushy-stunt virus, turnip yellow mosaic virus, cucumber virus, and a host of others.

Virus diseases of bacteria (bacteriophage) will be discussed in more detail later on.

Virus Diseases of Lower Animals Not Transmissible to Man

Numerous virus diseases of animals are known which are not infectious to man. The Rous sarcoma of fowls is perhaps the most highly fatal of all animal viruses, producing a cancerous growth in the inoculated chickens, and it approaches 100 per cent in its fatality rate. Myxomatosis of rabbits is likewise a highly fatal virus infection of these animals. Rabbit papilloma virus, on the other hand, produces a rather benign growth on the infected rabbit's skin. Many others of these types of virus diseases exist.

SPECIFIC VIRUS DISEASES OF MAN

Only a few representative viruses will be discussed. Since this field is rather extensive, a complete text covering only virus diseases would be necessary to adequately cover the enormous data available.

Viruses Attacking the Nervous Tissue

Poliomyelitis.—There is evidence that poliomyelitis (frequently called infantile paralysis) is a disease dating back to 3700 B.C. The disease was quite common during the time of Hippocrates. Between 1840 and 1900 several monographs were written on the clinical aspects of the disease, the data collected at that time being very accurate and to this day dependable. The causative agent of this disease was first isolated by Landsteiner and Popper in 1908 by inoculating monkeys with nervous tissue material from a fatal human case. The disease in monkeys corresponds almost identically to the disease in humans.

The virus causing this infection is one of the smallest known disease-producing agents, being 8 to 20 millimicrons in diameter. It is probably longer than it is wide. The virus is very hardy, withstanding saturated ether-water mixtures for at least twenty days. It will remain viable in sewage water for several months. It is, however, easily destroyed by heat (60°C. for 10-30 minutes) and chemicals (formaldehyde, bichloride of mercury, etc.).

The monkey (old world species) and chimpanzees are the most susceptible animals to experimental poliomyelitis infection. Rabbits, guinea pigs, sheep, goats, and a host of other animals are resistant to infection. One type (type 2) will infect mice and hamsters under certain conditions. Type 2 has also been propagated in embryo chick tissue, but thus far types 1 and 3 have not been successfully cultivated by this technique. All three types have been cultured in tissue culture using as the basic tissue monkey testes, monkey kidney, and certain human tissues, which proves that these viruses will reproduce in non-nervous tissue.

Three specific antigenic types are known: Type 1, Brunhilde; type 2, Lansing; and type 3, Leon. The name of type 1 is derived from the laboratory name of a chimpanzee in which the original virus was isolated. The name of type 2 is derived from the name of a city in Michigan, in which the patient resided from whom this virus type was isolated. The type 3 is named from a patient. All three types produce a type of specific immunity. For example, recovery from type 1 will not confer immunity for type 2 or 3, etc.

Poliomyelitis virus may be isolated from several sources: (1) from the nose and throat of acute cases; (2) from stool specimens of acute and convalescent cases; (3) from the spinal cord, medulla, and thalamus of fatal human cases; (4) from the blood of acute (prefebrile and beginning febrile) cases of the disease; (5) from sewage; (6) from several species of flies, including the common blow fly.

Poliomyelitis is a fairly common disease, particularly in children from the ages of 1 through 15, the highest incidence being from 4 to about 12. This varies, however, with different areas where epidemics occur. Adults are also susceptible, but most adults have had a very mild poliomyelitis infection during their childhood. It has been shown that about 75 per cent of the adult urban population have had the disease, probably in such a mild form that they did not realize they had had poliomyelitis. Blood samples to determine the level of immunity were previously tested in monkeys, but since 1951 when the tissue culture technique was perfected sufficiently, such tests are conducted by the latter technique. This has reduced the cost of single immunity tests from over a hundred dollars to only a few dollars each, provided laboratories have the proper equipment and trained personnel.

The disease, when definite symptoms are present, is usually ushered in, after an incubation period of 7-21 days, in a rather acute manner. The child (or adult) has a fever of 101-103°F.; headache, some nausea or vomiting, stiff neck and back muscles and stiffness of the muscles of the arms and legs. If the virus destroys nerve cells, there will be paralysis. In the *spinal* form, a variety of muscles may be affected: arms, legs, or both, bladder, etc. In the *bulbar* form, the most serious in relation to survival of the patient, the nerves of the base of the brain are affected. This paralysis makes swallowing difficult or impossible, also causing a change in vocal sounds, and frequently causes paralysis of the muscles of the breathing mechanism (diaphragm and intercostals). These bulbar cases may frequently require the aid of the so-called "iron lung" to aid the patient in his breathing during the acute phase of this disease. The death rate in the bulbar cases is usually about 10 per cent. The paralytic rate varies greatly with each epidemic. Since all three types of virus produce the identical clinical disease, no prognosis as to the outcome can be made on the basis of a specific virus type. Occasional cases may exhibit some mental disorientation, particularly in the acute stage of the disease, but no permanent mental damage is done, thus on recovery the intellect is equivalent to that which it was prior to the infection.

There are at the present time no known drugs or chemicals which will prevent or alter the course of the disease. Work is progressing along this line of attack. No vaccine is as yet available, but the financial aid and stimulation given to research workers by the National Foundation for Infantile Paralysis during many years will probably result, in the not too distant future, in a safe, adequate vaccine protecting against all three virus types.

Poliomyelitis is a seasonal disease, occurring during the late spring and summer months, decreasing in the fall. During epidemic periods, the greater number of cases are usually seen in August and September, with a decline beginning in October. In 1952 there were about 57,000 clinical cases of the disease in the United States, this being only a fraction of those actually infected; the latter were not, however, sufficiently ill to seek medical aid.

The disease is spread primarily by direct contact (coughing, sneezing, etc.), either from acute cases or from carriers of the virus who are immune and thus not ill. Contaminated food may, at times, aid in the spread of the infectious agent. Contaminated articles (pens, pencils, etc., and hands) may also play a role in the dissemination of the virus. Travel from one area to another is perhaps the most important means by which the virus is disseminated to widely separated areas.

There is no absolutely safe preventive measure. A number of suggestions appear to be helpful. Clean hands and clean food are essential. Keep children away from known infected persons. Do not allow children to become overtired or chilled during the poliomyelitis season.

Encephalitides.—Since there are several different specific types of this disease, only two will be discussed in order to present the essential principles.

Equine encephalomyelitis is a disease primarily infecting horses, rodents, and birds, but may be transmitted to man either directly from infected animals or by the bite of an infected mosquito. There are two specific types: (1) western equine encephalomyelitis and (2) eastern equine encephalomyelitis.

1. *Western Equine Encephalomyelitis.*—In 1931 the virus of western equine encephalomyelitis was isolated by Meyer and Howitt from the nervous tissue of a moribund horse. It was not until 1938 that the virus was isolated from a fatal human case. Since that time many thousands of human cases have been reported. Many hundreds of thousands of horses have succumbed from this disease anywhere from the state of Minnesota westward to California.

The virus is about 45 millimicrons in diameter and is disc shaped. It is easily cultured in the chick embryo and in tissue culture. Numerous animals such as guinea pigs, mice, rabbits, monkeys, hamsters, may be infected experimentally. The incubation period in humans is about 4 to 21 days.

After the incubation period the patient has a sudden fever, chills, severe headache, and frequently vomits. Drowsiness and mental stupor is frequent. Convulsions are common, with tremors and mental confusion. Paralysis is uncommon, but this disease may be confused with poliomyelitis. The mortality rate varies between 8 and 15 per cent. Most all of the recovered patients have no aftereffects, but a few may become mentally deficient due to brain damage.

No drugs or chemicals are suitable for treatment, but a highly effective vaccine is available for use on livestock only. An egg culture virus vaccine has been used on a few humans and appears to be effective in the prevention of the specific disease.

The virus is spread primarily by means of the bite of mosquitoes (*Culex tarsalis*, *Aedes aegypti*, and others) as well as by means of the tick *Dermacentor andersoni*. Many birds caught in nature have been found infected.   This includes quail, pheasants, blackbirds, owls, and a variety of others.   It has been suggested by some research workers that the disease is maintained in nature by bird mites—from the bird mite it is transferred to birds, then to mosquitoes, thence to man. There are a number of endemic areas in the United States, California's San Joaquin Valley being the most heavily infected area of the group. Also Yakima Valley in Washington, and various areas in the West and Southwestern United States are endemic areas.   In some of these endemic areas, inapparent infection (subclinical disease) accounts for an immunity of from 12 to 15 per cent in the adult population.

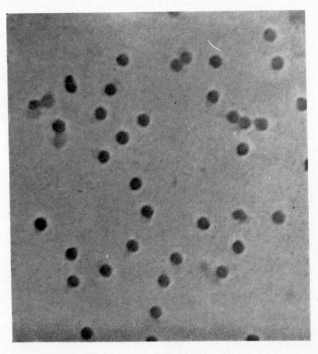

Fig. 35.—Equine encephalomyelitis virus, 110,000×.   Electron microphotograph with RCA Electron Microscope.   (Courtesy Dr. D. G. Sharp, Duke University.)

Prevention is to maintain adequate livestock immunity and to decrease mosquito populations in areas where the disease is endemic. Avoidance of direct contact with sick animals is a sound preventive measure.

2.   *Eastern Equine Encephalomyelitis.*—The virus of eastern equine encephalomyelitis was isolated in 1933 from infected animals.   It

was not, however, until 1938 that the virus was isolated from human nervous tissue. Like western equine encephalomyelitis (W.E.E.), horses are frequently affected with the eastern equine encephalomyelitis (E.E.E.) virus, the death rate being about 90 per cent in the latter disease. In humans, E.E.E. virus produces a fatality rate of about 90 per cent. Children appear to be more frequently attacked than adults.

The virus is about 45 to 50 millimicrons in diameter. In general, the virus behaves similarly to the W.E.E. virus. There is no cross immunity between the E.E.E. and W.E.E. viruses.

The disease in humans is more severe than the W.E.E. virus disease. There is a sudden onset with severe headache, nausea or vomiting, and a moderate fever which lasts from one to two days. There is then a recovery period of a day or so, then a relapse with a very high fever (105°F. or more) with convulsions, paralysis, and coma. In those who recover, there is a high rate of mental deterioration.

No specific treatment is available. Vaccines produced from chicken embryo infected tissue appears to be suitable for the vaccination of livestock only.

The disease is probably spread by the bite of infected mosquitoes, *Aedes sollicitans* and *Mansonia perturbans*. Chicken mites have also been shown to harbor the virus.

Rabies.—This disease has plagued mankind for many centuries' beginning probably before 500 B.C. From the twelfth to the nineteenth centuries, travel in both Europe and Asia was hazardous because of highwaymen and thieves and attacks by "mad" wolves. Later domestic animals became infected and epizootics occurred in many parts of Europe.

The disease was first transmitted experimentally to dogs in 1804. Pasteur, however, in 1881 established not only the viral nature of the disease, but also produced a preventive vaccine, using infected rabbit spinal cords as the vaccine source. He established the fact that, with continued passage of the virus in rabbits, it lost its "furious" properties for dogs and produced in rabbits a "dumb paralytic" disease. This "virus fixe" was then suitable as a human vaccine, since it had mutated from a virus (*street virus*) initially producing a "furious" or "mad" state in dogs to one that produced a paralytic disease in both dogs and rabbits without any furious state being present during the development of the disease in these animals.

The *street virus* varies in size from 160 to 240 millimicrons, while the *fixed* virus is somewhat smaller, 100-150 millimicrons in diameter. The virus is no more resistant to physical and chemical agents than the poliomyelitis or encephalitis viruses. It produces fairly regularly, variable sized bodies in the cytoplasm of the infected brain cells. These bodies, called *Negri bodies*, are sufficiently characteristic to be of diagnostic significance for this disease, rabies (hydrophobia).

The disease rabies is primarily a naturally occurring disease of wild animals (fox, coyote, wolf, skunk, etc.) and also domestic dogs. It is transmitted to man by the bite of an infected animal. In man the disease may vary in its incubation time from a week to as long as several months. The disease may resemble any one of a group of infections at first, with a slight fever, sore throat, and mild headache. After one to three days the patient may show signs of nervousness, skin itching, mild weakness of different muscle groups, change in vocal sounds and inability to swallow. The eyes become glassy and the patient may show fear of impending death. The disease is invariably fatal when symptoms begin.

Similar symptoms may be seen in dogs, but they usually wander aimlessly, biting or chewing at anything in their path—other dogs, cows, humans, etc. As a result of this indiscriminate biting, their stomachs may show a variety of indigestible articles such as stones, glass, wood, bottle caps, bits of cloth, buttons, nails, etc. They are glassy eyed. Frequently the muscles of the swallowing mechanism become paralyzed, resulting in the inability to swallow the collected saliva. This saliva becomes foamy, thus the reason for the "frothy or foamy" mouth of a mad dog (rabies or hydrophobia).

Dogs suspected of being rabid should not be killed, but restrained in a very strong pen. At the first indication of paralysis the animal should be killed by other means than by shooting in the head. The head should be carefully removed, placed in a large can, sealed, and labeled and ice packed around the outside of the can and then sent to the nearest State Health Laboratory or other special laboratory for diagnosis.

Any person bitten by a dog should be taken to a physician immediately for prompt treatment which may be lifesaving. Scrubbing the wound with soap, hot water and a brush is valuable first aid treatment. Immune serum is available and it appears to be of some value. The Pasteur type treatment, by daily injections of vaccine for about two weeks, is an effective preventive of this disease.

Prevention of the disease is best obtained by a completely vaccinated dog population. More rigid legislation is needed for antirabic vaccination of the dog population, particularly in endemic areas.

Viruses Attacking Primarily the Skin

1. **Measles (Red Measles)** is a common disease of childhood, seen in epidemic form in the late winter and spring. One exposure is usually sufficient to produce infection in a susceptible child. It is a disease spread by direct contact particularly communicated by sneezing and coughing of infected droplets.

The incubation period is about two weeks. The first symptoms are a slight sore throat, watery reddened eyes, runny nose, light cough and a slight fever. Small spots, *Koplik* spots, usually appear on the mucous membranes of the inner part of the cheek, near the premolar teeth. They are blue white with a slightly reddened periphery. Rash begins about one day or two after appearance of the *Koplik* spots. The rash appears back of the ears, on the neck and over the face, later spreading to the body, arms and legs. This rash is at first reddish, then as it spreads the rash areas coalesce, becoming blotchy, slightly purplish to brown in color.

The virus may injure the heart muscles and kidneys. However, the most serious effects are the secondarily invading microorganisms, frequently producing a bronchial pneumonia, inner ear infections, etc.

The course of the disease may be altered by an injection of gamma globulin a few days after exposure. No suitable vaccine is available. On recovery from the disease, the individual is usually immune for life.

2. **Rubella (German Measles)** is a rather common childhood disease which is highly communicable and is spread by direct contact. The incubation period is generally 16-18 days, but may vary in different children. Usually there is no fever, or only very slight. Common cold symptoms are seen. The rash begins on the face and spreads over the body—at first resembling red measles, then scarlet fever, but within about three days it will usually disappear. It is not a serious disease in children, but is rather serious disease in the pregnant woman.

If a woman contracts German measles during the first four months of pregnancy, severe damage may result in the fetus. Mental defects, cardiac malformations, deafness with some mutism, cataract producing blindness, and abnormalities of dentition may result.

Prevention is by using relatively large dosages of gamma globulin, preferably on or before the eighth day after exposure of a pregnant woman to German measles.   Some suggest that, if possible, all children before the age of puberty, particularly females, should be purposely exposed to this disease.   On recovery, immunity is usually lifelong.

3. **Smallpox** is a very old disease plaguing mankind since more than a thousand years before Christ.   It is still a common disease in the Orient, but in countries practicing preventive vaccination the disease is quite uncommon.   In endemic areas such as India, the death rate is near 40 per cent.

This is one of the larger viruses varying from 125-252 millimicrons. The virus is easily grown on tissue culture and in embryo eggs.   It also grows well on the shaved backs of rabbits and the abdomen of calves.

The naturally occurring virus is called "variola" virus, while the fixed (animal passaged) virus is called "vaccinia" virus.   The naturally occurring virus is transmitted directly to man by sputum droplets, touching contaminated articles or handling a patient ill with smallpox. After an incubation period of about 12 days the patient becomes suddenly and acutely ill, with generalized aching and high fever of 103°F. or more.   Two to four days later a rash is seen first over the face then spreads quickly over the rest of the body, including the soles and palms.   Recovery leaves the person with a life-long immunity.

Prevention is by prophylactic vaccination with vaccinia virus. This has proved successful since the successful experiments of Jenner in 1798 who applied the first vaccination or prevention against a virus disease.

The vaccine is a mutated variola virus, changed by animal passage, particularly by growing the virus on the shaved abdominal skin of the calf.   Vaccination against smallpox is recommended at an early age, eight months or less, and repeated every 4 to 6 years.   Some persons have a persistent immunity after their first vaccination, and each succeeding vaccination produces only an immune response.   The immune response is indicated by a slightly reddened, swollen area at the site of vaccination, appearing 2 to 4 days after vaccination and almost disappearing completely within 7-10 days.

Viruses Attacking Glandular Tissue

Mumps is a common childhood disease probably existing since severals centuries B.C.   The filterable virus nature of this disease was proved in 1934 by Johnson and Goodpasture.

The infectious agent of mumps (parotitis) varies from 90-135 millimicrons in diameter. It is easily cultivated on the allantoic membrane of the developing chick embryo.

The disease occurs primarily in children, but adults may and do contract the disease and in the latter group, after puberty, it may be a serious disease.

A single exposure is usually sufficient to infect a nonimmune individual. After an incubation period of two and a half to three weeks the patient complains of painful swallowing. This is not a true sore throat. The parotid glands at the angle of the jaw and just in front of the ear become infected and swollen. A single gland or both parotids may be affected. The jaws look puffy and flabby. Many other salivary glands may also be involved. Likewise other glands of the body, such as the pancreas, ovaries and testicles may be involved. When the glands of reproduction in postpuberty individuals are seriously damaged sterility may result. The disease, in uncomplicated cases, usually subsides within 10 days. Immunity is variable, and mumps infection of one parotid gland frequently fails to produce complete immunity, thus an infection later in the as yet uninfected gland may occur if re-exposed to this virus. Repeated exposures to the virus probably produce immunity in a fairly large percentage of the population without producing typical signs of the disease. A fairly simple test, the red cell agglutination test, can be used to detect virus or antibodies of immune individuals. Virus combines with the surface of certain species of red blood cells causing them to agglutinate. Specific antibodies will neutralize the virus, and in the above test, no agglutination of the red blood cells will occur.

Recent research indicates that an embryo egg vaccine will produce a fairly high state of immunity.

The disease is endemic during the entire year, but becomes epidemic during the winter and spring months.

Yellow fever is an endemic disease in Africa and South and Central America, existing in these areas for perhaps four hundred years or more. The French may have been able to complete the Panama Canal had it not been for this disease. The French engineer in charge of the project at the end of the 19th century predicted that this canal could never be built because all the workmen got sick and many died. Reed and co-workers discovered that the disease was transmitted by the mosquito *Aedes aegypti*, and by eradication of breeding places of these insects the disease incidence was materially decreased. The

United States Government took over the building of the Panama Canal and its dedication was honored by the world fair in San Francisco in 1915.

The virus of yellow fever varies in size between 12 and 29 milli-microns in diameter. It is easily cultivated in the developing chick embryo and easily infects mice, monkeys and hedgehogs, but rats and rabbits are resistant.

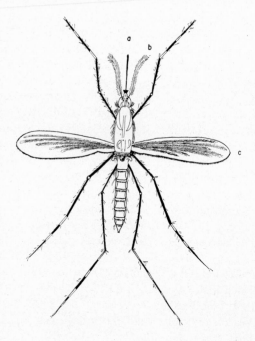

Fig. 36.—*Aedes aegypti* mosquito. *a*, Proboscis; longer than anophelene proboscis; *b*, Palpus (no antennae). *c*, Wing; no hairy spots as in anophelene.

After being bitten by an infected *A. aegypti* mosquito, there is an incubation period of about 3-4 days in a susceptible person. The disease strikes suddenly, with general aching, high fever (103-104°F.) and severe headache. After 3-4 days the patient becomes more acutely ill, with swollen eyes, bleeding from the mouth and other body spaces, vomiting (black vomitus), and a yellowing of the whites of the eyes and skin (jaundice). Severe damage to the liver is frequent.

Prevention is primarily by prophylactic vaccination, a highly effective vaccine being prepared from infected developing chick embryos. Eradication of mosquito breeding areas also has had a marked effect in lowering the incidence of this disease in endemic areas.

Viruses Affecting Pulmonary Tissue

The lung tissue is heir to many types of diseases, but virus infections are probably many times more common in humans than bacterial infections. The most common of this group of viruses is influenza.

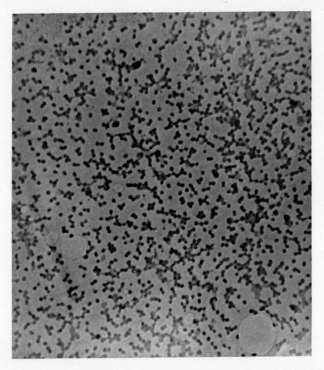

Fig. 37.—Swine influenza virus 15,400×. Electron microphotograph with RCA Electron Microscope. (Courtesy Dr. D. G. Sharp, Duke University.)

Influenza.—Influenza virus has probably been responsible for several pandemics (world wide epidemics) in the past several hundred years. One of the most severe pandemics occurred during and after World War I, at which time (1918-1919) over twenty million persons died of influenza.

The virus causing influenza was isolated in 1934 by Smith, Andrewes, and Laidlaw who also further established criteria for true influenzal infections. The size of the different virus types varies from 70 to 124 millimicrons. The virus is easily cultivated in the developing chick embryo. Ferrets and mice are highly susceptible. Three main antigenic types exist, Types A, B, and C, with several immunologically distinct subtypes of A. Immunity is specific for each type and subtype. The hemagglutination and hemagglutination inhibition tests are suitable for distinguishing the different virus types.

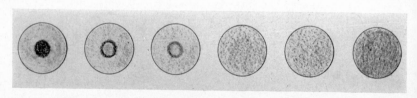

Fig. 38.—Hemagglutination test, influenza virus. Note in the first three tubes (from left to right) a finely granular deposit of agglutinated red blood cells (positive reaction). The fourth tube shows a weak positive reaction, indicating a small amount of virus present. The fifth tube shows a very weak positive reaction. The sixth tube served as control, no virus being added. This control tube shows a tightly packed button of nonagglutinated red cells.

The disease is a direct contact infection, the incubation period being short, 1 to 3 days. The disease strikes suddenly, with chills and fever, headache, soreness of the muscles and aching of the bones. Sneezing and coughing are frequent. The most serious sequela is the development of a bacterial pneumonia, which may prove fatal if prompt treatment is not instituted. Recovery leaves the person immune to the specific type of virus for probably several years, and in some cases lifelong immunity is developed. Prevention is by vaccination, using inactivated virus from allantoic fluid of chick embryos. Pooled virus types cannot be included in the vaccine mixture. Influenza may develop in vaccinated individuals, but the disease is comparatively mild.

The disease is more common in epidemic form in the late fall, winter and early spring, but endemically it exists throughout the year. Avoidance of direct contact with an ill person will decrease the chances of an infection.

Psittacosis is a naturally occurring disease of psittacine birds (parrots, parakeets, etc.) which may be transmitted directly to man. Other birds such as chickens, pigeons, seagulls, and many other species of birds may harbor the disease agent.

The disease was recognized initially in Europe in 1929, and since that time small epidemics have been reported from various parts of the world, including the United States.

The virus is one of the largest in size to come under the viral category. It varies between 220 and about 450 millimicrons. It is stainable with special stains and can be seen with the aid of the oil immersion lens of the light microscope.

The disease develops rather suddenly after an incubation period of one to two weeks. The general symptoms resemble influenza with more pronounced lung involvement. The disease may progress for two or more weeks, leaving the patient weak and debilitated for several weeks. The disease is effectively controlled by several antibiotics, penicillin, aureomycin and Terramycin.

Virus Pneumonia.—*Atypical virus pneumonia* (virus pneumonitis) resembles mild influenza, but the incubation period is longer (1 to 3 weeks) and the onset is gradual instead of sudden. The death rate approaches zero.

No complete agreement on the specific disease agent is available.

Pantropic Viruses

One rather recently discovered group of viruses called the Coxsackie group of "C" viruses was first isolated in 1948 by Dalldorf and Sickles from children with paralysis residing in Coxsackie, N. Y. The disease may be associated with poliomyelitis, but it is a distinct immunological group of virus, unrelated antigenically to the three poliomyelitis virus types.

The viruses of the "C" group are about 25-35 millimicrons in diameter. Initial isolation from humans is by means of day or two old suckling mice. A variety of tissues may be affected, including muscle, nervous tissue, liver and other areas. Several specifically antigenic strains have been isolated. Symptoms in humans are variable. Abdominal cramps, fever, and headache are fairly frequent symptoms. Gray-white areas with reddened base are seen in some cases. Apparently the fatality rate approaches zero.

Newcastle disease is primarily a disease of fowls, attacking the gastrointestinal tract and the respiratory and nervous systems. In man the virus produces damage to the eyes in a form of conjunctivitis, along with a slight fever, chills and headache. Recovery is inevitable if no secondary infections appear.

The disease is transmitted from ill fowls, including pheasants, chickens, and turkeys as well as other species of birds.

VIRUS DISEASES OF BACTERIA

The disease-producing bacterial viruses were discovered by Twort (1915) and studied extensively by d'Herelle (1917). These bacterial viruses may be isolated from sewage and fecal material of man, animals, and birds.

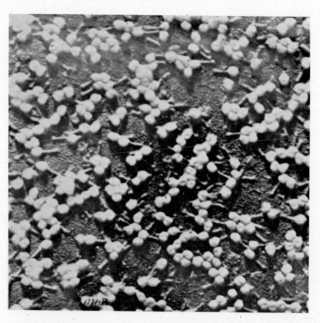

Fig. 39.—Group of spermlike elementary particles of colon bacillus bacteriophage, type T-4 (No. 1771 P), 46,500×. Electron microphotograph with RCA Electron Microscope. (Courtesy Dr. R. Wyckoff, National Institutes of Health.)

These filtrable disease agents of microorganisms are quite specific in their action, some limiting their activity to a particular strain of a given species. Thus they may be used as a means of identifying not only bacterial species, but also certain strains of a given bacterial species (e.g., *Sal. typhosa Vi* types). The T series of bacterial viruses (named bacteriophages by d'Herelle, now shortened to phage) have been more thoroughly studied than other types.

These bacteriophages vary from about 20 to over 100 millimicrons in diameter, with many showing a tail 120 millimicrons in length. Some may resemble spermatozoa in shape, with a head and tail. They attach to the specific bacterial surface and there penetrate and multiply within the bacterial cell. Under proper conditions the bacterial cell undergoes a lysis—thus a previously clouded culture media (clouded by bacterial growth) becomes as clear as the broth media originally seeded. If a mixture of bacteria and specific bacterial virus is plated on an agar plate, the phage (bacterial virus) will cause a destruction of the bacteria at those places where the virus is present. The lysed areas are clear centered with "moth-eaten" edges, varying in size depending on the particular virus used. These cleared areas are called plaques.

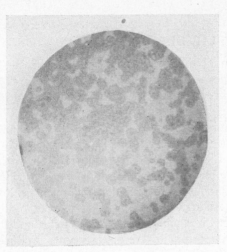

Fig. 40.—Bacterial virus (bacteriophage) plaque formation. The clear areas are the places where the bacterial virus has destroyed the bacteria. The dark areas are bacterial colonies and coalesced bacterial colonial growth unattacked by the bacterial virus.

The study of bacterial virus has opened new fields in the understanding of virus metabolism, virus variation, and mutation as well as a better understanding of microbic and viral genetics.

The Common Cold

Despite our advanced knowledge of many diseases, wonder drugs for treatment and vaccines for prevention, one glaring example of our total lack of knowledge is the "common cold." Many think it is due to a virus, others frankly admit they do not know. Nearly all scientists as well as the lay public admit it is a communicable disease—or a syndrome due to a multitude of inciting agents. Common colds occur endemically and epidemically during all months of the year, in all age groups. This disease affects each of our 65 million workers each year, with a loss of at least one day's work; at least half of those workers lose two day's work and one third lose three day's work. Likewise our 25 million school children between six and seventeen years of age lose at least a day's school work, thus a loss of money to the school. This disease therefore accounts for the actual loss of over *two billion* dollars annually.

More concentrated effort should be directed toward research in solving this problem.

2. RICKETTSIAL DISEASES

This group of disease agents, the "rickettsiae," was named in honor of Dr. Ricketts who in 1909 discovered the causative agent of Rocky Mountain spotted fever. In 1916, da Rocha Lima found similar organisms in lice from typhus fever patients.

These organisms appear to lie between bacteria and viruses in size, but like viruses they also require living tissue cells for growth and development. Like bacteria they are stainable with special stains. In size they are close to the smallest bacteria, varying from 0.25-0.3 micron wide by about 0.5 micron long. Some may, however, be 1.5-2 microns long. All appear to be gram-negative. Almost all are transmitted by insect bites such as lice, mites, ticks, and fleas.

The most important of the disease types produced by the genus *Rickettsia* (family, *Rickettsiaceae*) is classic typhus fever, a very old disease of mankind. This disease is one seen in famine, poverty, and in wars. It has been a determining factor in the outcome of great battles and great wars. Two forms are evident, epidemic (classic typhus) and endemic typhus fever, the latter one being a considerably less serious type of infection.

The next most important rickettsial disease is Rocky Mountain spotted fever, an unfortunate misnomer because the disease is found throughout the continental United States. It was, however, found rampant in the Rocky Mountain States, particularly in Montana, in the early 1900's. Thus, before the first quarter of the twentieth century had elapsed, a United States Public Health Service Laboratory

TABLE XVII
RICKETTSIAL DISEASES

DISEASE	RICKETTSIA	VECTOR	DISTRIBUTION	CARRIER	TYPE OF DISEASE
Epidemic or classic typhus fever	*R. prowazeki*	*Pediculus humanus* (body louse)	South America, Europe, Africa, Asia, Mexico, Russia, Japan	Man	Incubation 10-14 days; high fever, severe headache, aches, rash, prostration and mental depression
Endemic or murine typhus fever	*R. mooseri*	*Xynopsylla cheopis* (rat flea) and others	Ubiquitous	Rodents (particularly squirrels, rats)	Incubation 10-14 days; similar to classic typhus but less severe
Rocky Mountain spotted fever	*R. rickettsii*	*Dermacentor andersoni* (wood tick) *D. variabilis* (dog tick) and others	United States	Rodents (rabbits, squirrels, etc.)	Incubation 6-7 days; onset abrupt, high fever, chills, prostration, vomiting blood, rash appearing on ankles and spreading
Rickettsial pox	*R. akari*	*Allodermanyssus sanguineus* (mouse mite)	United States	Mice, cattle	Incubation 6-10 days; fever, headache, chills, rash producing black crusts later develop
Scrub typhus	*R. orientalis*	*Trombicula akumushi* (and others)	Asia, Sumatra	Rodents	Incubation 10-12 days; sudden chills, headache, bloodshot eyes, rash
Q fever	*R. burneti*	*Dermacentor* sp. and others	United States, Australia, Mediterranean, Africa, Europe	Rodents	Incubation 14-26 days; sudden chills, headache, cough, weakness

was established in Hamilton, Montana, for the study of Rocky Mountain spotted fever. This modernly equipped laboratory and its location is a scientist's dream for carrying out pure and applied scientific research.

Several other diseases of this group are infectious for man, including Q fever, rickettsial pox, etc., and will be listed along with the above two diseases in Table XVII.

A rather interesting fact concerning this group of microorganisms is their apparent relationship with the *Enterobacteriaceae, Proteus vulgaris* X strains. Serum from patients with rickettsial diseases will agglutinate these different proteus strains, OX-19, OX-K, and OX-2. This test is called the Weil-Felix test. Also, complement fixation tests, using extracts from the capsular material of specific rickettsiae as antigens, are excellent diagnostic tests.

Most of the rickettsial diseases may be prevented by specific vaccination, with formalinized rickettsiae which have been grown on chick embryos. Recovery from the disease produces a lasting immunity. Aureomycin and Terramycin are drugs of choice in treating rickettsial infections.

Selected Reading References

Dalldorf, G.: The Coxsackie Viruses, Bull. N. Y. Acad. Med. **26**:329, 1950.
Francis, T., Jr.: Mechanism of Infection and Immunity in Virus Diseases of Man, Bact. Rev. **11**:147, 1947.
Meyer, K. F., and Eddie, B.: A Review of Psittacosis for the Years 1948-1950, Bull. Hyg. **26**:1, 1951.
Rivers, Thomas, Editor: Viral and Rickettsial Infections of Man, ed. 2, Philadelphia, 1952, J. B. Lippincott Company.
Sabin, A. B.: Paralytic Consequences of Poliomyelitis Infection in Different Parts of the World and in Different Population Groups, Am. J. Pub. Health **41**:1215, 1951.
Smith, Kenneth M.: An Introduction to the Study of Viruses, London, 1950, Sir Isaac Pitman and Sons, Ltd.
Zinsser, H.: Rats, Lice and History, Boston, 1935, Little, Brown and Co.

CHAPTER 29

MOLDS AND YEASTS—FRIENDS AND ENEMIES

1. SAPROPHYTIC FUNGI

The distribution of the common molds and yeasts is as universal as the true bacteria. Likewise there are as many different kinds of yeasts and molds that inhabit the earth as there are true bacteria. Fortunately for us, nearly all species of molds are harmless. Actually many of these microorganisms are beneficial for our existence. Numerous fungi toil night and day aiding our soils to produce edible plant life. Dead plants and animals serve as food, and very little prodding is necessary to get these molds and yeasts to work. Others may need a little stimulation and training, and, after harnessing, they are put to work in our factories, milk plants, bakeries, and a variety of other endeavors which enables man as well as animals to obtain a better living. A species of termites (white ants) found in the Sudan in Africa cultivates, deep underground, certain fungi for a food supply. Yeasts, a by-product from the manufacturing of various products, furnishes an excellent protein supplement for animal feed. A few molds may be nuisances, causing spoilage of fruits and vegetables in the home, and some may cause spoilage of jellies and jams; however, the benefit of this great crew of workers far surpasses the damages they may do.

The actual foundation of the science of mycology (study of fungi) was initiated by Baukin, who in 1623 listed about one hundred species of fungi. The existence of these plants was, however, known long before Christ, as the Chinese advocated the use of certain molds for the treatment of local infections. The Romans and Greeks were able to identify edible nonpoisonous mushrooms. Many accounts in the Bible refer to crop diseases as cereal rusts being due to offensive acts against the gods.

From the work of Tourneforts (1694) and Nicheli in 1729 (*Nova Planatarim Genera*) on fungi, the science of mycology began to take roots. The foundation laid down by Linnaeus for classification of plants in his *Species Planatrum* in 1753 started a more comprehensive study of this group of microorganisms. The basis for our current classification was established by Fries in his *Systema Mycologieum*, published between 1821 and 1832 in three volumes. Contemporary mycologists, Emmons, Salvin, C. E. Smith, Conant, Almedia, Sparrow, Tatum, Henrici, Waksman, and others have contributed greatly to our better understanding of fungi.

Classification

No attempt will be made to present an extensive classification of this group of microorganisms, the fungi, but merely skeletonize the group into a framework to support the main concept of classification.

Kingdom   — Plant
Phylum   — Thallophyta
Subphylum — Fungi

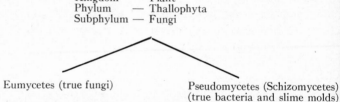

Eumycetes (true fungi)   Pseudomycetes (Schizomycetes)
(true bacteria and slime molds)

Class 1.   Phycomycetes (300 genera): Over 1500 species.
     (sexual and asexual reproduction)
         (a)   Nonseptate mycelia
             1.   Genus Mucor
             2.   Genus Rhizopus
         (b)   Septate mycelium
             1.   Genus Pythium

Class 2.   Ascomycetes: Sac Fungi (over 30,000 species)
     (sexual and asexual reproduction)
         (a)   True yeasts—multiply by budding
             1.   Genus Saccharomyces—wine yeasts
         (b)   Septate mycelium
             1.   Genus Penicillium—antibiotic mold
             2.   Genus Aspergillus—citric acid mold

Class 3.   Basidiomycetes (over 25,000 species)
     (sexual and asexual reproduction)
         (a)   Mushrooms, toadstools
             1.   Genus Amanita—poisonous mushrooms
         (b)   Smuts
             1.   Genus Ustilago—parasitize flowering plants
         (c)   Rusts (Uredinales)
             1.   Genus Puccinia—parasitize cereal grains

Class 4.   Fungi Imperfecti (Deuteromycetes) (over 20,000 species)
         Contains the imperfect stage or lacking sexual spores
         Practically all of the human pathogens belong to this "catchall"
            group of Fungi Imperfecti
         Genera: Blastomyces, Coccidioidomyces, Trichophyta, etc.

General Morphology of Fungi

All fungi are plants without roots or leaves and do not possess chlorophyll. A mass of intertwined spiderweb-like structures called *mycelium* are the "roots" which anchor the mold to the food supply.

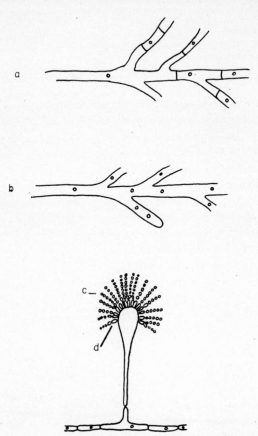

Fig. 41.—*a*, Septate mycelium of a mold.   *b*, Nonseptate mycelium.   *c*, Conidiospore of *Aspergillus* sp.   *d*, Conidiophore attached to conidial head.

This *mycelium* may be broken or jointed in segments, "septate," or it may be free flowing protoplasm, smooth and interrupted, "nonseptate," depending on the genus and species. Reproduction is primarily by spores. They are multicellular, except some of the true yeasts, which are unicellular and reproduce primarily by budding.

REPRODUCTION BY SPORES

A. Asexual spores
    1. Sporangiospores—spores inside sporangia or bulbous swellings.
    2. Conidiospores—develop on true conidiophores.
    3. Blastospores—true budding as in the yeasts.   Saccharomyces, etc.
    4. Arthrospores—mycelium swells and fragments.
    5. Chlamydospores—cells of mycelium swell and form spores.

B. Sexual spores
    1. Basidiospores—formed in a club-shaped structure with usually four basidiospores.
    2. Ascospores—these spores are formed in rounded sacs.
    3. Zygospores—formed by the joining of two similar cells.

Spores of fungi are usually quite resistant to physical and chemical agents and may cause spoilage of bacteriological culture media when not killed by the sterilization process.

The nutritional needs of most all the true fungi are so simple that they can manufacture their own nutritional requirements from only water, organic matter, and a few basic elements.   Moisture and warmth plus a food supply such as dead leaves, grass, or dead animal matter supports the growth of these organisms.   Some even thrive on insects which they are able to ensnare.   In soils and other habitats of molds, the pH can vary widely, from an acid pH of 1.6 to an alkaline pH of 10 or 11.   Thus in very acid or very alkaline soils the molds will survive while ordinary bacteria cannot endure these wide ranges of acidity and alkalinity.

Practically all the fungi will grow well at ordinary room temperature (68-75°F.) and incubator temperature (98.6°F.) may actually retard growth.

THE PHYCOMYCETES

Rhizopus

One of the most common species of the class phycomycetes is *Rhizopus nigricans*.   It is frequently found as a white to gray white cottony growth on bread.   The asexual spores develop within a walled structure called a sporangium which eventually bursts and the spores are disseminated into the surrounding area.   Thus, moldy bread will aid in seeding other bread or starchy foods in the household.   In the sexual state *zygospores* are produced by the union of the different molds of the same species, the contacting of the so-called "positive" and "negative" cells produces a heavy black growth of sexual spores. The mycelium of this mold is nonsegmented or nonseptate.

Mucor

The genus *Mucor* possesses many species but *M. mucedo* is perhaps the most commonly encountered one of this group. These molds are generally black. They are found primarily on decayed vegetable matter, on fruits, and manure. Their mode of reproduction is similar to that of the Rhizopus species.

Pythium

Most species of the genus *Pythium* are saprophytic soil inhabitants, but others may attack young seedling plants such as tobacco, cabbage, and tomato.

THE ASCOMYCETES

Penicillium

The most common genus of this group is *Penicillium*, the genus of fungi which produces the wonder drug "penicillin." They produce a greenish colored growth. They prefer room temperature, and many species fail to grow in the incubator at 37°C. Under the microscope the growth resembles a many fingered hand or a witch's dilapidated broom. There are about 600 species, two of which are used in penicillin production, *P. notatum* and *P. chrysogenum*. *P. glaucum* is a common species that causes fruit spoilage. Several species are used in the manufacturing and curing of cheeses; *P. camemberti* hydrolyzes casein, imparting the special flavor of Camembert cheese; *P. roqueforti* produces volatile acids from the butter fat, giving the Roquefort cheese its distinctive flavor.

Aspergillus

The species of the genus *Aspergillus* vary somewhat in color, some being black, others white, while others are green on the top surface with a yellowish color at the agar surface. The germinal part of this mold, the conidial head (vesicle), resembles the Giant Blunderboar's war club, the pointed spikes being the sterigmatas and conidiospores.

These fungi are found in decaying vegetable matter, hay, and grains. *A. niger* is used in commercial citric acid production; *A. niveus* is employed for the production of citrinin. Two species may be pathogenic, *A. nidulans*, and *A. fumigatus*, producing a pulmonary disease sometimes seen in grain workers and pigeon feeders. *A. glaucus* may cause leather spoilage and also spoilage of walnuts, chestnuts, and other edible foods.

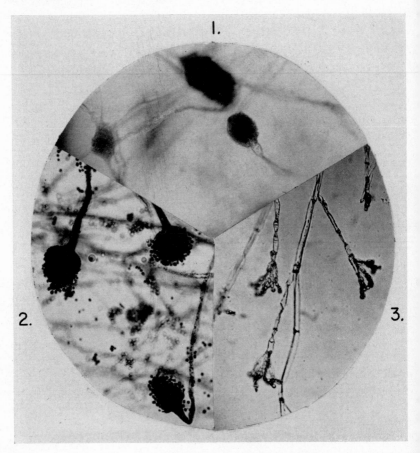

Fig. 42.—*1, Mucor mucedo.   2, Aspergillus niger.   3, Penicillium notatum.*

Yeasts

Without this group of fungi, our breadmaking would be a difficult task. The yeasts have played an important part in the development of civilization.

The genus *Saccharomyces* contains many useful species, all of which reproduce primarily by budding. When grown, they produce a typical "yeasty" odor. Yeasts are used as foodstuff for both animals and man. They are rich in protein enzymes, vitamins, and nucleoprotein; thus they are a readily available source of vital foodstuffs required by the animal body.

The common breadmaking yeast, *S. cerevisiae*, produces active growth in bread dough, evolving a great deal of carbon dioxide causing the bread to rise. Before baking, bits of the leaven (raw bread dough) can be stored in the ice box to prevent yeast multiplication and can be used as a yeast start. A recent innovation in bakeries or stores is unbaked rolls, etc. The dough is allowed to rise, then heated in an oven just enough to kill the yeast cells and destroy the enzymes. There is no further rising or even souring of the rolls. The purchaser merely completes the baking process in her own oven.

S. ellipsoideus is a wine yeast and is extensively used in fermenting grape juice to produce a variety of wines. *S. piriformis* is a yeast employed in the production of ginger-beer. *S. sake* is utilized to ferment rice starch; the resulting alcoholic beverage "sake" being the national Japanese beverage.

THE BASIDIOMYCETES

This class of fungi is best represented by the common mushroom. Puffballs, tree (bracket) fungi, smuts, and rusts are also included in this group.

The mushroom is composed of a stalk, which is a thick bundle of mycelia. These turn outward, forming the gill-like partitions and the umbrella-like top.

The most frequently eaten mushroom is of the genus *Agaricus*, while the most deadly species are those of the genus *Amanita*. In warm climates, after the first spring rains, the mushrooms may literally pop up all over from the ground. Commercially, mushrooms are cultivated in temperature-controlled cellars, caves, tunnels, or mushroom huts.

The Uredinales and mushroom poisoning are discussed in the following section, pathogenic fungi.

2. PATHOGENIC FUNGI

In Part 1 of this chapter it has been pointed out that the majority of the yeasts and molds are not only harmless to man, but also many are actually beneficial in food production, fertilization of soils, and manufacturing of a variety of chemicals. Some are "pests" in the laboratory, the spores floating freely in the air and on dust particles frequently causing contamination of laboratory media. Such a contamination led Fleming to the discovery of penicillin. Other fungi, especially in the tropics where it is always warm and moist, literally eat the leather shoes off one's feet, destroy clothing, destroy tents, and practically all nonmetallic objects.

The pathogenic as well as the nonpathogenic fungi require such simple foodstuff that they will grow on practically anything that is warm and damp—wet sacks, wet newspaper, clothes, shoes, etc.

Several days to several weeks may be required for adequate growth of the pathogenic fungi. Most of them grow well at room temperature, but a few grow better at incubator temperature. An acid medium enhances the growth of fungi while decreasing ordinary bacterial growth, thus a medium known as Sabouraud's culture medium is frequently used, since its nutrients favor the growth of fungi and its pH (5.2—5.6) being acid, discourages massive bacterial growth.

Many yeasts and molds are potentially dangerous, producing various types of disease, while others produce minor diseases. One of the major fungus infections is Blastomycosis, while a minor fungus infection is ringworm. Most of the disease-producing fungi are classified as *Fungi Imperfecti*.

DISEASE-PRODUCING YEASTS

Candida albicans

One of the very common yeast infections is caused by *C. albicans*. Several species of *Candida* are known, but only one appears to be pathogenic.

Three parts of the body frequently are attacked, skin, lungs, and nervous tissue. Lung infections as well as skin infections may spread to the nervous tissue, the most serious type of this disease. Generalized spread throughout the body is invariably fatal. No specific treatment is available.

This organism, in the yeast phase, is 2-4 microns in diameter, and shows prolific budding. It resembles the *Saccharomyces* (bread and beer yeast) in its morphology. It stains intensely gram-positive. If this organism is seeded on a corn meal agar plate by scraping deeply into the agar it will grow in the mycelial phase. The growth resembles a Pyracantha bush (*Coccinea pauciflora*) with large clusters of chlamydospores on the branched fungus resembling the berries of a Pyracantha. This microorganism ferments glucose and maltose with acid and gas production. Acid only is produced from sucrose, and lactose is not attacked. The organism grows easily in the yeast phase on ordinary nutrient agar, either at room temperature or in the incubator.

Candida albicans produces sores in the mouth, often referred to as thrush. Prolonged use of antibiotics may occasionally allow this yeast to initiate an infection. Apparently many of the common mouth bacteria are destroyed by the antibiotic, but the yeast is not affected. Thus the yeast cells are given a selective growth advantage over the bacteria. The disease is communicable from person to person and in nurseries this infection may be seen in epidemic form. Other parts of the body may be affected, particularly the hands and face and often the vagina. The skin becomes scaly, dry, cracked, and frequently becomes secondarily infected with bacteria.

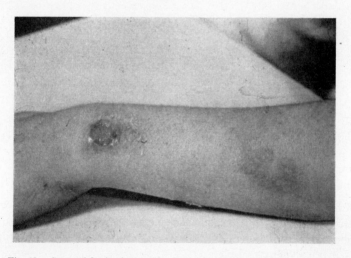

Fig. 43.—Sporotrichosis ulcer on the wrist of a 15-year-old boy. Note the lesion up the arm which has not ulcerated.

Sporotrichum Schenckii

Sporotrichum schenckii, the cause of the disease sporotrichosis, is, in the yeast phase, a rather large cigar-shaped or elliptical yeast. It exists in both the yeast phase and mycelial phase in infections and may grow on culture media as either the yeast phase (white colonies) or the mycelial phase (dark tan to brown colonies). The yeast phase is more frequently seen at incubator temperature. In the mycelial form, the delicate hyphae supports numerous conidiophores with clusters of conidia, resembling a bunch of cherries on a small branch. Biochemical reactions are undependable and identification is accomplished by its typical morphology.

The spores are widespread and infection is caused by introduction of the spores into the skin by means of scratches, thorn pricks, etc. The infection is at first localized at the site of the introduction of the spores, but after a week or more spreads to contiguous lymph nodes. The organisms then spread along the arm (more frequently the hand or wrist is initially infected) and attack and liquefy lymph nodes en route. Ulceration of the liquefied lymph nodes is frequent. Iodides are specific for the treatment of this disease.

Cryptococcus Neoformans

Torulosis, caused by *C. neoformans* is a rather uncommon disease, but sufficiently serious to warrant consideration.

The yeast, *C. neoformans*, is a fairly large, round yeast body reproducing by budding. It possesses a thick heavy capsule and is gram-positive. It does not form mycelia. Growth on ordinary media produces large, watery, slightly gray-brown colonies possessing buttery consistency. This yeast grows equally well at either room or incubator temperature.

Hormodendrum

Hormodendrum species all produce similar types of growth on laboratory media. The colonies are black, irregular, and spreading mycelia are raised from the surface of the agar. In tissue the spores reproduce by splitting and at times resemble four distinct window panes in an irregular circle.

The disease caused by *Hormodendrum pedrosoi* and *Phialophora verrucosa* is called chromoblastomycosis, a chronic disease of the skin. It may last for several years. The skin is warty, granulomatous, dry, and scaly. No specific treatment is known.

The organisms causing this disease inhabit the soil and apparently must be introduced by means of an injury to the skin. It is probably not communicable from person to person.

THE DIMORPHIC MOLDS

Three closely related fungi comprise this group, the genera being *Histoplasma, Blastomyces*, and *Coccidioides*. All produce, in the body, yeastlike bodies, but never mycelia. On culture the mycelial phase

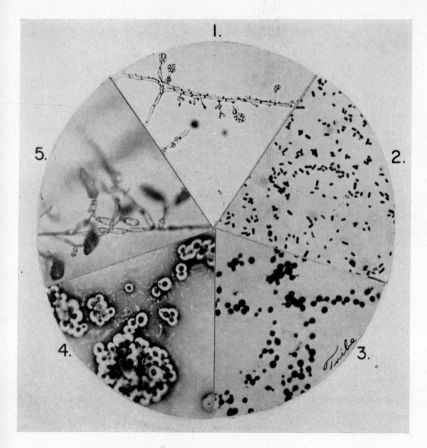

Fig. 44.—*1, Sporotrichum schenckii. 2, Candida albicans. 3, Cryptococcus neoformans* (Gram stain). *4, Cryptococcus neoformans* showing capsules (Congo red stain). *5, Hormodendrum pedrosi.*

is more frequent, but at incubator temperature, *Histoplasma* and *Blastomyces* will produce the yeast phase, while *Coccidioides* fails to produce a yeast phase on culture media.

Histoplasma Capsulatum

Histoplasma capsulatum is a small yeastlike body, 0.8 to 1.0 micron, usually exhibiting irregular staining. The colonies on plain or blood agar are similar to colonies of micrococci. The mycelial phase on culture media produces a velvety white, cottony colony. After several weeks, large tuberculate chlamydospores are formed, resembling the balance wheel of a watch. Identification is made by morphology and not by biochemical methods.

This organism is widespread, being found in soil, in caves, in infected cats, dogs, skunks, and other animals. In the south and southwestern states investigators have found, by the use of the skin test (histoplasmin extract of the mold), that about 12 per cent of the population were infected; however, the cases were so mild, that no infection was known to the patients.

The mold may produce infections of the lungs, resembling tuberculosis. In fatal infections, the mononuclear cells of the blood and cells of the bone marrow are infected, producing a disease resembling certain anemias as in other blood diseases. No specific treatment is known.

Blastomyces Dermatitidis

Blastomyces dermatitidis is a moderately large budding yeast in tissue (infections). On culture the yeast phase colonies are wrinkled and waxy. The mycelial colonies are coarse and cottony white. This organism grows either at room or incubator temperature on ordinary laboratory media, including Sabouraud medium. The mature mycelia contain thousands of single pyriform conidia on tiny hyphae.

This organism is widespread, but more commonly seen in large valley farming areas. The organism may infect primarily the skin or the lungs. When disseminated, all body tissues may be involved. Skin ulcerations are produced, and in the lungs, solid masses are seen, with some cavity formation, resembling tuberculosis. Certain drugs such as aromatic diamidines may be helpful in the treatment of blastomycosis.

Patients showing a positive skin test with blastomycin (extract of cultures of the mold) have a good prognosis. Patients showing a positive complement fixation test have a less favorable prognosis.

Coccidioides Immitis

Coccidioides immitis is a spherule-like yeast body with a double refractible membrane and inside this body is contained many tiny endospores. This form is seen in infectious processes. On artificial culture media, this fungus produces only the mycelial phase, the colonies being white and cottony. After two or more weeks of culture, the mycelia swell, producing a jointed appearance, the characteristic arthrospores being diagnostic. When dry, these arthrospores are easily dislodged and are discharged into the air by simply removing a cotton plug from such a culture tube. Thus this organism is dangerous to work with unless special precautions are taken.

The fungus is widespread, but is more endemic in the San Joaquin Valley of California, and in certain areas of Arizona and New Mexico. In areas of frequent dust storms, the infection rate is quite high. Certain rodents are natural hosts of this fungus.

In man, the lungs are primarily the infection site, although local skin infection is sometimes seen. Fever, malaise, and general weakness, with bronchitis are frequent symptoms, sometimes lasting several weeks to several months. The death rate is low, but when the organism disseminates throughout the body, draining abscesses are seen in many organs, including the brain. This form is invariably fatal. The disease frequently resembles tuberculosis.

Positive skin tests, using coccidioidin (extract of cultures of the mold) suggests a favorable prognosis, while a positive complement fixation test indicates a spreading infection and a less favorable prognosis. No specific treatment is available.

THE RINGWORM GROUP

The most common and exasperating of the pathogenic fungi are those that produce local skin infections, manifesting a variety of skin lesions, frequently called ringworms. Three genera are recognized, *Epidermophyton*, *Microsporum*, and *Trichophyton*. Several species of these genera are capable of producing infections of both man and animals. The genus *Trichophyton* contains about a hundred species, those previously classified under the genera *Achorion* and *Endodermophyton* have been placed in the genus *Trichophyton*.

These fungi grow readily on ordinary nutrient or Sabouraud agar. Colonies may be slow in developing, sometimes requiring a week or more. Room temperature favors growth. Colonies may be smooth or wrinkled, powdery, or cottony.   Many produce colored colonies, and a variety of colors may be seen, white, red, purple-violet, brown, yellow, etc., with variations in shades of the above colors.

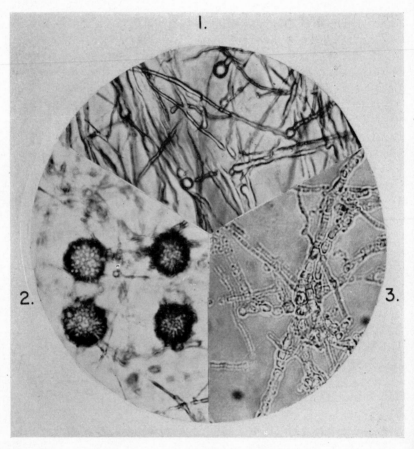

Fig. 45.—The dimorphic pathogenic fungi.   *1, Blastomyces dermatitidis.   2, Histoplasma capsulatum.   3, Coccidioides immitis.*

Three types of spores may be observed: (1) macroconidia (spindles and fuseaux), (2) microconidia, and (3) chlamydospores (thick walls).   The mycelia are complex, intertwined rootlike structures, in or on which the spores may be seen.   Some are coiled on the ends, resembling a spring.

The most common infection by this group of fungi is commonly known as "athletes foot" (tinea pedis), caused by many species of the genus *Trichophyton* and *Epidermophyton floccosum*.

The spores of these fungi are quite resistant to physical and chemical agents and when on shower floors, cement floors around swimming pools, or other areas where people walk in their bare feet the spores may be deposited on the feet. The moist areas between the toes are choice sites for these organisms to start an infection. The infection begins with a slight itching, then tiny blisterlike visicles that are brownish to cream colored. These tiny blisters break, the area beneath the blisters becomes dry, cracks, and becomes very painful, particularly if they are secondarily infected by various bacterial species. These infections in between the toes (athletes foot) may continue for several months to several years. Vigorous treatment is necessary, numerous remedies being available, but the serious types should be treated by a physician. Shoes contaminated with fungous spores can be sterilized by placing a piece of cotton soaked with formaldehyde in a closed container (air tight) in which the shoes are placed, leaving the shoes in this container for about forty-eight hours. Frequent change of shoes and boiled socks are effective preventive measures against spread on the feet.

Ringworm of the scalp (tinea capitis) is a fairly common childhood disease and may be seen in epidemic forms. Circular to irregular areas of transient baldness are common. Several species of *Trichophyton* may cause this disease, *T. schoenleini*, *T. tonsurans*, and *T. violaceum* being the most common. Another species, *Microsporum audouini*, which causes this disease is spread by direct contact not only from one area to another in the same child, but also infects playmates of the infected child. *M. canis* is primarily a ringworm infection of dogs and cats, but direct contact infection of children is possible from these diseased animals.

Different species of the dermatophytes may infect different parts of the body, for example, some prefer the scalp, others prefer the beard; fingernails and toenails may be so involved that they present a moth-eaten appearance. The hands may be so completely covered with ringworm infection that any type of work with the hands is impossible. Adults and children are equally susceptible to nearly all forms of ringworm infection.

Treatment is usually necessary for several weeks to several months to completely free the infected area of all spores.

MUSHROOM POISONING

Two species of the mushroom group (Basidiomycetes) are poisonous. The white *Amanita phalloides*, a common fungus (mushroom) in some areas, will produce severe symptoms within fifteen minutes after eating. Marked perspiration and salivation, moderate flow of tears, vomiting and severe watery diarrhea are the most important symptoms. Confusion is quite frequent. Death may ensue in forty-eight hours after the first symptoms.

Another mushroom of this genus, *A. muscaria* is also quite poisonous, but treatment of this type of poisoning is quite successful.

ERGOT POISONING

Another fungus that formerly caused a great deal of sickness in certain grain raising areas is *Claviceps purpurea*, a parasitic fungus of rye and other cereal grains. The poisonous substance, ergot, causes vascular damage, producing gangrene particularly of the fingers and toes.

THE UREDINALES

The rust fungi (*Uredinales*) comprise a group of about one hundred genera and about six thousand species. These fungi infect a variety of grains, particularly wheat. They also may produce woody galls on a variety of trees, especially conifers.

One of these fungi, the rust infection of wheat and other cereal grains, *Puccinia graminis*, has a rather interesting life cycle. From the wheat stalk the spores of this rust are carried by wind or bills of birds to a barberry bush where the spore continues to a new developmental cycle. After a series of cyclic changes, the new spore may then be again transferred to wheat. If barberry bushes are near wheat fields this cycle is annual. When no barberry bushes are available, wheat serves to complete the germinal cycle, but a shorter and different cycle is completed in the uredineal stage.

ACTINOMYCES AND NOCARDIA

Although the microorganisms of the genera *Actinomyces*, *Nocardia*, and *Streptomyces* are classified as true bacteria or *Schizomycetes*, they frequently produce diseases quite similar to the fungi, thus they will be discussed briefly in this chapter because of the latter character-

· istics. These microorganisms belong to the class *Schizomycetes* (fission fungi) and the order *Actinomycetales*.

Actinomyces bovis and *A. israeli* are anaerobic, gram-positive, branching rod forms of microorganisms. In anaerobic broth media the organisms grow in cottony ball-like structures, a stained specimen showing long, intertwined, bacterial cells, resembling a long piece of fine wire haphazardly wound into a ball.

The most common disease produced is "lumpy jaw" in cattle and actinomycosis of the jaw of man. The infection, may, however, spread from the local jaw area to various organs of the body, including the brain and heart. Primary lung infections also may be seen. In areas where the skin is dead and broken (necrotic), due to this infection, pus is present and in this pus tiny cream to yellowish colored granules are seen called "sulphur granules."

The *Nocardia* is the genus representing the *aerobic* actinomyces-like organisms. About five species of this genus are pathogenic for man. Their morphology is similar to the true actinomyces, being gram-positive, long and frequently branched rods. Some of the species of *Nocardia*, are, however, acid-fast. All species produce colored colonies. Only one species, *N. asteroides*, fails to coagulate milk. This same species also fails to liquefy gelatin, while the other species are proteolytic.

The disease produced by these microorganisms is called *Nocardiosis*, which is a chronic disease of the skin or bones, with tumorlike growths, frequently warty and draining pus. *N. asteroides* may produce a generalized systemic disease, often seen first in the lungs.

Streptomyces

The organisms of the genus *Streptomyces* are similar to the *Actinomyces* and *Nocardia* morphologically. The *Streptomyces* species are soil and water forms that aid greatly in soil fertilization as well as certain species being excellent antibiotic producers. Some of these organisms, when growing in mud of shallow portions of a lake, produce off-odors and musty smells in water.

Selected Reading References

Ainsworth, G. C.: Medical Mycology, London, 1952, Pitman Publishing Corporation.
Alexopoulos, J. C.: Introductory Mycology, New York, 1952, John Wiley & Sons, Inc.

Bunnell, I. L., and Furculow, M. L.:   A Report of Ten Cases of Histoplasmosis, U. S. Public Health Rep. **63:**299, 1948.

Henrici's Molds, Yeasts and Actinomycetes, ed. 2, revised by Skinner, C. E., Emmons, C. W., and Tsuchiya, H. M., New York, 1947, John Wiley & Sons, Inc.

Smith, C. E., Beard, R. R., Rosenberger, H. G., and Whiting, E. G.:   Effect of Season and Dust Control on Coccidioidomycosis, J. A. M. A. **132:**833, 1946.

Wolf, F. A., and Wolf, F. T.:   The Fungi, New York, 1947, John Wiley & Sons, Inc., Vols. 1 and 2.

CHAPTER 30

MICROSCOPIC ANIMALS

Most of the protozoa are unicellular free-living forms of the anima kingdom found principally in warm, stagnant water. The most common one is the *Amoeba proteus*, a simple unicellular animal about 0.25 mm. in diameter. It contains a nucleus, food vacuoles in the cytoplasm, and various other constituents usually found in animal cells. It is motile by extending its protoplasm into fingerlike projections, the so-called pseudopodia. Another common protozoan is the *Paramecium caudatum*, a water form, possessing hair like structures around its body, called cilia. Many thousands of species of the non-pathogenic types exist, most of which may be easily seen by means of the low power lens of an ordinary microscope.

These forms of microscopic animals derive their food from available substances in water such as other small animal cells, bacteria and bits of utilizable plant life. The single cell is therefore a complete animal, procuring its food, metabolizing the food, utilizing oxygen, excreting waste products, and reproducing its kind.

PATHOGENIC PROTOZOOAN PARASITES

Disease-producing protozoa are fairly numerous, particularly in Asia, the Tropics, Africa, and other places where the climate is warm to temperate. Some, however, may be found in colder climates, but as a rule are not too commonly found there.

Intestinal Parasites—Class Rhizopoda

The most serious one of this group is *Endamoeba histolytica*, a parasitic protozoa affecting the lower bowel of man and animals. Two forms may be seen, the ameboid form (vegetative) and the cyst form. The vegetative form is motile by means of protoplasmic pseudopodia and these may be observed under the microscope in fresh, warm, stool specimens. This form is about 20-30 microns in diameter. The cytoplasm of fresh vegetative forms may contain whole, intact red blood cells and ingested bacteria.   This vegetative (trophozoite) stage has a single nucleus.   Under unfavorable conditions the vegetative stage reverts to the *cyst* stage, becoming a small rounded organism 7-15 microns in diameter, with four well-defined nuclei.

Endamoeba histolytica produces a world wide distributed disease, amoebic dysentery, a chronic ulcerating disease of the lower bowel. From three to ten bowel movements per day are common.   The disease tends to be chronic.   *Endamoeba histolytica* burrow through the bowel and by a sinus tract tunnel make their way to the liver where they produce liver abscesses.

The cyst stage is the one capable of producing a new infection, since the acidity of the stomach and pepsin therein have no effect on the cyst.   After it reaches the small bowel, the trophozoite (vegetative) form develops.   Ingestion of contaminated food particularly fresh vegetables contaminated with sewage or feces, and water are the principal sources of infection.

Certain antibiotics and drugs are effective curative measures.

Endamoeba coli, a nonpathogenic protozoan parasite is similar in morphology to the true *Endamoeba histolytica*, but it is generally larger and not ameboid.   The cysts are larger (10-30 microns) and have eight nuclei.

Balantidium coli is a very large protozoa, egg-shaped, with cilia around the periphery.   It varies in size, being from 50 to 80 microns in length and circumference.

Its principal hosts are the monkey and pig, but man may become infected.   Ingested cysts will start the disease in susceptible animals.

Mastigophora (Intestinal flagellated protozoa)

Although several genera and species are known to inhabit the intestinal tract of man, they do not seem to be sufficiently hardy to produce a prolonged serious infection.   *Giardia lamblia, Chilomastix mesnili* and *Trichomonas hominis* are found in the human intestine, but the disease produced by any of these is relatively benign.   Another species, *Trichomonas vaginalis*, is frequently found in chronic vaginitis, producing a chronic discharge.   The organism is highly motile at body temperature, but quickly becomes nonmotile if the temperature decreases slightly.

Several similar organisms of the class *Mastigophora*, belonging to the family *Trypanosomidae* are of definite pathogenic importance.

These are called hemoflagellates because they parasitize the blood not only of man, but also of lower animals as well as some invertebrates. They resemble a small minnow without a fish type tail, but instead have a rather long single whiplike undulating membrane extending from the head to the tail. Some are gently curved. In the nonflagellated phase, they resemble a protozoan parasite, not unlike *Endamoeba coli.*

The genus *Trypanosoma* contains several species, the *T. gambiense* being the most pathogenic of the group. The *T. gambiense* is the cause of African sleeping sickness. After the bite of a tsetse fly which is the carrier, the patient may not show any ill effects for several weeks to several months. The trypanasoma invades the blood stream, then infects most all organs of the body. The climax of this disease is the invasion of the brain. Death ensues within a few months after the nervous tissue is attacked.

Several species are capable of producing disease in man, *T. cruzi* being equally as common as *T. gambiense.* Another species, *T. hippicum*, is transmitted by the bite of an infected vampire bat, but domestic animals are the more frequent victims. *T. equiperdum* is the cause of "dourine," a disease of horses resembling the disease syphilis in humans. *T. lewisi* and *T. duttoni* are species that infect rodents. In invertebrates (fleas, lice, etc.) these trypanosomes multiply in the hindgut, their feces carry the microorganism responsible for the spread of the trypanosoma. *T. lewisi* is especially important as a laboratory animal trypanasoma infection, research being easily carried out with this species. Nearly all species of trypanosoma are transmitted by the bite of infected insects.

THE SPOROZOA

One of the remaining great scourges of mankind still rampant in many areas of the world is malaria. Although the parasite causing this disease was not observed until 1880 by Laveran, the term malaria was coined by an Italian in 1753 (*mal*-bad; *aria*-air) since he was of the opinion that bad air was responsible for the disease because those living near swamps, or areas of stagnant smelly water, more frequently developed the disease.

The disease malaria is caused by four species of the genus *Plasmodium*: *P. vivax, P. malariae, P. falciparum*, and *P. ovale.* These parasites are transmitted to man by the bite of several species of the female *Anopheles* mosquito. The female requires blood for egg fertilization, and man, birds, and animals are their blood source.

Before the mosquito can transmit the disease to man, a series of morphological and physiological changes takes place in the plasmodium, termed the life cycle (*sporogony* in the mosquito and *schizogony* in man).

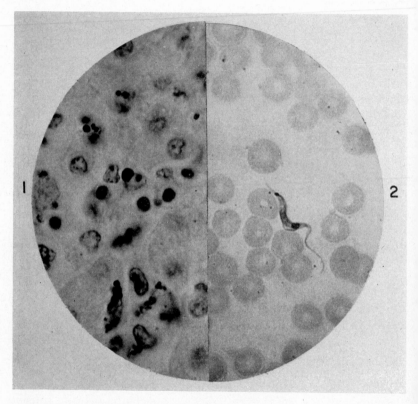

Fig. 46.—Blood parasites. *1, Plasmodium* sp. in liver cells. *2, Trypanosoma lewisi* in blood smear. Note the size of this parasite in comparison to the red blood cells.

In the sexual cycle the *microgametocyte* (male) and *macrogametocyte* (female) of the plasmodium are taken into the mosquito when infected blood is ingested. In the mosquito's stomach the *macrogamete* is formed. *Microgametes* develop with flagella, these then fertilize the *macrogamete*. The resulting *ookinetes* penetrate the stomach wall. Here they undergo growth and nuclear proliferation, becoming *oocysts*.

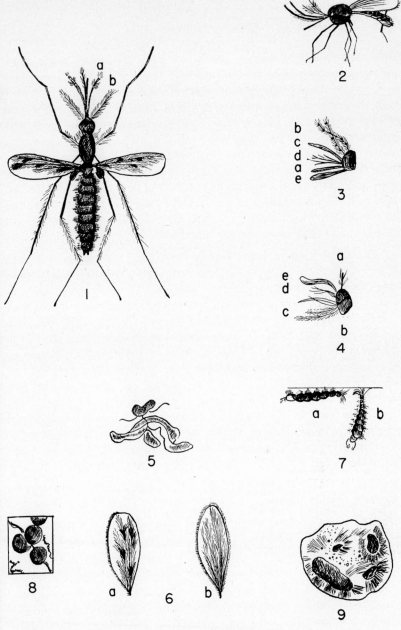

Fig. 47.—*1*, Male *Anopheles* mosquito; *a*, palpus; *b*, antennae. *2*, *Culex pipiens* mosquito. *3*, Head and biting apparatus of *Anopheles* female mosquito; *a*, palpus; *b*, antennae; *c*, proboscis; *d*, stylet. *4*, Head and biting apparatus of female *Culex* mosquito; *a*, palpus; *b*, antennae; *c*, proboscis; *d*, stylet. *5*, Anophelene salivary gland (25×); gland ducts join the common duct. *6*, *a*, Anophelene wing, distinguished by its fine spots; *b*, *Culex pipiens* wing, without spots. *7*, *a*, *Anophelene* larvae; note characteristic position parallel to water surface. *b*, *Culex* larvae; characteristic position vertical to water surface. *8*, Oocysts of plasmodia in stomach of Anopheles mosquito. *9*, Mature sporozoites of plasmodia within the mosquito.

The *oocysts* rupture, loosing many hundreds of *sporozoites* (7-9 microns long) which scatter about in the anatomy of the mosquito. Some find their way to the salivary glands of the mosquito, and in this *sporozoite* stage they are injected into the body of man by the biting female *Anopheles* mosquito. Here they penetrate the red blood cells and undergo several further morphological and physiological stages. In the red blood cell the stages are: (1) ring stage, (2) trophozoite stage, (3) schizont stage, (4) segmenter stage, (5) micro- and (6) macrogameto-cytes, and lastly (7) the merazoite stage. As they multiply in the red blood cell, they eventually cause the cell to swell and burst, liberating the *merazoites*, which then penetrate other blood cells. The asexual cycle is then repeated in the red blood cells.

The incubation periods of each vary, generally being from 7-10 days to 7-20 days. Severe fever occurs regularly, the elapsed time being dependent on the different species. With *P. ovale* and *P. vivax*, the febrile paroxysms occur each 48 hours; those infected with *P. malaria* have severe fever and chills every 72 hours; infections with *P. falciparum* frequently have a continuous fever, but may have 24 to 48-hour afebrile periods. These attacks of fever and chills are at times so severe that the patient may actually shake himself out of bed unless he is either restrained or "sideboards" are placed on the bed. Those who recover may have relapses or they may remain carriers of the plasmodium and thus are a hazard to their fellow men. Blood from a malarial suspect should be carefully examined before using it for transfusions. The plasmodium may infect many organs of the body, the spleen, liver, and brain being areas frequently affected.

Special stains (Giemsa and others) of blood smears are used for diagnosis of this disease.

A variety of drugs have been used with success in treating as well as preventing this disease. Quinine has been a long-used remedy. Newer drugs, chloroquine diphosphate, quinacrine (Atabrine), Plas-mochin and other chemical agents are also used successfully.

Prevention is best realized by mosquito eradication. Drainage of swamps, potholes, and stagnant pools discourage mosquito breeding. Oiling water surfaces, spraying with DDT and other methods have been quite successful in reducing mosquito populations. In many areas, mosquito abatement districts have been formed with teams of workers going out into adjacent as well as remote areas, spraying water holes and draining swampy areas.

The disease is still prevalent in the United States, about 10,000 infections taking place in 1948, with 155 deaths.

Toxoplasmosis, a disease caused by a plasmodium-like parasite, *Toxoplasma*, produces an encephalitic disease, particularly in children. Toxoplasmosis is a rather uncommon disease.

PARASITIC HELMINTHS

Many of the flukes and tapeworms are macroscopic (one can see them without the aid of a microscope) but some are microscopic. The eggs are all microscopic, thus falling into the category of microbiology.

Flukes

Most of these parasites are uncommon in the United States, but are fairly common in various parts of Asia. They are rather specific in their parasitism of certain organs of the body. A number of species of liver flukes may cause disease in man: *Clonorchis sinensis, Fasciola hepatica;* lung fluke (*Paragonimus westermani*), intestinal fluke (*Fasciolopsis buski*) and blood flukes (*Schistosoma mansoni, Schistosoma japonicum*).

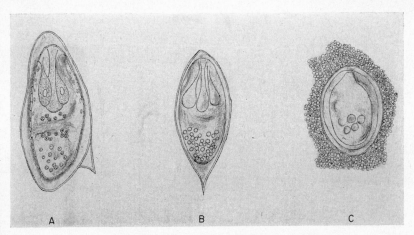

Fig. 48.   Ova of blood flukes.   *A, Schistosoma mansoni.   B, Schistosoma haematobium.   C, Schistosoma japonicum.*

The adult flukes are small (0.5-2.0 centimeters) wormlike animals. Some are transmitted by eating improperly cooked, infected fish, others may have a snail as an intermediary host (the flukes growing in the feelers of these snails) (*P. westermani*).   Most of these are diagnosed by finding the parasite egg in the stool.

Tapeworms

Only a few examples will be given since this phase of parasitology is rather extended and involved.

The swine tapeworm, *Taenia solium*, is a fairly common infection of man.   Eating improperly cooked infected pork is the method of infection.   Stools are examined for the eggs.

Echinococcus granulosus is the cause of hydatid disease or a cyst-like disease of the lungs and liver.   Eggs of this tapeworm are ingested by man, and a series of developmental changes take place.   After lodging in the liver or lung they form multiple cysts, resembling a group of white grapes in clusters.   These hydatid cysts contain the scolices (hydatid sand) of these worms, and if ruptured will propagate, producing new cysts in the tissue.

The disease is common in Alaska, Australia, and South America, but fairly uncommon in the United States.   Dogs, wolves, coyotes, cattle, and sheep are common hosts.   In the past four years, nine cases of this disease have been found in Utah, the youngest patient was a two-year-old girl and the oldest one a man near seventy.   The eggs may be also transmitted by recently contaminated water supplies, the feces of an infected animal gaining entrance to streams or ponds.

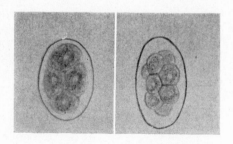

Fig. 49.—Ova of hookworms.   *A, Ancylostoma duodenale* (four-cell stage).   *B, Necator americanus* (eight-cell stage).

Hookworm Infestations

Hookworm infestations are found throughout the world.   In China, India, and other countries using human feces for soil fertilization the disease is very common.

The *Ancylostoma duodenale* and *Necator americanus* are two common hookworm species. Eggs of these parasites are passed in the feces, and if deposited in moist soil they will hatch into a larval form which will survive in the soil. The worm larvae can be ingested or it may "hook" onto the bare feet and penetrate the skin and finally migrate into the blood vessels. It then migrates to the lungs, gains access to the trachea, and is coughed up and swallowed and then attaches to the wall of the intestine. Anemia and marked weight loss are common symptoms of hookworm infestations.

Adequate treatment is available for this disease. Microscopic examination of the feces is the method of diagnosis.

Trichinosis

The parasite of this infestation was first observed in 1822 in Germany by Teidemann. The tiny worms (1.5-4 mm.) *Trichinella spiralis* infest the small intestine, and after the male fertilizes the female, the male leaves the body in the fecal waste of the body. The fertile female penetrates the intestinal wall, and after several days gives birth to many embryo worms. Some of these embryos enter the blood circulation and are distributed to various muscle areas, particularly to the diaphragm (large sheet muscle of respiration separating the lungs from the stomach, liver, and intestines). The embryos grow into the larvae stage in the muscle in about 3-4 weeks, at which time they are very painful and sometimes the toxicity produced may be fatal. The larvae become encysted, frequently calcifying after several years.

The rat is a natural host of this parasitic worm; pigs eat the infested rats and if we eat improperly cooked infested pork we will develop the disease. Wild bear meat is frequently infested with this parasite. Deep freezing meat will kill the larvae or proper cooking to "well done" of all pork are adequate measures for prevention of this infestation.

Selected Reading References

Belding, D. L.: Textbook of Clinical Parasitology, ed. 2, New York, 1952, Appleton-Century-Crofts, Inc.

Calkins, G. N.: The Biology of the Protozoa, ed. 2, Philadelphia, Pa., 1933, Lea & Febiger.

Herms, W. B.: Medical Entomology, ed. 3, New York, 1944, The Macmillan Company.

Most, H.: Parasitic Infections of Man, Symposium, New York Academy of Medicine, New York, 1951, Columbia University Press.

Neal, R. A.: Experimental Production of Pure Mixed Strains of Endamoeba Histolytica, Parasitology **42**:40, 1952.

CHAPTER 31

HOW DISEASES ARE SPREAD

Epidemiology and Prevention of Diseases

Man is frequently his own worst enemy in both carrying and disseminating disease microbes to his fellow man. Isolated, man is relatively free of microbial infections, but contact with strangers will frequently introduce other unseen enemies into his previously tranquil camp. Communities isolated during the winter months are frequently free of diseases, but on arrival of the first supply ship in the spring, great numbers of respiratory infections quickly begin among the previously disease free, isolated inhabitants.

As man has continually exposed himself to a variety of disease agents in the past several hundred generations he has built up a general resistance to many microbes. Certain diseases are now comparatively benign, such as measles, chicken pox, mumps, etc; however, the introduction of such a disease as measles into a formerly unexposed populace, as happened a few centuries ago, by the white man visiting certain Polynesian and Melanesian populated islands, nearly decimated many islands of their populace. The disease was staggeringly virulent, causing death to many thousands.

The seasonal occurrence of many diseases is another important factor to consider. The exact reason for the seasonal preference of a microorganism to attack its victims is not known. Many seasons are referred to as certain "disease seasons"—measles season, poliomyelitis season, mumps season, etc. Measles cases, for example, begin in earnest in the winter months, reaching the epidemic peak in March, April, or May, then decrease rapidly. The whooping cough season usually begins in November and December, producing the greatest number of cases in the spring months. Poliomyelitis, on the other hand, is a summer and early fall disease, reaching its peak in numbers of cases in August or September. In South America, however, the peak number of poliomyelitis cases is in December, January, and Feb-

ruary, their summer months.   Insect borne diseases are seen only
during the period of time when these insects are active.

With knowledge gained by the tireless research workers on the
causative agents of disease, how they are spread, how they may be
prevented, and the more recent discoveries of the antibiotics for treatment, our world has been made a happier and more pleasant place in
which to live.   The mysterious unknown concerning disease has been
stripped of its fear, no longer producing panic among the informed
people.   A few of our diseases are still unsolved riddles, but scientific
research will eventually lead us to the conquering of the most stubborn
of our now unsolved diseases.

Three factors must be present before an infection can be initiated.
First, an infectious agent must be present.   Second, a susceptible host
must be in the environment.   Third, the susceptible host and the
infecting agent must collide.

A variety of factors influence the presence or absence of a parasite
in a given environment.   The habitat and environmental patterns of
some parasites influence their existence or absence in a given area.
An example of this concept is the disease yellow fever.   If we have
an environment in which the yellow fever mosquito (*A. aegypti*) exists,
it is not necessarily the only criteria for the disease yellow fever to
likewise coexist.   The virus must be introduced into the mosquito.
The picture is still not framed.   In order for the virus to develop
adequately in the mosquito for transmission to man, the infected
mosquito must have a suitable temperature—usually not below 68-
70°F. or not above 80-85°F. for a period of about 12 continuous days.
Thus areas in which the *A. aegypti* exists may not be yellow fever areas
if the temperature gradients are unfavorable for virus growth in the
mosquito.

Likewise the condition of the parasite may influence its effective
disease-producing ability.   Since man's defense mechanisms (antibodies, enzymes, leucocytes, etc.) has continually fought back all
invasions by microorganisms many of these disease agents have been
"softened" and the resulting microorganism may produce only mild
symptoms in the host.   Changes or mutations reflecting different
grades of invasiveness of microorganism are frequent occurrences.

HOST DEFENSE

Immunity as used here designates a previous contact with a
disease-producing agent.   Having had a disease and then recovered,

the host is frequently immune to a second attack with the identical parasite. There are, however, exceptions, and second attacks by certain microorganisms may occur. Some persons as well as animals will develop such a very mild, undetectable disease that hosts have no idea they have had a certain disease, yet they develop a solid immunity and resist infection with re-exposure to the same microorganism. Many people and animals are artifically immunized by the use of vaccines. Thus, we have in a large segment of our population many individuals who, despite an epidemic of a given disease, are immune to this infection, the presence of antibodies in the blood and other immune mechanisms such as tissue immunity being responsible.

Some immune individuals may, however, turn the tables and pass along, to unsuspecting susceptible hosts, a virulent microorganism. These immunes are called *carriers* because they carry pathogenic microorganisms in their noses, throats, stools, etc. Such diseases as typhoid fever, scarlet fever, diphtheria, poliomyelitis, and a host of others may be spread by "immune carriers."

In a given family unit, the normal flora of microorganisms of each member of a family group is so frequently passed back and forth that each member is used to the others' microbes and no illness is produced. However, this family group may pass some of its home grown microorganisms on to other families, and cause disease in its neighbors. Children frequently borrow microorganisms from their schoolmates and innocently pass these foreign microbes around to family members, causing a disease process in most of the family group. Common colds and measles are good examples of this type of transfer of infection within a family unit.

A resistant state of a person or animal is used here to designate the indifference of a given host or his cells to a microorganism. There is no answer to this "natural resistance" to certain disease agents, but many animals as well as man fail to develop certain diseases. Man, for example, is completely resistant to microorganisms that readily cause diseases in plants; conversely, measles of trees is possible only in story books. A rabbit or guinea pig may be injected with thousands of monkey infecting doses of poliomyelitis virus, yet these two species of rodents show no ill effects whatsoever.

Susceptibility of a host to a parasite must also be reckoned with. Some people are "born" more susceptible to disease than others, usually because their immunity-producing mechanism is poorly constituted.

Besides inherited predisposition to diseases there are many other factors that aid in producing disease susceptibility of an individual.

Proper diet plays a significant role in susceptibility to diseases. Lack of certain vitamins may decrease general tissue resistance to a disease. Shortage of certain amino acids invites infection. This is easily shown by feeding rats a diet lacking in a certain amino acid. The rats voluntarily eat one meal of this diet, then refuse to eat more, preferring to starve. By force feeding with a stomach tube, all rats develop a fatal bacterial infection within 7 to 10 days.

Hormonal imbalance frequently increases susceptibility to infections. The uncontrolled diabetic readily develops boils, abscesses, and other infections processes. Cortisone has been found to increase the susceptibility of hamsters to the Lansing (type 2) poliomyelitis virus.

PARASITE VIRULENCE

Virulence is used here to denote the ability of a microorganism to gain entrance into the body of the host, overpower the resistance of the host, and produce a disease. Various phases of virulence may be observed. Some organisms are so feebly virulent that they produce inapparent or nonrecognizable infections; others may produce moderate infection with mild symptoms; still others produce a severe disease with a high death rate. The resistance or grades of immunity of the host also must be taken into consideration when we discuss an actual infection of a single person, or when we consider an infectious process in epidemic form.

During some epidemic periods, a given microorganism will be highly virulent, producing a severe disease, at other times the disease agent (apparently the same type) becomes less virulent and produces a mild disease. The influenza virus and the diphtheria organisms are good examples of this variation in virulence. In the 1952-1953 influenza pandemic, the virulence was mild, the death rate of this subtype A_1 was practically nil, while in the 1917-1918 influenza pandemic the death rate was very high. In Europe a few years ago a diphtheria epidemic was extremely virulent, while normally the disease is fairly easily cured with prompt use of antitoxin.

The many factors involved in the virulence of a microorganism, such as capsules, exotoxin production, toxic bacterial body cells, mutations, etc., have been discussed in Chapters VIII and XI.

Frequently as an epidemic progresses, with microorganisms being passed fairly rapidly from one susceptible person to another, the virulence increases, and generally a more severe disease is seen in the mid-part of an epidemic period. As food for these microorganisms becomes scarce (population becoming immune) the epidemic dies out —or literally, the microorganism eats itself out of house and home.

HOW DISEASES ARE SPREAD

Direct contact infections are usually spread by sneezing or coughing so that a susceptible host breathes these microorganisms into his nose and throat. Such diseases as diphtheria, scarlet fever, poliomyelitis, mumps, measles, influenza, etc., are examples of "contagious" or direct contact infections. The common cold, although the exact cause of this disease is unknown, is undoubtedly a direct contact disease. Tuberculosis is also a direct contact disease.

Indirect contact infections are spread by more or less indirect means. Johnny who has a sore throat (perhaps streptococcus, diphtheria, etc.) chews on a pencil. Willie, a few minutes later chews on this same pencil and the microorganisms are transferred from the pencil to the unsuspecting Willie. Eating contaminated food, drinking water or milk that has been loaded with microorganisms, are all indirect means of taking into one's body disease producing parasites.

Another frequent unsuspecting method of indirect transfer of microorganisms, particularly in children, is by means of their hands. Coughing in one's hand, then holding hands in playing games transfers microorganisms from one hand to another. The fingers are then transferred to the mouth and again the chain is completed. Adults are also guilty of "finger and hand" transfer of disease microbes. A typhoid carrier who discharges *Sal. typhosa* in their stools frequently "soil" their fingers and if improperly washed, these fingers may transfer these microbes to food or water. Even such "carriers" may contaminate the uncovered lip of a bottle containing properly pasteurized milk and by pouring the milk over the lip of the bottle, transferring the typhoid organisms to the glass of milk. Another commonly "finger-hand" transfer of a microorganism is by individuals who may have an infected wound. If it is a *Micrococcus pyogenes* var. *aureus*, and the individual mixes salads, or other foods to be eaten uncooked, the organisms may multiply rapidly and produce an enterotoxin in the food. Many small epidemics are caused in this manner, particularly in the warm summer months at picnics where food may be kept many hours without refrigeration.

The *insect* spread of disease-producing microorganisms is a fairly frequent method. Flies are not at all fastidious in their eating and walking habits. Diseases such as the diarrheas, dysenteries, cholera, as well as others, are frequently spread by such insects. Biting insects such as mosquitoes spread diseases—(malaria, yellow fever, certain encephalitides) as well as other diseases. Biting flies, such as the tsetse fly, the deer fly (*Chrysops discalis*) and others also transmit disease-producing microorganisms. Ticks, body lice, mites, and various other types of insects are capable of transmitting a variety of bacterial, rickettsial, and virus diseases. A rather interesting insect, *Triatoma sanguisuga*, has a unique way of transmitting disease. It first takes its fill of blood, them moves up and defecates in the wound, then backs up and scratches the feces into the wound. Likewise diseases of plants may be spread by insects—the leaf hopper frequently spreading the curly top virus disease of beets.

Wind and *dust* storms may be responsible for the spread of certain disease microorganisms. In the San Joaquin Valley, dust and winds carry the spores of *Coccidioides immitis* resulting in an increased number of infections during these periods. Dust particles in hospitals may carry infectious microorganisms—thus oil mopping is helpful in preventing such spread.

Animal reservoirs may be responsible for the spread of disease. Rats and mice frequently harbor certain *Salmonella* species and contaminate food and water. Rats also harbor the organism of rat-bite fever, the *Leptospira* causing jaundice, and the microorganism causing plague. Rabbits are frequently infected with *P. tularense* and handling such infected rabbits may transfer the organism to humans. The bite of a rabid dog, skunk, or coyote may transfer the virus of rabies to other animals as well as to man.

Thus the spread of disease agents depends on many factors, and if a link in the chain of spread is broken, no disease will exist.

EPIDEMICS

A disease is said to be *epidemic* when more than the normally occurring number of cases develops in a community, state, or nation. When a disease is prevalent in a given community, more or less continually, but only a few cases appear during the year, it is said to be *endemic.* When a given disease "breaks out all over," or is world-wide in distribution and in epidemic form in many countries of the world at the same time it is called a *pandemic.* The influenza pandemics of 1917-1918 and 1952-1953 as previously mentioned are examples.

Epidemics in the making may be forecast or predicted with fair accuracy if good epidemiological records are kept. A slight, but constant, continued weekly rise in the number of cases of a given disease, over and above a normal or mean number of cases of the identical disease for the same period of time, acts as a barometer for epidemic prediction.

Most all states have laws requiring the physician to report certain reportable diseases to the state health department. The state epidemiologist then analyzes the weekly reports and keeps accurate tab of the "health pulse" of his state.

PREVENTION OF DISEASE

Actual quarantine of a family unit that has a case of illness is probably not too effective in preventing community spread. Isolation of the ill patient in a separate room or in a hospital is much more effective. However, during epidemics of disease or seasonal periods of disease increase (such as measles in the spring) many children are carriers of the virus, and spread of the virus is easily accomplished from one child to another during the latter phase of the incubation period. Detection of such carriers and isolation is an impossible task. However, detection of carriers of typhoid organisms, diphtheria organisms, meningitis organisms, etc., is possible and isolation of carriers does decrease the disease spread.

Children as well as adults should not visit other children or adults with a communicable disease. This simple precaution frequently decreases community spread.

Closing of schools seems to be not too effective in the prevention of spread of many communicable diseases. When children are out of school they have as many contacts and frequently more intimate contact with other children in their neighborhood by playing at each other's houses. When children are ill, they should not be sent to school or allowed to have visitors until they are free of infection. These simple precautions would immeasurably decrease communicable disease—even the common cold. Coughing, sneezing, and expectorating are the "jet planes" of communicable disease microorganisms.

The prevention of diseases spread by food can be accomplished by sanitary preparation and proper refrigeration. Water-borne diseases can be prevented by proper sanitary control of water sheds, water filtration, and chlorination and effective sewage handling. Milk-borne diseases are easily prevented by adequate pasteurization and disease-free cattle and sanitary milking parlors. Control of insects such as mosquitoes, lice, flies, etc., reduces the type of diseases spread by these insects. Likewise, rodent control effectively reduces diseases spread by rodents.

Prophylactic prevention of certain diseases can be accomplished by the use of gamma globulin, if this immune substance is given shortly after known exposure. Red measles and German measles are examples of such prophylactic prevention of disease.

Certain sulfonamide drugs may be used as preventive measures; other drugs such as quinine for the prevention of malaria have been used for many years.

Vaccination

Many diseases are completely prevented by adequate vaccination. In adequately informed countries, smallpox vaccination has almost completely eradicated this disease. Diphtheria is still a menace, but vaccination has decreased this dread disease to a shadow of its former killing orgies. Whooping cough is not nearly so common as it was a few years ago, due to effective vaccination programs. Yellow fever, a previous scourge in many tropical and semi-tropical countries, has been effectively strangled by widespread vaccination. Rocky Mountain spotted fever is now a comparatively rare disease, due primarily to effective vaccination. Typhoid fever has been reduced to a point where cases are so rare in some areas that medical students see only one or two cases of this disease in their total medical school training period. This disease has been brought under control by effective sanitary control of water and sewage supplies and vaccination. In the past few years another vaccine has been added to our armament against disease—mumps vaccine. Vaccines against influenza are partially successful, but due to the bizzare antigenic mutants, a completely effective vaccine is not yet available.

Future preventive measures for several other diseases are just across the horizon, and one of these, a vaccine against all three antigenic types of poliomyelitis virus, seems destined to be a reality within the next decade.

Selected Reading References

Dubos, Rene J. (Ed.): Bacterial and Mycotic Infections of Man, ed. 2, Philadelphia, 1952, J. B. Lippincott Company.
Herms, W. B.:   Medical Entomology, New York, 1944, The Macmillan Company.
Meyer, K. F.:   The Animal Kingdom, Reservoir of Human Disease, Ann. Int. Med. **29**:326, 1948.
Simmons, J. S., and Collaborators: Global Epidemiology, Philadelphia, 1944, J. B. Lippincott, Company.

APPENDIX

AUDIO-VISUAL AND VISUAL EDUCATION

All too frequently teachers fail to have at their fingertips the addresses or data on where they may obtain available films, slides, etc., which they wish to show to their students. Many such films, etc., may be rented at a nominal fee. Some may be purchased. The following list is therefore presented as a ready reference of names and addresses from which teaching films, slides, etc., may be secured.

1. Society of American Bacteriologists, Dr. Harry E. Morton, University of Pennsylvania, School of Medicine, Philadelphia 4, Pa.

2. Medical Film Guild, Ltd., 167 West 57th Street, New York, N. Y.

3. Clay, Adams Co., Inc., 141 East 25th Street, New York 10, N. Y.

4. Iodine Educational Bureau, Inc., 120 Broadway, New York 5, N. Y.

5. Movie Film Rental Library, Akin and Bagshaw, 2023 East Colfax Avenue, Denver 6, Colo.

6. Lederle Laboratories, Film Library, 643 South Olive Street, Los Angeles 14, Calif.

7. Academy International of Medicine, Film Catalogue Library, 214 West Sixth St., Topeka, Kan.

8. American College of Surgeons, Medical Motion Picture Films, 40 East Erie St., Chicago 11, Ill.

9. Department of the Army, Washington, D. C., For area address nearest to you write to the above address. Also: Gillette Pictorial Center, Building 603, Presidio of San Francisco, Calif.

10. General Electric Company, Motion Pictures Catalogue, Schenectady, N. Y.

11. U. S. Department of Agriculture, Office of Information, Motion Picture Service, Washington 25, D. C.

12. Encyclopedia Britannica Films, Wilmette, Ill.

13. United World Films, Inc., 445 Park Ave., New York 22, N. Y.

14. A Directory of 16 mm. Film Libraries, Bulletin No. 10, 1949, Department of Health, Education and Welfare, Office of Education, Washington, D. C.

15. American Medical Association, Motion Picture Library, 535 North Dearborn St., Chicago 10, Ill.

16. Educational Motion Pictures, University of Utah Extension Division, Audio-Visual Bureau, University of Utah, Salt Lake City 1, Utah.

IMMUNOLOGICAL TESTS USED IN DIAGNOSIS

TEST	TYPE OF TEST	DISEASE DIAGNOSED
Agglutination	Antigen-antibody (bacterial cells and specific antibody)	Most bacterial diseases
Allergic	Allergens-extracts of pollens, meats, grains, dust, etc., injected intradermally	Allergic diseases
Ascoli	Precipitation (extract of infected tissue and known antibody)	Anthrax
Blastomycin	Intradermal injection of extract of *Blastomyces* sp.	Blastomycosis
Brucellergin	Intradermal injection of extracts of *Brucella* sp.	Brucellosis
Coccidioidomycin	Intradermal injection of extracts of *C. immitis*	Coccidioidomycosis
Cold agglutination	Agglutination of Group "O" human or homologous red cells in the ice box	Atypical virus pneumonitis
Complement fixation	Specific antigen and antibody and complement	Most all diseases
Dick	Streptococcus toxin injected intradermally (1/50 STD)	Scarlet fever susceptibility or immunity
Foshay	Intradermal injection of antigen or antibody	Tularemia
Hemagglutination	Red cells (human or chicken) plus virus	Influenza, mumps, and other virus diseases
Heterophile	Agglutination of sheep red blood cells	Infectious mononucleosis
Histoplasmin	Intradermal injection of extracts of *H. capsulatum*	Histoplasmosis
Kahn, Hinton, Eagle, Mazzini	Flocculation	Syphilis
Pfeiffer	*Vibrio cholera* injected intraperitoneally into immune guinea pig or mouse; organisms lysed	Cholera
Quellung	Capsular swelling with specific antibody	Pneumococcus, Hemophilus, and Klebsiella infections

IMMUNOLOGICAL TESTS USED IN DIAGNOSIS (CONTINUED)

TEST	TYPE OF TEST	DISEASE DIAGNOSED
Schick	Diphtheria toxin injected intradermally (1/50 MLD)	Diphtheria immunity or susceptibility
Trichophytin	Intradermal injection of extracts of *Trichophyton* sp.	Ringworm
Tuberculin	Extracts of tubercle organism injected intradermally	Tuberculosis
Virucidal	Virus plus antibody mixture injected into animal	Viral and rickettsial diseases
Vollmer patch	Tuberculins applied to skin on a patch of gauze	Tuberculosis
Wassermann (complement fixation)	Antigen-antibody and complement	Syphilis and other diseases
Weil-Felix	Agglutination (antibody from rickettsial diseases plus OX strains of proteus)	Rickettsial diseases

TEMPERATURE CONVERSION

CENTIGRADE TO FAHRENHEIT

° CENTIGRADE	° FAHRENHEIT	
−100	−148	*Freezing point* (sea level)
−90	−130	Centigrade = zero (0)
−80	−112	Fahrenheit = 32°
−70	−94	
−60	−76	*Boiling point* (sea level)
−50	−58	Centigrade = 100°
−40	−40	Fahrenheit = 212°
−30	−22	
−20	−4	Calculations for conversion of tem-
−10	+14	peratures of one scale to another:
0	+32	
+10	50	(1) Centigrade to Fahrenheit:
20	68	degrees C. $\times$ 9/5
30	86	plus 32 = ° F
40	104	
50	122	(2) Fahrenheit to Centigrade:
60	140	degrees F. minus
70	158	32 $\times$ 5/9 = ° C.
80	176	
90	194	
100	212	
110	230	
120	248	
130	266	

ACIDITY AND ALKALINITY

The taste of vinegar imparts a sharp, sour taste which denotes acidity. The taste of washing soda has a slippery, caustic taste which denotes alkalinity. Acidity and alkalinity are denoted by the term pH or an actual measure of the strength of the particular fluid in relation to acid (H^+) or alkali (OH^-). When a solution is neutral to either acid or base the pH is 7.0. If on the pH scale it is below 7.0, then the fluid is acid. If the number is above 7, the fluid is alkaline. The pH scale is from 1 to 14. The pH of our blood is about 7.35 which is on the alkaline side of the scale; the pH of our stomach fluids is about 1.8 or quite acid. The pH may be measured electrometrically or by means of dyes which change color at different pH values.

INDICATOR DYES FOR DETERMINING pH

NAME OF CHEMICAL	pH RANGE	COLOR CHANGE
Acid cresol purple	0.2–1.8	Red-yellow
Meta cresol purple	1.2–2.8	Red-yellow
Bromphenol blue	3.0–4.6	Yellow-blue
Bromcresol green	3.8–5.4	Yellow-blue
Methyl red	4.4–6.0	Red-yellow
Bromcresol purple	5.2–6.8	Yellow-purple
Bromthymol blue	6.0–7.6	Yellow-blue
Phenol red	6.8–8.4	Yellow-red
Cresol red	7.2–8.8	Yellow-red
Thymol blue	8.0–9.6	Yellow-blue

All bacteriological media must be adjusted for pH, usually in the range of 6.8 to 7.6. Too acid a media frequently inhibits bacterial growth. Also, indicator dyes added to media enables one to determine if an organism produces an acid or alkaline reaction.

GLOSSARY

A

Abiogenesis: Spontaneous generation; the presumed belief of generation of life from nonliving matter.

Aerobe: An organism that requires free atmospheric oxygen for respiration.

Agar (agar agar): A polysaccharide extracted from sea weed and used to solidify cultural media.

Agglutination: The clumping of cells in a suspension by action of a specific antibody.

Allergy: A state of hypersensitivity to a foreign substance usually a protein.

Anabolism: The building processes of living cells.

Anaerobe: An organism that is capable of living in the absence of free oxygen.

Antagonism: The inhibiting effect of one organism on another.

Antibiotics: Metabolic substances produced by other forms of plant life.

Antibody: A specific substance formed in the animal body in response to the introduction of antigen.

Antigen: A substance, usually a foreign protein, which when introduced into the animal body stimulates the production of specific antibodies.

Antiseptic: A substance which will prevent the growth of vegetative pathogenic microorganisms without necessarily destroying them.

Antiserum: A serum that contains antibodies.

Antitoxin: Antibody found in blood serum which is capable of neutralizing or destroying its homologous toxin.

Aseptic: The absence of living microorganisms.

Attenuate: To decrease the virulence of a microorganism.

Autogenous vaccine: A vaccine prepared with organisms from the same patient who is to be treated with it.

Autolysis: The spontaneous disintegration of cells by the action of their own enzymes.

Autotrophs: A group of organisms capable of utilizing inorganic compounds for their nutrition.

B

Bacteremia: The presence of bacteria in the blood because of a spill-over or leakage from an infected area.

Bactericidal: Destructive to bacteria.

Bacteriolysis: The disintegration of bacterial cells.

Bacteriophage: A specific virus which attacks bacteria.

Bacteriostasis: The prevention of growth and multiplication of bacteria without killing the organism.

390

Binary fission: Reproduction by division into two equal parts.
Bipolar: At both ends of the cell.

C

Capsule: An envelope of polysaccharide or protein which surrounds certain microorganisms.
Carrier: An individual who harbors and may disseminate pathogenic organisms but does not show symptoms of the disease.
Catabolism: The destructive metabolism of living cells.
Catalyst: A substance which changes the velocity of a reaction usually by increasing it.
Cell: The unit of specific protoplasmic organization.
Chemotherapy: The treatment of disease with chemical substances or drugs.
Chlorination: The addition of chlorine, primarily for killing microorganisms, as in water treatment.
Chromogenesis: The formation of pigments or colors.
Coagulation: The process of clotting or curdling of a fluid, usually protein.
Commensalism: A relationship of organisms in which one of the associates is benefited without either benefit or harm to the other.
Communicable disease: A disease which is transmitted from one individual to another by direct contact.
Complement: A heat-labile substance occurring in the blood which is necessary for the lytic action of certain antibodies.
Contagious: Capable of being transmitted by direct contact from one individual to another.
Culture: A population of microorganisms.
Cytoplasm: The protoplasm of a cell exclusive of the nucleus.

D

Denitrification: The reduction of nitrates with the liberation of free nitrogen.
Disinfection: The process of destroying infectious microorganisms.

E

Ecology: The study of the relationship between organisms and their environment.
Endemic: Pertaining to the constant presence of a number of cases of a disease in a given area.
Endospore: A heavy-walled, resistant body formed within the microorganism.
Enteric: Pertaining to the intestinal tract.
Endotoxin: A poisonous substance produced within and not liberated until dissolution of the cell.
Enzyme: A substance produced by the living cell which changes the velocity of a reaction.
Epidemic: An unusual or above average number of cases of a disease in a given area.
Etiology: The science of the causes of disease.
Exotoxin: A poisonous substance produced within the cell and then excreted int o the surrounding medium.

F

Facultative: Not obligatory; capable of assuming a different state.
Fauna: The animal life of a region.
Fermentation: The incomplete breakdown of carbohydrate resulting in the production of acids, alcohols, carbon dioxide, and water.
Filamentous: Composed of delicate threadlike structures.
Flagellum: A delicate whiplike appendage of bacterial cells.
Flora: The plant life of a region.
Fomites: Articles, such as eating utensils, pencils, etc., which may serve to transmit an infectious disease.
Fumigation: Destruction of living agents by gaseous fumes.
Fungus: A multicellular thallophyte that lacks chlorophyll.

G

Genus: A classification embracing related species.
Germs: Microorganisms—here, meaning bacteria.
Germicide: An agent which kills pathogenic microorganisms.

H

Hapten: An incomplete or partial antigen which may or may not react with its antibody and is frequently unable to bring about the production of antibody except when combined with protein.
Hemolysin: A substance which will cause the destruction of red blood corpuscles, especially antibodies or enzymelike substances produced by bacteria.
Hemolysis: The destruction of red blood corpuscles with the liberation of hemoglobin.
Heterogeneous: Different in form or function.
Heterotrophs: Organisms which require organic substances for growth.
Homologous: The same form or function.
Host: Any animal or plant which harbors a parasite.
Hydrolysis: A chemical reaction involving the splitting of a molecule with the addition of water.
Hypersensitiveness: A state of abnormal response of the body to a subsequent introduction of substances which produce little or no reactivity when first introduced.
Hypha: A single strand or filament of a fungus composing the mycelium.

I

Immunity: Resistance to disease; it may be natural or acquired.
Immunology: The science of the study of resistance and susceptibility to disease and reactions of disease agents.
Incubate: To maintain a culture under proper conditions suitable for growth.
Infection: The invasion of the body with the production of disease by virulent organisms.
Inflammation: The reaction of the tissue to injury characterized by pain, heat, redness, and swelling.
Inoculation: The introduction of material into an animal or other medium.

Intracellular: Within the cell.
Intramuscular: Within the muscle.
Intravenous: Within the vein.
In vitro: Within a glass, observation in a test tube.
In vivo: Within the living body.
Isolation: Keeping a patient with a disease from contacting other people.
-Itis: Suffix denoting inflammation of.

L

Lesion: An injured area of tissue.
Lethal: Deadly, fatal.
Leucocyte: A white blood corpuscle.
Leucocytosis: A condition of an increased number of white corpuscles in the blood.
Leucopenia: A reduction in the number of white blood corpuscles in the blood.
Lysis: The destruction of cells.

M

Medium: The substrate for the cultivation of microorganisms.
Metabolism: The physical and chemical reactions occurring in living cells by means of which life is maintained.
Metachromatic granules: Internal granules which differ in staining properties from the rest of the cell.
Micron: A unit of measurement having a value of one one-thousandth of a millimeter or approximately one twenty-five thousandth of an inch.
Microorganisms: A minute (microscopic) animal or plant.
Morbidity: The frequency of occurrence of cases of a disease.
Morphology: The study of size, shape, structure, and arrangement of organisms.
Mortality: The percentage of deaths from a given cause.
Mutation: A change from some parental character occurring in the offspring; more permanent than variation.
Mycelium: A mass of hyphae or fungous filaments.
Mycology: The study of fungi.

N

Nitrification: The oxidation of ammonium compounds to nitrates.
Nitrogen fixation: The formation of nitrogen compounds from free nitrogen.
Nucleus: A structure within the cell which controls cell functions.
Nutrient: A substance for the growth of an organism.

O

Opsonin: An antibody which enhances phagocytosis.
Osmosis: Diffusion through a semipermeable membrane.

P

Pandemic: A world-wide spread of an epidemic of a given disease.
Parasite: An organism which lives upon or within another living organism at whose expense it obtains some advantage.

Pasteurization: The heating of fluid for a short time to destroy undesirable microorganisms.

Pathogen: A disease-producing microorganism.

Pathology: The science dealing with tissue changes and processes in disease.

Peptones: Partially broken-down proteins.

Peptonization: The process of converting proteins into peptones.

Phagocyte: A white blood cell which is capable of ingesting solid particles.

Phagocytosis: The process of engulfing solid particles by cells.

Photosynthesis: The formation of carbohydrate from carbon dioxide and water using light as the source of energy.

Plasma: The fluid portion of the blood less blood cells; it contains fibrinogen or its precursor.

Plasmolysis: The shrinking of the protoplasm of a cell due to the loss of water.

Pleomorphism: Having several forms or shapes of the same species.

Precipitin: An antibody causing the specific soluble antigen to precipitate or go out of solution.

Proliferation: The multiplication of cells.

Prognosis: A prediction as to the outcome of a case of disease.

Proteins: Organic compounds of high molecular weight which contain nitrogen and yield amino acids on hydrolysis (egg white, body cells, etc.).

Proteolysis: The hydrolysis or decomposition of proteins.

Purulent: Containing pus.

Pus: Dead leucocytes.

Putrefaction: The decomposition of animal or plant proteins by microorganisms with the production of foul odors.

Pyemia: Septicemia with the production of secondary foci of infection.

Q

Quarantine: A period of isolation of exposed or infected groups of persons from contact with unexposed persons.

R

Resistance: The ability of the body to ward off disease.

Respiration: Any oxidative change whereby energy is released for metabolism of the cell.

S

Sanitation: The maintenance of a healthful environment.

Saprophytes: Organisms that obtain food from dead organic matter.

Sensitize: To render more responsive to a substance.

Septicemia: The multiplication of bacteria in the blood stream often erroneously called "blood poisoning."

Serum: The cell free and fibrin free straw-colored fluid of coagulated blood.

Smear: A thin layer of material spread on a glass slide.

Spore: A thick-walled resistant cell.

Sterile: Free from all living matter.

Sterilization: The process of complete destruction of all living matter.

Susceptible: Capable of being influenced or easily affected.

Synergism: The ability of two or more species of microorganisms to accomplish a change that neither can bring about alone.

T

Therapy: The treatment of a disease.

Thermolabile: Susceptible to heat.

Thermophile: Organism which grows best at above body temperatures.

Toxemia: The presence of poisons or toxic substances in the blood or body.

Toxin: A potent poison produced by living cells which is capable of stimulating the production of antibody called antitoxin.

Toxoid: A modified toxin which has lost its toxic properties but has retained its ability to stimulate antitoxin formation.

V

Vaccine: Usually nontoxic toxins, killed or avirulent microorganisms used for injection into the body to produce antibodies.

Vectors: Animals and insects that carry infectious agents from one host to another.

Virulence: The disease-producing ability of an organism.

Virus: An ultramicroscopic infectious agent which passes through the ordinary bacterial retaining filters and reproduces only at the expense of living cells.

Y

Yeast: Unicellular plants, usually spherical or ovoid in shape, which multiply asexually by budding.

INDEX